AF556784

SECURITY FOR TELECOMMUNICATIONS NETWORK MANAGEMENT

IEEE Press
445 Hoes Lane, P.O. Box 1331
Piscataway, NJ 08855-1331

Technical Reviewers

Robert L. Barker, *Bell Atlantic, Silver Spring, MD*
Jan Bates, *Tellium, Inc., Oceanport, NJ*
Gerald D. Chandler, *Sterling Software Corporation, Tinton Falls, NJ*
Richard Graveman, *Telcordia Technologies, Inc., Morristown, NJ*
Lakshmi Raman, *Telcordia Technologies, Inc., Red Bank, NJ*

SECURITY FOR TELECOMMUNICATIONS NETWORK MANAGEMENT

Moshe Rozenblit
Telcordia Technologies, Inc.
New York, NY

IEEE Communications Society, *Sponsor*

Salah Aidarous and Thomas Plevyak, *Series Editors*

The Institute of Electrical and Electronics Engineers, Inc., New York

This book and other books may be purchased at a discount
from the publisher when ordered in bulk quantities. Contact:

IEEE Press Marketing
Attn: Special Sales
445 Hoes Lane, P.O. Box 1331
Piscataway, NJ 08855-1331
Fax: 1 732 981 9334

For more information about IEEE Press products, visit the
IEEE Press Home Page: http://www.ieee.org/press

Printed in the United States of America

10 9 8 7 6 5 4 3 2 1

ISBN 0-7803-3490-6
IEEE Order Number PC5393

Library of Congress Cataloging-in-Publication Data

Rozenblit, Moshe, 1944–
Security for telecommunications network management / Moshe Rozenblit.
p. cm. — (IEEE Press series on network management)
Includes bibliographical references and index.
ISBN 0-7803-3490-6
1. Telecommunication — Security measures. 2. Computer networks — Security measures. I. Title. II. Series.
TK5102.85.R69 1999 99-38975
621.382—dc21 CIP

To my Parents —
Rachel and Israel

Contents

Preface xvii

Acknowledgments xix

List of Figures xxi

List of Tables xxiii

PART I TMN OVERVIEW 1

CHAPTER 1 A Brief History of TMN 3

CHAPTER 2 Architectural Views of the TMN 7

2.1 Functional Architecture 7
2.2 Physical Architecture 9
2.3 Communications/Information Architecture 11
2.3.1 Physical Layer 13
2.3.2 Data Link Layer 13
2.3.3 Network Layer 14
2.3.4 Transport Layer 14
2.3.5 Session Layer 15
2.3.6 Presentation Layer 15
2.3.6.1 Distinguished Encoding Rules 15
2.3.7 Application Layer 16
2.3.7.1 ACSE (Association Control Service Element) 18
2.3.7.2 ROSE (Remote Operations Service Element) 19
2.3.7.3 STASE-ROSE 19
2.3.7.4 FTAM (File Transfer Administration and Maintenance) 20
2.3.7.5 X.500 Directory User Agent 20
2.3.7.6 CMISE (Common Management Information Service Element) 20
2.3.7.6.1 Containment 20
2.3.7.6.2 Addressing MOs 21
2.3.7.6.3 Inheritance 22
2.3.7.6.4 CMISE Services 22
2.3.7.6.5 Managers and Agents 23
2.3.7.7 SMASE (System Management Application Service Element) 24
2.3.7.8 Proper Naming 25

2.3.7.8.1 OBJECT IDENTIFIER 26
2.3.7.8.2 DistinguishedName 26
2.3.7.8.3 Security Audit Trail—An Example 27
2.3.7.9 EDI 28
2.3.7.10 CORBA 29
2.3.7.11 SNMP—Keeping It Simple 30
2.3.7.11.1 A Simple Structure of Management Information 30
2.3.7.11.2 SNMP PDUs 32
2.3.7.11.3 Version 2—Bigger and Better 33
2.3.7.11.4 SNMPv3—A Version for All Seasons 35
2.3.8 Security-Related Components of the TMN Stack 37

PART II SECURITY OVERVIEW 39

CHAPTER 3 TMN Attacks and Defenses 41

3.1 TMN General Threats and Vulnerabilities 41
3.1.1 Potential Security Attackers 41
3.1.2 Potential Security Threats 42
3.1.2.1 Unauthorized Access 42
3.1.2.2 Eavesdropping 42
3.1.2.3 Masquerade 43
3.1.2.4 Modification of Information 43
3.1.2.5 Repudiation 43
3.1.2.6 Replay, Reroute, Misroute, Delete Messages 43
3.1.2.7 Network Flooding 43
3.1.3 Potential Security Risks 43
3.1.3.1 Theft of Information 43
3.1.3.2 Unauthorized Use of Resources 44
3.1.3.3 Theft of Service 44
3.1.3.4 Denial of Service 44
3.1.3.4.1 Single-User Denial of Service 44
3.1.3.4.2 Networkwide Denial of Service 44
3.1.4 Impacts of Security Risks on the TMN 45
3.1.4.1 Configuration Management—Provisioning 45
3.1.4.2 Performance Monitoring 45
3.1.4.3 Fault Management—Trouble Administration 46
3.1.4.4 Accounting Management 46
3.1.4.4.1 Usage Measurement 46
3.1.4.4.2 Tariffing/Pricing 46
3.2 Threats Unique to the TMN 46
3.2.1 Generic TMN Vulnerabilities 47
3.2.2 TMN Domain-Specific Vulnerabilities 47
3.2.2.1 Generic ICEC Threats 48
3.2.2.2 Application Specific Threats 48
3.2.2.2.1 Service Ordering 48
3.2.2.2.2 Trouble Administration 49
3.2.2.2.3 Testing 49
3.2.2.2.4 Performance Management 49
3.2.2.2.5 Alarm Notification 49
3.3 Security Services 50

3.3.1 Connection Access Control 50
3.3.2 Peer Entity Authentication 50
3.3.3 Data Origin Authentication 50
3.3.4 Integrity 51
3.3.4.1 Selective Field Integrity 51
3.3.4.2 Whole Message Integrity 51
3.3.4.3 Session Integrity 51
3.3.5 Confidentiality 51
3.3.5.1 Selective Field Confidentiality 51
3.3.5.2 Whole Message Confidentiality 52
3.3.5.3 Traffic Flow Confidentiality 52
3.3.6 Non-Repudiation 52
3.3.6.1 Non-Repudiation of Origin 52
3.3.6.2 Non-Repudiation of Receipt 52
3.3.7 Access Control 53
3.3.8 Security Alarm 53
3.3.9 Security Audit Trail 55

CHAPTER 4 Basic Security Mechanisms 57

4.1 Hashing 57
4.1.1 Keyed Hashing 58
4.1.2 S-key 60
4.2 Encryption 60
4.2.1 Symmetric Key Encryption 61
4.2.1.1 Padding 61
4.2.1.2 IV Selection 63
4.2.1.3 Error Propagation 64
4.2.1.4 Triple DES 64
4.2.1.5 Digital Seals 65
4.2.2 Asymmetric Encryption 65
4.3 Digital Signatures 66
4.4 Certificates 68
4.5 Access Control Mechanisms 69
4.5.1 Rules 70
4.5.2 Initiator ACI 71
4.5.2.1 Authenticated Identity Initiator 71
4.5.2.2 Anonymous Authenticated Initiator 71
4.5.2.3 Anonymous Unauthenticated Initiator 72
4.5.3 Request ACI 72
4.5.4 Target ACI 73
4.5.4.1 Access Control Lists 73
4.5.4.2 Capabilities 73
4.5.4.3 Security Labels 73
4.6 Diffie–Hellman Key Exchange 73
4.6.1 Ephemeral Diffie–Hellman Key Exchange 74
4.6.2 Certified Diffie–Hellman Parameters 74
4.7 Authentication Protocols 74
4.7.1 Challenge-Response Authentication 75
4.7.2 Stateful Authentication 76
4.8 Mapping Security Services to Security Mechanisms 78

CHAPTER 5 **Support Mechanisms 81**

5.1 Security Alarms 81
5.2 Security Audit Log 82
5.3 Key Distribution 82
5.3.1 Key Lists 82
5.3.2 Public Key Distribution 83
5.4 Directory 83
5.4.1 Automatic Registration 83
5.4.2 Directory Access Control 85
5.4.3 Multiple Security Domains 86
5.5 Protocols for Security 86
5.5.1 Anatomy of Secure Communication Protocols 87
5.5.2 Security In Layers 88
5.6 GSS-API—Security In a Box 89
5.6.1 GSS Handshaking 90
5.6.2 GSS Secure Transfer 91
5.6.3 GSS Administrative Interfaces 93
5.6.4 Status Reporting 94
5.7 GULS—Soaring Security 94
5.8 SSL3—Safe Surf 95
5.8.1 A Firm Handshake 95
5.8.2 Secure Transfer 99
5.8.3 Alerts 100
5.9 IPsec 100
5.9.1 A Discrete Handshake 100
5.9.2 Secure Transfer 101
5.10 Security Engineering 102

PART III SECURING THE TMN 105

CHAPTER 6 **Security of OSI-Based TMN Protocols 107**

6.1 In the Beginning—ACSE Security 107
6.1.1 Identification 107
6.1.2 Authentication 108
6.1.3 ASE-Specific Security 109
6.2 CMISE Security 109
6.2.1 Electronic Bonding Authenticator—Homegrown Security 110
6.2.1.1 Historical Background 110
6.2.1.2 The Authenticator 111
6.2.1.2.1 Vulnerabilities 112
6.2.1.2.2 Protocol Implications 113
6.2.1.2.3 Operations Implications 113
6.2.1.2.4 DES Padding 114
6.2.1.2.5 Key Management 114
6.2.1.2.6 Future Proofing 115
6.2.2 Selective Field Protection—If ABSolutely Necessary 115
6.2.2.1 Data Representation 117
6.2.2.1.1 Character Strings 117
6.2.2.1.2 Time 117
6.2.2.1.3 Octet Strings 117

6.2.2.1.4 Bit Strings 117
6.2.2.1.5 Boolean 117
6.2.2.1.6 Integers 118
6.2.2.1.7 Real Numbers 118
6.2.2.1.8 OBJECT IDENTIFIER 118
6.2.2.1.9 OBJECT DESCRIPTOR 118
6.2.2.2 Syntax for Security Transformations 118
6.2.2.3 DES Padding For ABS 124
6.3 STASE-ROSE 124
6.3.1 Security Transformations on ROSE PDUs 125
6.3.2 Peer Entity Authentication 126
6.3.3 Negotiation of Security Algorithms 127
6.3.3.1 Defaults 127
6.3.3.2 Negotiation 129
6.3.4 Dynamic Update of Security Parameters 132
6.3.5 STASE-ROSE Services 132
6.3.5.1 SR-TRANSFER Parameters 132
6.3.5.1.1 ROSE-PDU 133
6.3.5.1.2 Encryption-Type 133
6.3.5.1.3 Encryption-Parameters 133
6.3.6 Interaction between Application Service Elements 135
6.3.6.1 Association Establishment 135
6.3.6.1.1 Association Initiator 135
6.3.6.1.2 Association Responder 136
6.3.6.2 Association Release 137
6.3.6.2.1 Sender 137
6.3.6.2.2 Receiver 138
6.3.6.3 Association Abort 138
6.3.6.3.1 Sender 138
6.3.6.3.2 Receiver 139
6.3.6.4 Data Transfer 139
6.3.6.4.1 Sender 139
6.3.6.4.2 Receiver 139
6.3.7 STASE-ROSE Protocol 140
6.3.7.1 Abstract Syntax Definition of APDUs 140
6.3.8 Use of GSS API with STASE-ROSE 145
6.3.8.1 Security Context Negotiation 146
6.3.8.2 Data Transfer Phase 147
6.3.9 STASE-ROSE Current Status and Future Developments 148
6.4 Q3 Security 149
6.5 X.500 149
6.6 X.25 150

CHAPTER 7 EDI-Based TMN Security 153

7.1 TLS1 for EDI 153
7.2 Interactive Agent 154
7.2.1 Message Formatting 154
7.2.1.1 IA Status Message Detail Format 157
7.2.1.2 Optional Message Receipts 157
7.2.2 Message Syntax Definitions 158
7.2.2.1 ASN.1 Syntax for Basic EDI Messages 159

7.2.2.2 ASN.1 Syntax for EDI with Message Integrity 159
7.2.2.3 ASN.1 Syntax for EDI with Non-Repudiation 160
7.2.2.4 ASN.1 Syntax for IA Status 162
7.2.2.5 ASN.1 Syntax for Optional IA Receipts 162
7.2.3 Client Specifications 163
7.2.4 Server Specifications 165
7.2.4.1 SSL Read Processing 167
7.2.4.2 Route Data to Translator 167
7.2.4.3 Receipt Logging 167
7.2.4.4 Message Validation 167
7.2.4.4.1 Message Integrity 167
7.2.4.4.2 Non-Repudiation 168
7.2.4.5 Server Disconnect 168
7.2.4.6 Example of Parsing the Received Message 168
7.2.4.6.1 Basic EDI Message 169
7.2.4.6.2 Message Integrity 169
7.2.4.6.3 Non-Repudiation 171
7.2.4.6.4 IAstatus 174
7.2.4.6.5 IAreceipt 174
7.2.5 Interfaces 175
7.2.5.1 Data Communications Protocol 175
7.2.5.2 EDI Translators 175
7.2.6 Design Considerations 176
7.2.6.1 Multiprocessing/Multithreading 176
7.2.6.2 Non-Persistent Connections 176
7.2.6.3 Resumable SSL3 Sessions 176
7.2.6.4 Connectivity 176
7.2.6.5 Message Priority 176
7.2.7 Operational Concerns 176
7.2.7.1 Security 176
7.2.7.2 Flow Control 176
7.2.7.3 Logging 176
7.2.7.3.1 Logging Levels 177
7.2.7.3.2 Log Files 177
7.2.7.4 Routing 177
7.2.7.5 Firewalls 177
7.2.7.6 Digital Certificates 177
7.2.8 Error Handling/Recovery 177
7.2.9 Implementation Issues 178
7.2.9.1 Interoperability 178
7.2.9.2 Port Assignments 178
7.2.9.3 Partner Responsibilities 178

CHAPTER 8 CORBA-Based TMN Security 179

8.1 Overview of General Inter Orb Protocol (GIOP) 180
8.2 Telecom Non-Repudiation Inter-Orb Protocol (TeNoRIOP) 180
8.2.1 Non-Repudiation for Request 181
8.2.1.1 Digest for Request 182
8.2.1.2 Digital Signature for Request 184
8.2.1.3 Strict Correlation 184
8.2.2 Non-Repudiation for Reply 185

8.2.2.1 Digest for Reply 186
8.2.2.2 Digital Signature for Reply 187
8.3 IDL Syntax for Non-Repudiation Evidence 188
8.4 Local API Interface Specification 191
8.5 Timing of Non-Repudiation Evidence 193
8.5.1 Timing Agreements 193
8.5.1.1 Non-Repudiation of Origin 193
8.5.1.2 Non-Repudiation of Receipt 193
8.5.1.3 Behavior While Waiting 194
8.5.1.3.1 Trusting Behavior 194
8.5.1.3.2 Cautious Behavior 194
8.5.1.3.3 Suspicious Behavior 194
8.5.1.4 Notifications 195
8.6 Non-Repudiation Protocol Machine 195
8.6.1 Message Sender 195
8.6.2 Message Receiver 196

CHAPTER 9 SNMP-Based TMN Security 197

9.1 SNMPv1 Security 197
9.2 SNMPv2 Security 198
9.2.1 Proper ID Required 198
9.2.2 On the Relativity of Time 199
9.2.3 Secure PDUs 199
9.3 SNMPv3 Security 200
9.3.1 User-Based Security Model 201
9.3.1.1 Simple Times 201
9.3.1.2 Key Items 201
9.3.1.3 USM PDUs 202
9.3.1.4 USM APIs 204
9.3.2 View-Based Access Control Model (VACM) 206
9.3.2.1 Who's Calling 206
9.3.2.2 Domain of Discourse 207
9.3.2.3 VACM Services 208

CHAPTER 10 Portraits Gallery 209

PART IV SECURITY MANAGEMENT 221

CHAPTER 11 Management of Security Information 223

11.1 Security Administration Functions 224
11.1.1 Login Management 224
11.1.2 Notification Management 225
11.1.3 Access Control Management 225
11.1.4 Management of Encryption Keys 225
11.2 Information Model Description 226
11.2.1 Login Management 226
11.2.1.1 Pre- and Post-Login Messages 226
11.2.1.2 User Management 227
11.2.1.3 Password Management 228

11.2.1.4 Channel Management 230
11.2.1.5 Session Management 231
11.2.1.6 Security MOs Management 232
11.2.2 Notification Management 232
11.2.3 Access Control 236
11.2.3.1 Targets 236
11.2.3.2 Rules 237
11.2.3.3 Initiators 237
11.2.3.3.1 Access Control Lists 237
11.2.3.3.2 Capabilities 237
11.2.3.3.3 Security Labels 237
11.2.3.4 Authentication for Access Control 238
11.2.3.5 MOs for Access Control 238
11.2.4 Key Management 240

PART V SECURITY OPERATIONS 243

CHAPTER 12 Security Functions and Operations 245

12.1 Prevention Services 245
12.2 Detection Security Services 246
12.3 Illustrative Scenarios 246
12.3.1 Authentication and Access Control Scenario 246
12.3.2 NEL Alarm Detection, Containment, and Recovery 248

CHAPTER 13 Security Management Functions and Operations 251

13.1 Prevention 252
13.1.1 BML 252
13.1.2 SML 254
13.1.3 NML 254
13.1.4 EML 254
13.1.5 NEL 254
13.2 Detection 254
13.2.1 BML 254
13.2.2 SML 254
13.2.3 NML 255
13.2.4 EML 255
13.2.5 NEL 255
13.3 Containment and Recovery 255
13.3.1 BML 255
13.3.2 SML 256
13.3.3 NML 256
13.3.4 EML 256
13.3.5 NEL 257
13.4 Security Administration 257
13.4.1 BML 257
13.4.2 SML 257
13.4.3 NML 258
13.4.4 EML 258
13.4.5 NEL 259

13.5 Security Management Scenarios 259
13.5.1 Establish/Change Privileges 259
13.5.2 Audit Detection of a Security Violation, Containment, and Recovery 262

CHAPTER 14 Future Enhancements to the TMN Security 267

14.1 Secure Interworking 267
14.1.1 Application-Based Security 267
14.1.2 CMIP/CORBA 268
14.2 Public Key Infrastructure 268
14.3 Internal Certification of External Entities 268
14.4 External Certification Authorities 268
14.5 Security Alarm Management 268
14.6 Security Audit Trail Management 269
14.7 X Interface Security 269
14.8 F Interface Security 269
14.9 Update of Related Standards 269

Abbreviations 271

Suggested Reading 277

References 279

Index 285

About the Author 296

Preface

The Telecommunications Management Network (TMN) controls the underlying Telecommunications Network (TN). Therefore, whoever controls the TMN controls the TN. The TMN needs to be protected, but protecting it presents unique challenges that have only been addressed in the last couple of years. The current state of the art of TMN security is documented in numerous, but different and arcane, national and international standards and implementation agreements. The aim of this book is to tie this multitude of components into a coherent whole. It covers all aspects of TMN as defined in international (ITU-T) and North American (ANSI-T1) standards as well as in implementation agreements published by major industry forums. It does not cover proprietary protocols and implementations.

INTENDED AUDIENCE

This book seeks to assist network architects, operations planners, system designers, software engineers, and Operations Systems (OSs) developers in understanding and addressing the security implications of the TMN. While targeted primarily at the telecommunications industry, the book is also relevant to other industries with stakes in both security and telecommunications: finance, large manufacturing, military, and government.

For planners and architects, it may obviate the need to read impenetrable standards by providing a clear view of the essentials. For systems designers and software developers, it provides an easier access to such standards.

I assume the reader is acquainted with rudimentary concepts of object-oriented programming and has some basic familiarity with telecommunications networks. Knowledge of the TMN structure and security techniques is helpful but not required. Indeed, the first part offers an introductory tutorial on TMN for security buffs, whereas the second part introduces network managers to the amazing world of electronic security. The third part combines the two in addressing the security of TMN, while the fourth part shows how TMN can be used to manage security-related information. Finally, the fifth part addresses security operations—both operations that support the security of the TMN and operations needed to manage security.

Moshe Rozenblit
Telcordia Technologies, Inc.
New York, NY

Acknowledgments

It has been my great privilege to work with many bright and stimulating people on topics covered in this book. Foremost is Dr. Lakshmi Raman (from Telcordia Technologies, Inc.) whose contribution to TMN is monumental. In particular, she has greatly improved the quality of all Open System Interconnection (OSI)-related TMN security standards. Venkat Rao (from GTE) is my cochair of the Security Subcommittee of the Electronic Communications Implementation Committee (ECIC); he has provided the seminal ideas of the Application-Based Security standard (ANSI T1.254) and pioneered the use of Secure Socket Layer version 3 in the TMN. The Security Transformations Application Service Element for Remote Operations Service Element (STASE-ROSE) standard (ANSI T1.259 and ITU-T Recommendation Q.813) is based in part on work done by Saikumar Dubagunta (from DSET); I had the pleasure of serving with Lakshmi and Saikumar as a coeditor of T1.259. Pal Kristiansen (from Telenor) has contributed the section on Generic Security Service Application Programming Interface of the international version of STASE-ROSE—ITU-T Q.813. Bernhard Nauer (from Siemens) was the main contributor to TMN Security Overview (ITU-T Recommendation M.3016). Brian Bearden (from SBC) has been my coeditor on the proposal for the Telecommunications Non-Repudiation Inter ORB (Object Request Broker) Protocol. John Murphy (from MCI-Worldcom) is the editor of the standard on non-repudiation and integrity for Electronic Data Interchange (EDI)-based TMN transactions (ITU-T Recommendation Q.815). Much of the North American standard on Q-interface security (ANSI T1.261) is based on discussions in the Distributed Network Management Architecture of the SONET Interoperability Forum (SIF), chaired by Connie Hunt (from SBC). Lakshmi has guided my work on the information model for security management in Bellcore's Generic Requirements 1253. Frederic Cherbonnier (from France Telecom) discovered a subtle flaw in the Electronic Bonding authenticator (T1 Technical Report No. 40), which resulted in a change in the key update schedule. Knut Laag chaired the TMN security group of the European Telecommunications Standards Institute (ETSI) and promoted fruitful trans-Atlantic cooperation in this area. Tim Bauman and my colleagues at Telcordia Technologies provided an encouraging and stimulating environment for writing this book.

TMN security owes much to many more people who have contributed through their activities in T1M1, ECIC, Study Group 4 of the International Telecommunications Union

(ITU), ETSI, SIF, the Object Management Group (OMG), and the Internet Engineering Task Force (IETF). Last but not least, I am most grateful to the reviewers of this book. In particular, Lakshmi's abundant and detailed technical comments have contributed greatly to the accuracy of many sections in this book; Dr. Gerald Chandler's (from Sterling Software) comments did much to enhance the structure of the book; Bob Barker (from Bell Atlantic) provided valuable insight; and Ms. Jan Bates's (from Tellium) review added much clarity and linguistic correctness.

I am honored that Salah Aidarous and Tom Plevyak have invited me to write this book; I hope they are not disappointed with the result. I am grateful to Linda Matarazzo, Karen Hawkins, Betty Pessagno, Tony VenGraitis, and the IEEE Press team for their role in the production of this book.

You can e-mail me at mrozenbl@telcordia.com

Happy reading.

Moshe Rozenblit
Telcordia Technologies, Inc.
New York, NY

List of Figures

Figure 2.1: TMN Management Functional Areas and Layers—illustrative entries 8
Figure 2.2: Functional components of an Operation System Function Block 9
Figure 2.3: Example of a TMN physical architecture 10
Figure 2.4: The lower layer of the ISO model 11
Figure 2.5: OSI layering 13
Figure 2.6: Seven layer OSI stack 14
Figure 2.7: Makeup of an Application Service Element 16
Figure 2.8: ASE service primitives 17
Figure 2.9: Application layer constituents 18
Figure 2.10: Interleaved replies to multiple requests 19
Figure 2.11: A security audit trail MO 21
Figure 2.12: Example of a management information tree 21
Figure 2.13: Example of a containment relationship 22
Figure 2.14: Example of an inheritance tree 23
Figure 2.15: Manager and agent systems 24
Figure 2.16: Symmetric CMIP interface 24
Figure 2.17: Cascading management 24
Figure 2.18: Basic setup for systems using CMISE 25
Figure 2.19: Minimum configuration for systems using SMASE 25
Figure 2:20: Example of a directory schema for a people directory 27
Figure 2.21: SNMPv3 architecture 35
Figure 2.22: Main interactions of security modules 38
Figure 4.1: Structure of MD5 for one 512-bit block 59
Figure 4.2: MD5 operating on a whole message 59
Figure 4.3: DES encryption and decryption in the ECB mode 62
Figure 4.4: DES encryption in the CBC mode 62
Figure 4.5: DES decryption in the CBC mode 62
Figure 4.6: Example of the RSA procedure 66
Figure 4.7: Signing and verifying 67
Figure 4.8: Partial order of security services 80

Figure 5.1: Life cycle of a secure communications protocol 88
Figure 5.2: In-stack secure communication protocol 90
Figure 5.3: GSS out-of-stack secure communications protocol 90
Figure 5.4: GSS-API handshaking sequence 91
Figure 5.5: GSS-API secure transfer phase 92
Figure 5.6: SSL3 handshake protocol 96
Figure 5.7: Example of tunnel mode authentication and transport mode encryption 101
Figure 6.1: Encryption in the application 116
Figure 6.2: Negotiation of security algorithms 129
Figure 6.3: SR-TRANSFER Service Primitives 133
Figure 6.4: Interaction during association establishment 136
Figure 6.5: Interaction during association release 137
Figure 6.6: Interaction during association abort 138
Figure 6.7: Interaction during data transfer 139
Figure 6.8: Use of GSS-API with STASE-ROSE at association establishment time 146
Figure 6.9: Use of GSS-API with STASE-ROSE during data transfer 147
Figure 7.1: IA Client-server interaction 155
Figure 7.2: Message format architecture for mandatory messages 156
Figure 8.1: Non-Repudiation for request 182
Figure 8.2: Non-Repudiation for reply 185
Figure 8.3: Message sender protocol machine 195
Figure 8.4: Message receiver protocol machine 196
Figure 9.1: Secure Gateways (SG) protecting TMN entities 197
Figure 9.2: Using hashing for encryption/decryption 203
Figure 9.3: A view tree family 207
Figure 10.1: TMN security communications protocols 209
Figure 10.2: GULS—for all OSI-compliant protocols 210
Figure 10.3: SSL3/TLS1 for TCP/IP only 211
Figure 10.4: IPsec—for IP only 212
Figure 10.5: TR40—for ACSE-using protocols 213
Figure 10.6: Application-Based Security—for CMIP only 214
Figure 10.7: STASE-ROSE for ROSE-based protocols 215
Figure 10.8: GSS-API 216
Figure 10.9: Interactive Agent—for EDI only 217
Figure 10.10: TeNorIOP—for CORBA only 218
Figure 10.11: SNMPv2 and SNMPv3 security 219
Figure 11.1: Example of security manager and managed systems 223
Figure 11.2: Possible name bindings (containment) for login management MOs 227
Figure 11.3: Notification management interactions 235
Figure 11.4: Inheritance of notification management MO classes 235
Figure 11.5: Possible name binding for notification management MOs 236
Figure 11.6: Access control MO class inheritance hierarchy 239
Figure 11.7: Relationship of access control MOs 239
Figure 12.1: Authentication and access control 247
Figure 12.2: NE alarm detection of a security violation, containment, and recovery 249
Figure 13.1: Partitioning of TMN security functionality 252
Figure 13.2: Establish/change privileges 260
Figure 13.3: Audit detection of a security violation, containment, and recovery 263

List of Tables

Table 3.1: Summary of Security Threats and Risks 45
Table 3.2: Security Services against Security Threats 54
Table 3.3: Security Services Correlated with Security Risks 55
Table 4.1: Security Mechanisms to Provide Security Services 78
Table 4.2: Security Mechanisms to Protect against Security Threats 79
Table 5.1: Context-Level Calls 92
Table 5.2: Per-Message Calls 93
Table 5.3: Credential Management Calls 93
Table 5.4: Support Calls 93
Table 5.5: Fatal Error Codes 94
Table 5.6: Informatory Status Codes 94
Table 6.1: Negotiated Algorithms 128
Table 6.2: Encryption-Type Values 133
Table 6.3: Components of EncryptionParameters 134
Table 9.1: IN and OUT Parameters of USM Service Primitives 204
Table 9.2: IN and OUT Parameters for USM Authentication Primitives 205
Table 9.3: IN and OUT Parameters for Privacy Primitives 206
Table 13.1: Security Management 253

PART I
TMN OVERVIEW

This brief introduction to TMN is intended for security professionals who are interested in how the TMN can be secured and in how they can benefit from it for managing security information. TMN mavens can skip it. The goal of this section is not to provide an in-depth understanding of TMN, but rather to offer just enough insight into it to permit a thorough understanding of its security implications. As such, the coverage is quite uneven: aspects of TMN that are relevant to security are covered in glorious detail, and others are barely acknowledged. Furthermore, most of the examples used to illustrate general principles of the TMN are related to security applications.

This overview is constructed as answers to three questions:

1. Why TMN?
2. What is TMN?
3. How does it work?

The first question is answered by reconstructing a plausible (rather than accurate) history of the TMN's origins. The second question is addressed by presenting different, complementary architectural views of TMN. The answer to the last question explains the mechanisms of TMN, or at least those that are relevant to security.

1

A Brief History of TMN

A straight plunge into the minutiae of TMN can have a chilling effect on the unprepared reader. Perhaps the best warmup exercise is to gain some appreciation for its motivation and history. A basic understanding of the goals and reasoning behind the final (or rather, current) product will help in viewing the TMN's countless details in a coherent, logical perspective. This section, however, does not seek to provide an accurate historical chronology. Rather, it reverse-engineers what might have been the history of TMN and what might have been the rationale behind some of the major decisions in its development.

The fictional history starts with one of the most significant events in modern telecommunications. Tellingly, the event in question was a local, legalistic event rather than a global, technological breakthrough: the breakup of the Bell System in the United States on January 1, 1984. Suddenly, Local Exchange Carriers (LECs) responsible for some 100 million telephone lines were allowed and encouraged to procure their equipment from a large array of competing suppliers. Those LECs, which were used largely to obtain their equipment from a single source (AT&T's Western Electric, forebear of Lucent Technologies), enthusiastically embraced the new opportunity. While switching and transmission components from different vendors could interoperate through use of the same signaling and transmission protocols, they sadly did not have the same interfaces to Operation Systems (OSs). The new LECs had to acquire different OSs to manage Network Elements (NEs) from different suppliers. The increased cost of duplicate systems was compounded by decreasing flexibility: OSs from different suppliers could not exchange information or communicate with the same set of higher level OSs and Work Stations (WSs). It soon became clear that the cloud of noninteroperability was looming large over the silver lining of multivendor procurement. TMN was conceived to diffuse this cloud.

A band of TMN visionaries soon gathered under the auspices of the newly formed Alliance for Telecommunications Industry Solutions (ATIS) in an effort to turn dream into reality. Their objective was plug and play: allowing an OS from any supplier to plug into NEs, OSs, and WSs from any other supplier in perfect harmony. This said, the only remaining task was to define the salutary plug. The search was guided by the following criteria:

1. Interoperability
2. Full network management functionality

3. Complete freedom of local implementation
4. Broad industry acceptance

Each of these criteria has had such a profound impact that it deserves some clarification.

Interoperability means that two communicating systems need not use the same hardware platform (for example, they can use different processors), the same operating system (for example, one system can be based on Unix while the other uses Windows NT), or the same conventions for representing data (for example, one system can use ASCII while the other uses EBCDIC). Furthermore, one system does not even need to know what hardware, operating system, or data representation is used by the other. A considerably more challenging requirement is that an OS that manages a NE need not know how that NE is built. (That information may be a trade secret that the NE supplier will not share with the OS supplier.)

Network management functionality mandates that any interaction required for the purpose of efficient network management be supported over TMN-defined interfaces. This became the opportunity for network managers and operations planners to dust off and augment their wish lists of functions for effective network management. Here are two examples of wanted functionality, one dealing with switch configuration management, and the other, closer to security, dealing with access control management:

- Send a message to a Central Office (CO) requesting that all the line cards that were installed after January 15, 1995 and in which feature A is active should have feature B inactive and report the change, if any, to the accounting system.
- Specify that any SONET technician from the North District can reconfigure any SONET NE in that District, can test any SONET NE in the North and Central Districts, and can check the status of any SONET NE.

Local implementation freedom precludes any constraints on how a system is built, as long as it interfaces with other TMN systems. This requirement allows suppliers to use their unique expertise and creativity to develop better products, rather than mandating a commodity market for NEs and OSs.

Industry acceptance is the all-important reality check: the TMN specifications must gain broad acceptance from competing suppliers and competing service providers in order to ensure the availability and deployment of TMN-based products.

Such criteria may have seemed foolhardy in 1984. They still do today, even though numerous TMN-based applications are now up and running.

The first order of the day for the TMN founders was to look for any ready solution that might be out there. Three candidates were rounded up:

1. **TL1** (Transaction Language 1) from Bellcore. Several suppliers had implemented TL1. However, as a Bellcore specification, rather than an open standard created by a public body, its broad industry acceptance was in doubt. Furthermore, it was not clear that TL1 could conveniently support the wish lists for network management functionality that had been accumulating.
2. **SNMP** (Simple Network Management Protocol) from the Internet Engineering Task Force (IETF). Actually, SNMP was developed later, but this fictional history does not worry about such details. SNMP was originally designed to manage fairly small networks (by large LECs' standards) of fairly simple elements (for example, modems) with rudimentary processing capabilities (for example, 8-bit processors). It was not deemed capable of providing the functionality needed to manage large networks of complex elements (for example, COs).

3. **OSI** (Open System Interconnection) from the International Standards Organization (ISO). OSI was recognized as the emerging framework that satisfied all the criteria listed earlier and was therefore adopted as the foundation of TMN. At that time, however, ISO had no readily implementable standards specifying, for example, how an OS could remotely reconfigure a CO. And so along with its adoption came the realization that embracing OSI was going to be the first step of a long journey.

In order to ensure that TMN had a broad, international constituency and that its focus was on telecommunications, it was entrusted to the International Consultative Committee for Telephone and Telegraph (CCITT), which eventually became the International Telecommunications Union–Telecommunications standardization sector (ITU-T). In order to promote more intensive local efforts and customized specifications, regional standards organizations undertook to assist and complement the ITU-T efforts. These included most notably: Standards Committee T1, under ATIS and with accreditation by the American National Standards Institute (ANSI) in North America, and the European Telecommunications Standards Institute (ETSI) in Europe. The Tele Management Forum (TMF), formerly Network Management Forum (NMF), an industry consortium, soon integrated the emerging TMN standards, each addressing a small piece of the puzzle, into coherent network management solutions, with broad, international industry participation. In the United States, the Electronic Communications Implementation Committee (ECIC) was created, under ATIS, to provide complete and detailed implementation agreements for specific TMN applications between LECs and Interexchange Carriers (ICs).

An awesome library of TMN standards and implementation agreements has now been made available for supporting a broad spectrum of network management applications. Quite a few applications have already been implemented and deployed. Yet a lot more remains to be accomplished before TMN can be declared a success. The task is made all the more challenging by evolving business needs that require ongoing readjustment and fine-tuning of the TMN.

As early TMN applications were being deployed, it became clear that the impressive power and flexibility of OSI-based TMN carried an equally impressive price tag. Two lower-cost alternatives were therefore devised for applications with more modest requirements:

- Electronic Data Interchange (EDI) was inducted into the TMN to support applications that call for simple document exchanges (for example, placing a service order).
- Common Object Request Broker Architecture (CORBA) was added to the TMN, only for some applications at present, on the assumption that it would soon be freely available on every computing platform.

As we extrapolate our story into the future, all the preceding may become history. The reason is the explosion of data communications. In the early 1980s, data communications was riding on the voice network like a starling hitching a ride on the back of an elephant. Since then data communications, especially of the TCP/IP variety, has grown by leaps and bounds. Many industry gurus are predicting that by the year 20xy there will be only one network protocol and it will be IP. (As of this writing the value of xy is very much guru-dependent.) IP networks are being managed with the Simple Network Management Protocol (SNMP). As IP networks are poised to overtake all of telephony, calls are being heard to incorporate SNMP into TMN.

Although network planners crave the simplicity of one IP fits all, operation planners shudder at the prospects of having to deal with the rudimentary SNMP (it really is simple) for their growing charges. When IP networks were tiny (compared to telephony networks),

SNMP was ideal. As SNMP is applied to monumentally larger tasks, it may prove a bit too simple. Most likely, when SNMP becomes part and parcel of TMN, the ITU-T (as keeper of TMN) and the IETF (as keeper of IP and SNMP) will join forces to fortify SNMP for its higher calling. At this point, our futuristic history of TMN is overtly speculative. Nevertheless, since SNMP in TMN is a virtual reality, it is addressed in this book.

A glutton for protocols may claim that if one network management protocol is good, many are better. Eventually, however, the riches become downright embarrassing. Indeed, the proliferation of interface protocols detracts from interoperability. It is expected, however, that different protocols will be used for different applications, thus allowing for a better match between business needs and underlying technology.

2

Architectural Views of the TMN

Amalgamating all the details of a large, complex building in a single blueprint would result in a useless mess. Much clarity is achieved by providing a simple floor plan for the masons, a separate wiring diagram for the electricians, a plumbing layout for the plumbers, and a heating, ventilation, and air conditioning schematic for the duct workers. Similarly, TMN can be most clearly understood in terms of three complementary architectural views:

- Functional architecture
- Physical architecture
- Communications/information architecture

These architectures address what, where and how, respectively, the TMN's mission gets done.

While the floor plan, wiring diagram, plumbing layout, and duct scheme can be depicted on separate sheets of paper, their compilation is highly interactive: the layout of the plumbing pipes, the routing of electrical wires, and the placement of the venting ducts impose obvious physical constraints on each other. Similarly, the construction of the TMN architectures has been proceeding along simultaneous, interacting threads. Thus, while the final architectures can be represented independently of each other (well, almost), understanding some of the decisions leading to any one of them involves unavoidable cross references to the others.

2.1 FUNCTIONAL ARCHITECTURE

Functional architecture is the starting point of TMN. It enumerates all the application functions that the TMN supports. Indeed, it is prudent to specify what we are building before we go ahead and build it. Those functions are enumerated in ITU-T Rec. M.3400 [M.3400] as standardized **Management Application Functions (MAFs).** At the highest level, such enumeration would consist of a single entry: manage a telecom network. This is quite correct and concise, but not the least illuminating; a much more detailed breakdown is needed. The exact level of detail that is needed will become clear when we explore the information architecture. It turns out that the rather voluminous M.3400 is only a preliminary, though crucial, step toward what would be an exhaustive specification of TMN functionality. Yet, M.3400 already contains hundreds of MAFs. Rather than list those MAFs in random or alphabetical order, the functional architecture groups them into function sets of related MAFs. Since the number of function sets is still fairly high, these are partitioned into five standard **Management Functional Areas (MFAs): Configuration Management (CM), Fault Management (FM), Performance Management (PM), Accounting Management (AM),** and **Security Management (SM).** The five TMN Management Functional Areas are distinguished by the kinds of activities they perform.

Bellcore's GR-2869 [GR2869] introduces another, complementary classification of the function sets into five logical layers known as the **Logical Layer Architecture (LLA).** The layers are differentiated by the kinds of things the activities affect. The five functional layers are: **Business Management Layer (BML), Service Management Layer (SML), Network Management Layer (NML), Element Management Layer (EML),** and the **Network Element Layer (NEL).**

Figure 2.1 depicts the five Management Functional Areas and five functional layers of a TMN with a few illustrative entries.

The differences, and complementarity of the two classifications can be illustrated with a few simple examples: The decision to offer a new service is part of BML CM. Similarly, specifying the performance parameters and the price of the service is within the BML's PM and AM, respectively. Handling a customer service request for the service is part of the SML CM. Informing the customer of the service performance and rendering a bill are within the SML's PM and AM, respectively. Configuring the network to support the service (for example, by allocating certain capacity to some routes) is part of NML CM.

This structured enumeration of TMN functions is intended to help a telecommunications network manager identify the management needs of new services. It also serves as a means of conveying those needs to its telecommunications equipment and operations systems

Layers	Management Functional Areas				
	Configuration Management	Fault Management	Performance Management	Accounting Management	Security Management
BML	Decide on a new service	Decide on repair priorities	Decide on service parameters	Decide on service pricing	Formulate security policy
SML	Process service request	Process customer trouble reports	Monitor service performance	Negotiate price with customer	Manage certification paths
NML	Map the service into network nodes	Root cause analysis	Monitor network performance	Collect and correlate usage data	Internal key distribution
EML	Configure network elements to support the service	Manage NE testing	Instruct NEs on reporting of performance data	Instruct NEs on collecting usage data	Manage security audit trails in network elements
NEL	Respond to EML configuration requests	Respond to EML testing requests	Respond to EML requests for reporting performance data	Respond to EML requests for collecting usage data	Respond to EML requests for security audit trail changes

Figure 2.1 TMN Management Functional Areas and Layers—illustrative entries

providers, either external or in-house. The TMN framework also emphasizes the use of standard interfaces between the telecommunications network and operations systems and among operations systems.

Although the MAFs proclaim what the TMN does for network management, doing it requires additional functionality, such as data communications networks, database management systems, and secure communications. Bellcore's GR-2869 defines this functionality as Common Operations Management (COM). It refers to generic functions that provide common support for all Management Functional Areas across all TMN layers. In addition, it includes some functionality that appears on the boundary of the TMN. Figure 2.2 depicts this functionality. *(F, X,* and *Q* in the picture correspond to interfaces to a WS, to an entity outside the TMN, and to an OS or an NE, respectively.) Much of this functionality corresponds to that of the functional components of a TMN Operation System Function (OSF).

The dichotomy between MFAs and COM is especially helpful in clarifying the role of security in the TMN. The Security Management MFA represents the TMN's role in managing security information, including security information for the Telecommunications Network—for example, security for wireless communications and security for network signaling. The Security Function COM refers to the capabilities needed to ensure the security of the TMN.

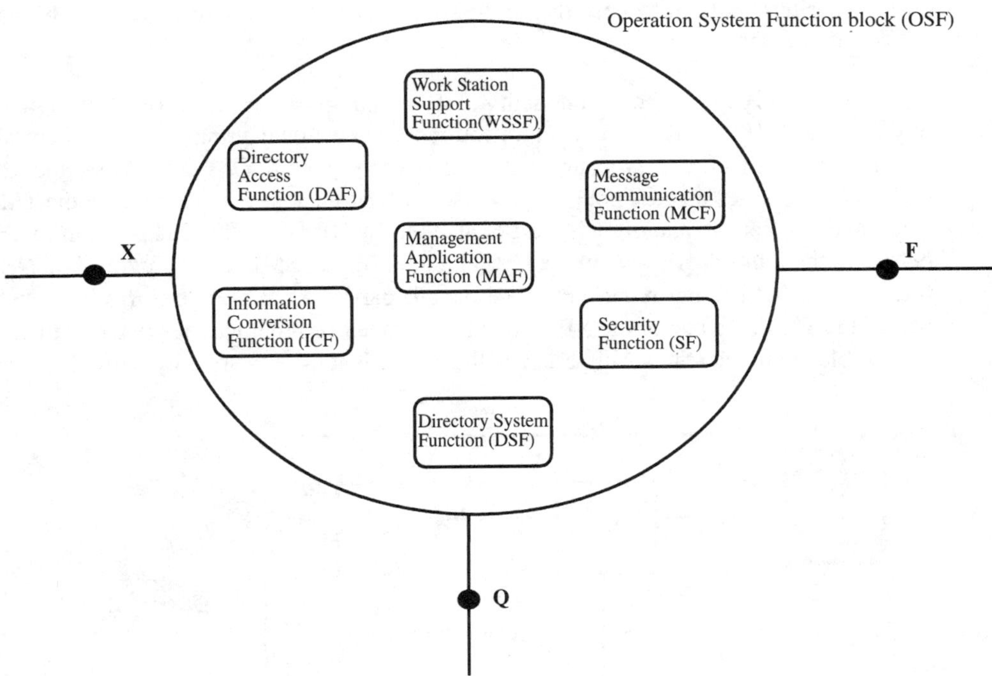

Figure 2.2 Functional components of an Operation System Function block

2.2 PHYSICAL ARCHITECTURE

Interoperability is the TMN's major objective. Interoperability between two systems is achieved by specifying all the messages that those systems can exchange, along with the semantics of those messages. This allows independent developers to produce those systems, using whatever means, tools, and technology each sees fit. As long as each system is capable of sending and receiving the specified set of messages, those systems will interoperate. Of course, we need to know what those two systems are up to and why they need to communicate before we can specify the messages that they can exchange. Thus, an OS that helps reconfigure COs will use a different set of messages from an OS that manages access control privileges. Still, some aspects of the communications between various systems are quite similar: both OSs in our example need some transmission facilities to communicate with their interlocutors, and both need to do so securely.

Although some needs are common to many systems, some classes of systems may have their special requirements. For instance, in securing the communications between two systems, we can easily distinguish the following cases:

- When an OS communicates with a NE or another OS, the security of the communications can be simplified by requiring that all OSs and NEs within the TMN support the same set of encryption algorithms and recognize the same, local Certification Authority (CA). Furthermore, part of the security can be based on the fact that OSs and NEs don't wander around, so they keep their network addresses.
- When an OS communicates with a WS, it cannot rely on the WS's physical address for security since WSs can be quite mobile.
- When an OS communicates with an external entity, it may need to resort to a broader range of encryption algorithms in order to accommodate the external entity's peculiarities, and it may need the ability to recognize foreign CAs or to process certification paths.

In order to capture the commonality of the communications needs of TMN systems, the physical [M.3010, T1.210] architecture introduces functional groupings that share similar communications facilities. Not surprisingly, they correspond to OSs, NEs, WSs, and external entities (i.e., entities that are not part of TMN, but that interact with OSs inside the TMN; an external entity can be, and often is, part of another TMN). Typically, OSs perform MAFs that belong in the four upper layers in the logical model (BML, SML, NML, and EML). NEs perform MAFs that belong in the NEL. The (small) part of a NE that performs NEL MAFs is part of the TMN; the rest of the NE, which provides telecommunications functionality, is part of the Telecommunications Network, not the TMN. Just as NEs are only partially within the

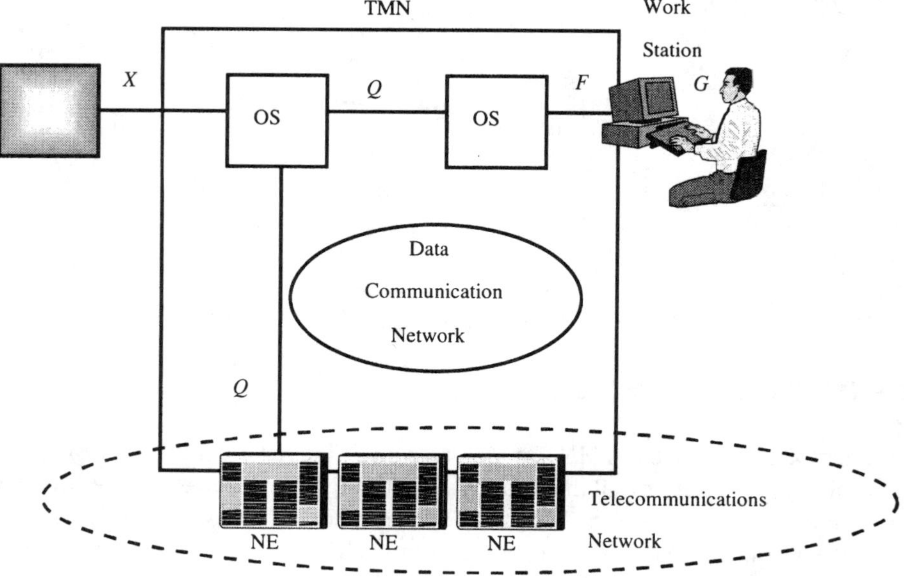

Figure 2.3 Example of a TMN physical architecture

TMN, so are WSs. The part of the WS that provides network management functionality belongs in the TMN; the part of the WS that interfaces with the user—for example, providing a GUI (Graphical User Interface)—is outside the scope of the TMN.

To complete the picture, the physical architecture also includes a Data Communications Network (DCN) that allows the TMN elements to interact. The resulting physical architecture is exemplified in Figure 2.3.

Figure 2.3 shows the various interfaces that are defined by the physical architecture:

- The *Q* interface between an OS and another OS or a NE, inside the same TMN.
- The *F* interface between an OS and a WS, inside the same TMN.
- The *X* interface between an OS and an external entity.

The figure also illustrates the *G* interface between a WS and a human user. The *G* interface, however, is outside the scope of the TMN.

2.3 COMMUNICATIONS/INFORMATION ARCHITECTURE

The TMN communications architecture [Q.811, Q.812, M.3010, and T1.210] is based on the OSI reference model [X.200]. The primary goal of the model is to define standard interfaces between systems with different internal architectures in order to promote interoperability. A straightforward specification of such interfaces can result in an instant dinosaur: if it were to specify, for example, X.25 [X.25] packets over DS0 over copper wires, it would be useless for TCP/IP packets, or DS1 transport, or optical fiber physical medium. (This example is based on the lower layers of the OSI model, illustrated in Figure 2.4.)

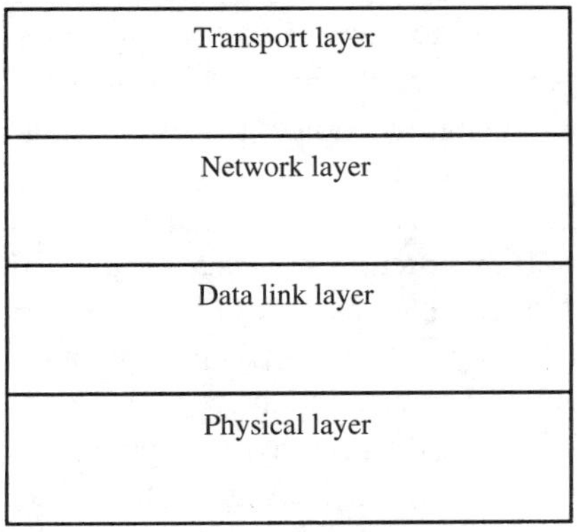

Transport layer
Network layer
Data link layer
Physical layer

Figure 2.4 The lower layers of the ISO model

In order to accommodate the competing needs of interoperability and flexibility, the ISO model introduces the concept of layering. Since this concept will be frequently reused in TMN security, a brief discussion is provided here.

In the simple illustration above, we want, for example, to send IP packets over a network, connecting two end points of that network. To that end, the ISO model defines a network layer, responsible for getting IP packets across the network. The network layer uses services provided by a link layer. The link layer carries the IP packets across a link between two nodes in the network. At each intermediate node, the network layer decides which link shall be used next and again trusts the corresponding link layer with transport over that link. The service the link layer provides to the network layer is independent of the bit rate on that link. Thus, the network layer need not know if a DS0 or a DS1 link is used between the two nodes. The OSI model specifies the services provided by the link layer to the network layer, though it does not specify the syntax of the messages—called **Interface Data Units (IDUs)**—exchanged between the two layers. Indeed, this is a local implementation matter. The system developer may even combine the functionality of the two layers in a single module (though software mavens counsel against it). The payload of the IDU, called a **Protocol Data Unit (PDU),** consists of the bits that the network layer wants moved to the next node, along with some additional, optional, **Interface Control Information (ICI)** that tells the link layer which services are needed (error correction, for example).

Similarly, the link layer uses the services of the physical layer to encode the bits and move them from one node to the next. The link layer does not know or care if the bits are encoded as amplitudes of a microwave signal, electrons in a wire, or photons in a fiber. So it is that the link layer uses the services provided by the layer beneath it to provide services to the layer above it.

The whole process needs to be replicated, in reverse, at the next node. The physical layer receives microwaves, electrons, or photons and converts them into a stream of bits that it delivers to the link layer. The link layer may check for any transmission errors (if such service is supported) and deliver its payload to the network layer above it, similar to the payload entrusted by the network layer to the link layer at the originating node.

When the receiving network layer gets its PDU, it must know what to do with it, for example, where to route it. This information is contained in the header portion of the PDU it receives, called **Protocol Control Information (PCI).** The remainder of the PDU constitutes the **Service Data Unit (SDU);** it is the payload that the network layer carries on behalf of the transport layer, the layer immediately above the network layer.

IP specification provides a complete and detailed definition of its PDUs. In particular, it fully defines the PCI for each PDU. Similarly, X.25, another network layer packet switching protocol, specifies its PDUs and PCIs. The link layer does not need to know if the network layer to which it offers its services uses IP, or X.25, or any other protocol. Thus, each layer provides services [X.210] to the layer above it without knowing how or why that layer uses those services. Each layer (except the physical layer) uses the services of the layer below without knowing how those services are provided. This remarkable decoupling between layers allows the introduction of new technologies in one layer without affecting any of the other layers.

The salient features of the layering concept are illustrated in Figure 2.5. It shows that layer N + 1 in system A wishes to send a SDU to its peer layer in system B. It appends a PCI (header) to the SDU (payload) to form a N + 1 PDU destined for the remote N + 1 layer. It appends ICI to the N + 1 PDU to form an IDU that it passes to layer N in the same system. It is instructive to note that layer N + 1 adds two headers to the SDU: the PCI is appended for the benefit of the peer layer on the remote system, while the ICI contains instructions for the underlying layer in the same system. Layer N removes and processes the ICI portion of the IDU in order to decide what to do with it. The remaining portion of the IDU (i.e., the N + 1 PDU) is the SDU (payload) that layer N needs to send to its peer in system B; and the whole process repeats. The standards that define each layer specify the semantics (but not the syntax) of the ICI; that is, they specify the services that the layer provides and what kind of information it exchanges with the layer above it. The standards further specify the syntax and semantics of the PDUs that can be exchanged between peer layers in different systems. A crucial aspect of layering is that it defines the syntax of the PCI, which is processed by the peer layer in the remote system, but not the syntax of the ICI, which is processed by the adjacent layer in the same system.

Since layering promotes flexibility, it may seem that the more the merrier: each layer would provide only limited functionality; therefore, improvements could be made by tweaking only a small bit at a time. True. However, each layer adds its own PCI (header), thereby increasing transmission overhead. A proliferation of layers would also impose a higher processing cost. Balancing flexibility against efficiency, ISO produced the seven-layer OSI com-

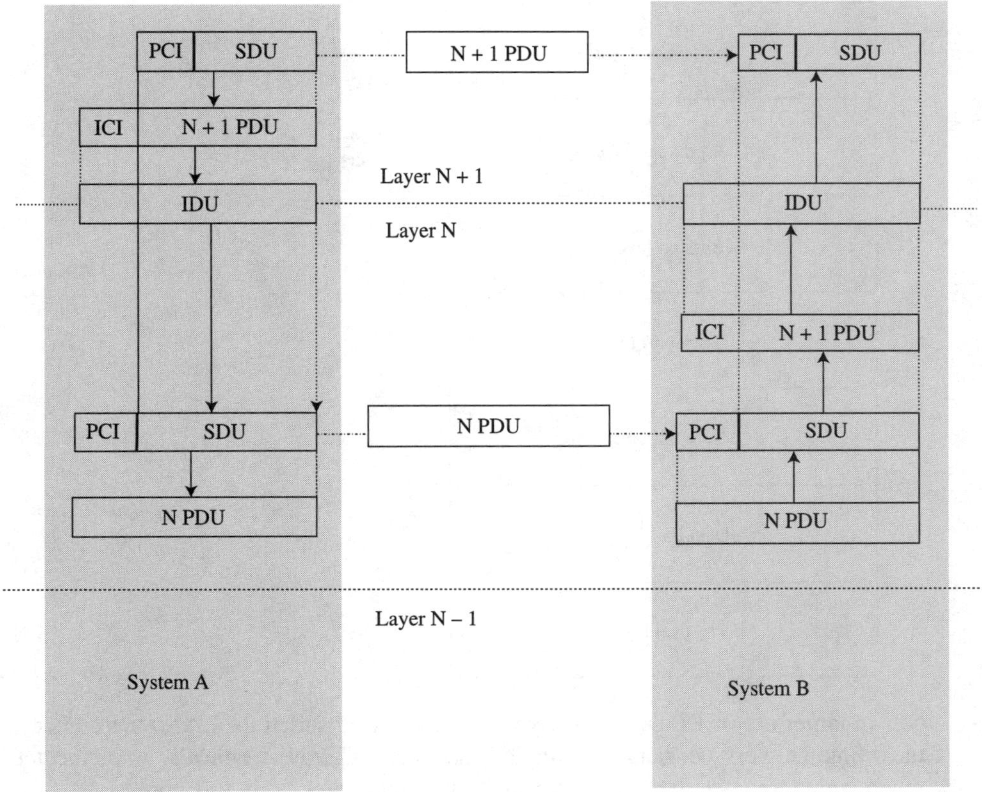

Figure 2.5 OSI layering

munications protocol stack. It also allows further fine-tuning: if the services of one layer are not needed, they can be reduced to an almost symbolic minimum. At the other extreme, if the services needed from a given layer proliferate, the layering principle can be applied within that layer. For instance, "sublayers" have been defined within some layers to provide security services. The seven layers of the OSI model are depicted in Figure 2.6.

The four **lower layers** of the OSI protocol stack ensure reliable end-to-end data transport. They are discussed first.

2.3.1 Physical Layer

The physical layer describes the actual medium (i.e., electrical or optical signals) used to transmit information between two nodes. Examples of physical layer protocols are T1 electrical connections or optical OC-3 SONET connections. Physical layer security is provided through hardware encryptors at each end of the link. Because there are at present no generally accepted standards for such hardware encryptors, and encryptors from different suppliers do not usually cooperate, there is no further discussion in this book of physical layer security.

2.3.2 Data Link Layer

The data link layer is responsible for the transmission, framing, and error control needed over a single communication link. Examples of link layer protocols are: Ethernet or Token Ring in a LAN (Local Area Network) environment or the LAP-B and LAP-D proto-

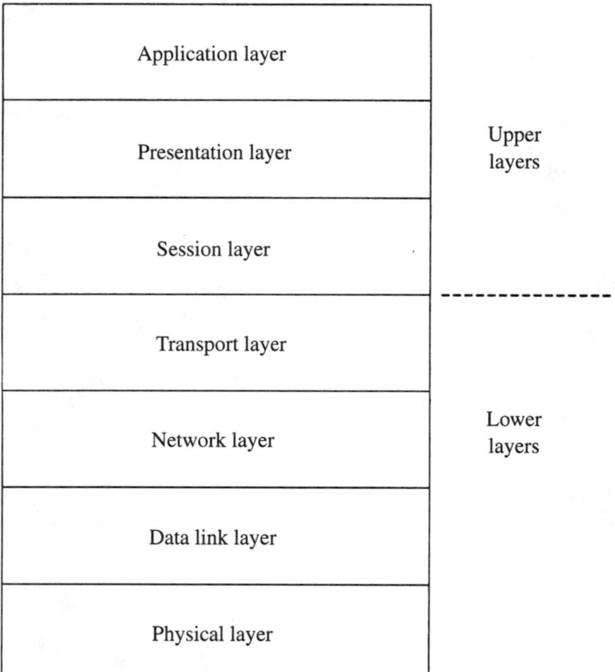

Figure 2.6 Seven-layer OSI stack

cols in larger networks. Data link layer security is well suited for LAN environments, where the whole network consists of a single link. Since a TMN is typically implemented over a much larger area, the data link layer security is not addressed in this book.

2.3.3 Network Layer

The network layer provides data transmission over a network and is independent of both the media and topology of the underlying subnetworks. At each intermediate node within the network, the network layer is responsible for routing: picking the next link over which the message should be carried, and using the services of the data link layer over the selected link. A secure network layer provides security across the network. The most common examples of network layer protocols are X.25 and the Internet Protocol (IP). Network layer security includes controlling access to the network through Closed User Groups (for X.25) and firewalls (for IP). Both are addressed in Part 3 of this book. Network layer security can also protect information as it is carried across the network through the Network Layer Security Protocol (for OSI) and IPsec (for IP). However, such network layer security can protect only whole data streams between end systems. In general, TMN applications require finer granularity of protection, for example, providing different types of protection for messages related to different applications on the same end systems.

2.3.4 Transport Layer

The network layer ensures that each packet reaches its destination. However, it can route different packets along different routes (depending, for example, on changing network traffic loads). Therefore packets may arrive at the destination in a different order from the order in which they were originally sent. One of the services of the transport layer is to turn chaos into order and ensure that the packets are delivered to the end system in the correct sequence. More generally, the transport layer provides reliable transfer of data between end systems across the network. It further allows data from different applications on a system to be multiplexed into a single connection. Transport layer services ensure that data was not corrupted in transit and that messages are delivered to the receiving system in the same order in which they were submitted by the sending system. The most common examples of transport layer protocols are the Transport Control Protocol (TCP), which along with IP forms the TCP/IP protocol suite that is the foundation of today's Internet, and the TP0-TP4 protocols defined by ISO for OSI. Several essential security services (authentication, data integrity, and data confidentiality) can be provided within the transport layer, as will be seen later in this book.

The three **upper layers** of the OSI stack deal with communication functions carried out by the end systems, independently of the interconnecting network.

2.3.5 Session Layer

When two people talk over the phone (or even face to face), communication is much improved if one person listens while the other talks. In general, people can tell when the other person has finished talking (as opposed to pausing for breathing) and when it is okay to interrupt. The session layer enforces similar etiquette for computer communications: it keeps track of when it is okay to transmit data. For many TMN applications, both communicating systems can send and transmit simultaneously. For such applications there is not much left for the session layer to do. The session layer does not support any security services.

2.3.6 Presentation Layer

One of the explicit goals of TMN and the OSI model is to support communications between open systems with different internal representations of data (e.g., ASCII and EBCDIC). The presentation layer supports this goal by converting any local representation into a common, agreed-upon representation. It also allows the two communicating systems to negotiate what kind of data they intend to exchange in the course of the connection and which common representation they will use. The presentation layer does this through presentation context management and syntax matching. **Context management** ensures that the parties agree on what information will be exchanged (for example, customers' telephone numbers) and how it shall be represented (for example, sequences of 10 decimal digits). The transfer syntax used in TMN [X.208, X.680, X.681, X.682, X.683, and X.690] represents each field as a triplet (type, length, value); for example, a customer name may be transmitted as:

- Type: string of printable characters
- Length: 21 octets
- Value: encoding of the customer name's 21 characters

Syntax matching is an agreement between the parties on the transfer syntax used to encode the messages (for example, that each decimal digit shall be encoded as two's complement). The presentation layer takes data from any application and structures the data to be sent over the network in a format that can be understood by the receiving system.

2.3.6.1 Distinguished Encoding Rules. One aspect of common data representation that is especially important to security has to do with the uniqueness of the transfer syntax. It is best introduced with an example of lists encoding.

A SEQUENCE is an ordered list. When a new element is added to a SEQUENCE, it must be inserted in a precise location, typically in a lexicographical order. If the order of the elements in a list has no special meaning, then it may be simpler to organize the data as a SET.

Thus, a system may keep a SET OF user IDs of all users currently logged in. One system may actually store that list in lexicographical order, whereas another system may store such information in the chronological order of the logins. In either case, the order is a matter of local convenience with no significance.

The most commonly used encoding rules, the Basic Encoding Rules (BER), explicitly support the freedom of ordering elements in a SET. When an information element of type SET is encoded in BER, the elements in the SET can be arranged in any order. This laudable flexibility causes a serious security problem. Indeed, if a digital signature follows a SET in order to provide a proof of non-repudiation for the SET, then the ordering of the elements within the SET is crucial. The digital signature will be valid only if the elements in the SET are kept in the same order as when the signature was computed. To remedy this, and similar shortcomings, a few constraints can be added to the BER in order to ensure unique encoding. The result is the set of Distinguished Encoding Rules (DER). It is common practice to use DER, rather than BER, when security transformations are applied to PDUs.

2.3.7 Application Layer

The application layer interfaces directly with the user's application. It allows an application (for example, a switch configuration management application) residing on one system to form an association with a corresponding application on another system, so that the two applications can communicate.

The application layer is best introduced through some examples of services it provides to the applications that use it:

- Association setup and termination
- Peer entity authentication during association setup
- Protection for messages
- Matching multiple, interleaved responses to their respective queries
- Directory access
- Remote file transfer and management

Since the application layer supports a rich and expanding functionality, it is practical to decompose it into several **Application Service Elements (ASEs).** ASEs are self-contained modules of software, which reside in the application layer. Each ASE performs specific functions needed in communications between two systems. The general structure of an ASE, as shown in Figure 2.7, reuses the basic concepts of the layering architecture. Each ASE has a mode of operation that can be represented as a finite state machine; it offers services to its user(s), and it communicates with a peer ASE, on a remote open system, using a protocol spe-

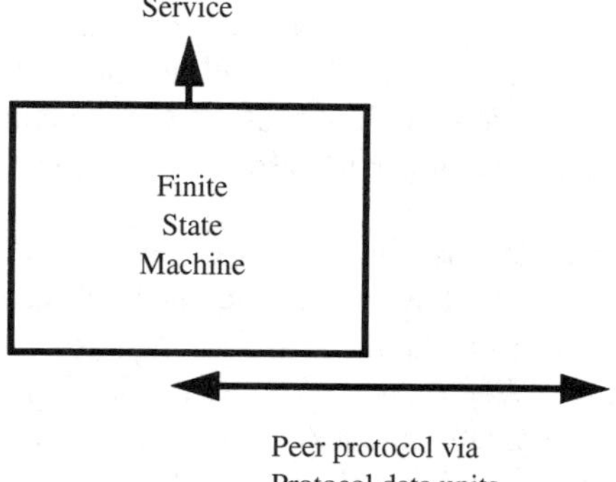

Figure 2.7 Makeup of an Application Service Element

cific to that ASE. As in the case of layering, the OSI standards specify the ASE-to-ASE protocol and the services that an ASE offers, but not the messages that it exchanges with its users.

The services that an ASE offers to its users are described as service primitives. Typically, a service primitive requested by a user results in a PDU transmitted from the originating ASE to its peer ASE on the remote system. The receiving ASE then presents an "indication" to its user. The indication contains the information that the originating ASE user intended for the terminating ASE user, in a format that is usable by the receiving ASE's user. If the protocol between the two peer ASEs calls for a reply, then the user receiving the indication requests its ASE to send a response. Again, this request results in a PDU to the remote ASE. Upon receiving the response PDU, the ASE presents a "confirmation" to its user. The exchange is often represented with a time sequence diagram as illustrated in Figure 2.8.

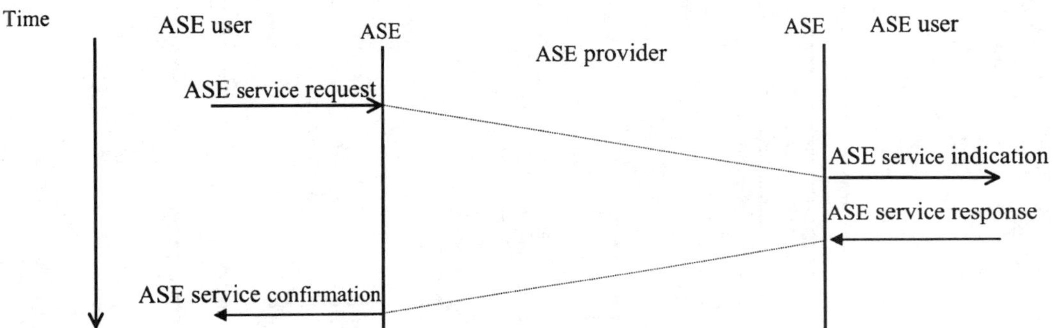

Figure 2.8 ASE service primitives

An application may use several ASEs to provide the communications needs of that application. For instance, one ASE may be used to establish an association, whereas another ASE may be used to access a directory over the established association. A list of such ASEs defines an Application Context.

A mere collection of ASEs without a leader is like a group of musicians without a conductor. The role of conductor for ASEs within an Application Context is provided by a **Control Function (CF).** The CF specifies how the ASEs interact with each other (i.e., which ASE uses the services of another ASE) and the order of their invocation. For instance, first the ASE responsible for association control is invoked to set up an association (this is always the case); then a directory access ASE may be invoked to query a directory; and finally the association control ASE is invoked again in order to terminate the association (this is the case for every normal association termination). Unlike the ASEs, the CF does not exchange information with any module in a remote system; it is strictly a local function.

An ASE's peer entity protocol may encompass several options. For instance, the ASE responsible for association control may or may not exchange information needed for mutual authentication. In order to interoperate, two communicating systems must agree on the options that will be supported by the ASEs that they intend to use. The CF and the ASEs that support an application, along with the communication protocols of the ASEs, constitute an **Application Entity (AE).**

An AE, along with the processing of the information exchanged through the ASEs, constitutes an **Application Process (AP).** The generic constituents of the application layer are illustrated in Figure 2.9.

In principle, a single AP can have several AEs. For instance, an application may need to exchange authentication information when interacting with some directories but not when

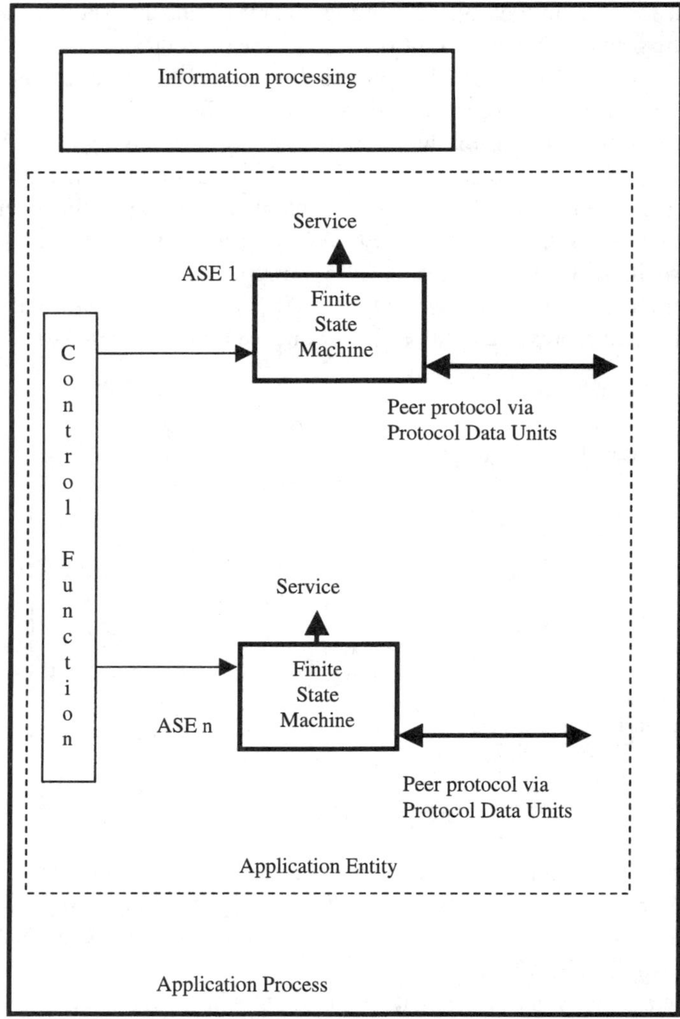

Figure 2.9 Application layer constituents

interacting with others. In practice, however, each AP has only a single AE. In such cases, it may not be necessary to specify the AE when an association is established.

The purpose of an AE is, of course, to support the exchange of information needed by some application. For instance, one of the ASEs in an AE may collect information regarding network troubles; the processing of that information analyzes such data and pinpoints the underlying cause. In general, however, the software that performs the information processing (i.e., the actual network management application) can be provided independently of the underlying communications protocol stack. The application uses the communications stack through a well-defined **Application Programming Interface (API)** and need not be aware of how the information is being exchanged. For this reason, this book views network management applications as residing above the seven-layer protocol stack and not as part of any AEs that are inside the stack in the application layer.

The ASEs currently used in the TMN are introduced below.

2.3.7.1 ACSE (Association Control Service Element). ACSE [X.217, ISO8650] supports the establishment and tear-down of associations between applications. Its primary purpose is to allow the two applications to identify themselves and to negotiate which set of ASEs they will use in the course of the association. The identification can be accomplished by exchanging the names of the calling and called AP and AE, as well as the specific invocation of each AP and AE (useful if two systems establish different associations between different invocations of the same APs and AEs). The exchange of AP- and AE-related information is, however, optional. The negotiation is accomplished by exchanging the name of the Application Context (i.e., collection of ASEs) that will be used during the application. The association initiator starts by proposing an Application Context, along with (optionally) a list of alternates. The responder can respond with the proposed Application Context, with an Application Context from the list of alternates, or (if no list of alternates is provided by the initiator) any other Application Context. If any one of the communicating systems does not like what it gets, it aborts the association.

ACSE also allows the ASEs in the Application Context to exchange initialization information. It simply includes a field for each of the ASEs that can carry whatever information that ASE needs to convey to its remote peer at association setup time. A security ASE can use this facility to negotiate security parameters (encryption algorithms, acceptable Certification Authorities, . . .) with its remote peer.

ACSE has an optional authentication Functional Unit that allows the communicating entities to exchange authentication information.

2.3.7.2 ROSE (Remote Operations Service Element). A complex query may result in multiple replies. For instance, a request for the names of all the users who have updated routing tables in any Network Elements during the last month may result in many replies as the names of those users are being collected. While those replies are arriving, the requesting entity may issue other complex queries (for example, a list of all the users currently logged in to Network Elements). The responses to the various queries may be randomly interleaved, as illustrated in Figure 2.10. Such chaos could drive an ASE to distraction. ROSE [X.219-88, X.219-93, X.229] helps to reestablish order.

ROSE appends an Invoke-ID field (supplied by the ROSE user) to each request; the same Invoke-ID is appended to all the resulting responses. The ROSE user uses those Invoke-IDs to match requests with responses.

2.3.7.3 STASE-ROSE. Security Transformations Application Service Element for Remote Operations Service Element (STASE-ROSE) [T1.259, Q.813] provides security for

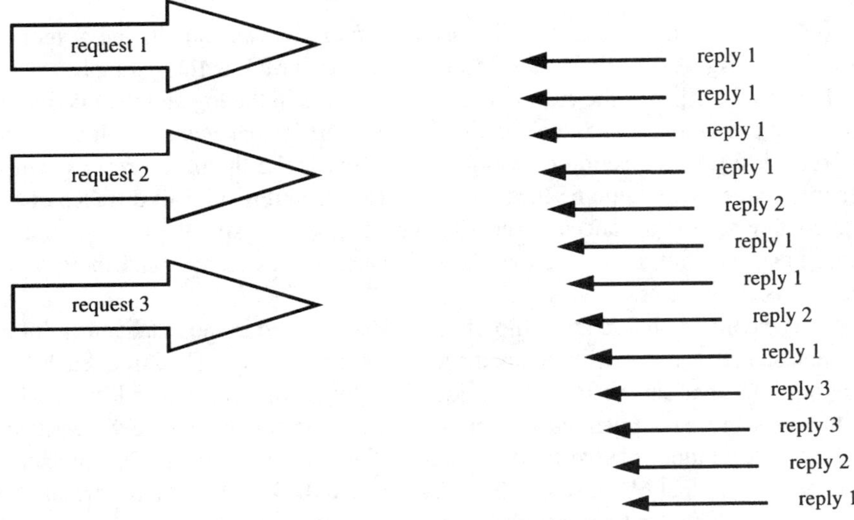

Figure 2.10 Interleaved replies to multiple requests

ROSE PDUs. STASE-ROSE supports a complete set of security transformations on ROSE PDUs, providing integrity, confidentiality, and non-repudiation. STASE-ROSE further supports strong peer entity authentication and negotiation of security parameters.

2.3.7.4 FTAM (File Transfer Administration and Maintenance). Reliable transfer is typically provided by appending a checksum to a message and verifying that the checksum at the receiving end is the same. In the case of a large file transfer, which may consist of hundreds of megabytes, this may not be practical. Indeed, if a single transmission error has occurred, the whole file will need to be retransmitted. It is more efficient to provide numerous checksums throughout the file and retransmit only the portions that have been corrupted in transit. FTAM [ISO8571] provides such facilities for file manipulations, as well as offering the remote user an unambiguous view of the file, regardless of local representation and storage mechanisms.

2.3.7.5 X.500 Directory User Agent. The X.500 Directory User Agent (DUA) [X.500] allows an application to access an X.500 Directory (i.e., a general-purpose directory as defined in ITU-T Rec. X.500) in order to retrieve information (such as naming and addressing information) for systems with which the application wishes to communicate.

2.3.7.6 CMISE (Common Management Information Service Element). CMISE [X.710, X.711] provides the communications services to be used for network management (such as retrieving or setting a parameter). In order for two systems to cooperate successfully on the management of a complex NE, such as a Central Office (CO), they must have a common, shared view of the NE, its components, and the kind of management activities that are pertinent for each component. The shared management knowledge is represented in an object-oriented method [X.720, X.721] through an information model. A Management Information Model (MIM) consists of a set of Managed Object (MO) classes. An instance of a MO class can represent some managed resource. A MO typically includes attributes that describe the resource (for example, serial number, date installed, status), notifications that it can issue (for example, a security alarm if there is an unauthorized attempt to change the value of an attribute), and actions that can be requested from that MO (for example, test the transmission link with a 2-kHz carrier and report the result). MO classes are defined using the standard Guidelines for Definition of MO (GDMO). Their salient features are described next.

2.3.7.6.1 CONTAINMENT. An instance of an MO can contain one or more instances of other MOs. Figure 2.11 illustrates a security audit trail MO. In this example, the security audit trail has two attributes: the current number of records in the log and the maximum number of records it can hold. It can issue a notification to alert its manager when it is full, and it can be instructed to take the action of wrapping (i.e., overwrite the oldest records with the newest records) rather than dumping the newest records when it is full. An instance of this MO class can contain several instances of security audit trail record MO class. Each security audit trail record can have attributes such as the time when it was created and the type of event that caused its creation.

The containment relationship of MO instances used in the management of a system can be represented as a tree, with the managed system at the root of the tree. Each MO must have one attribute that can be used for uniquely distinguishing it from all other MOs contained within the same MO. Each instance of an MO can therefore be uniquely identified through the sequence of unique relative (with respect to the containing MOs) distinguishing identifiers from the root to that MO instance. The tree is called the Management Information Tree (MIT). Figure 2.12 illustrates such a tree. The management information tree illustrated in this figure is based on a set of rules, specifying, for example, that instances of a security audit trail record

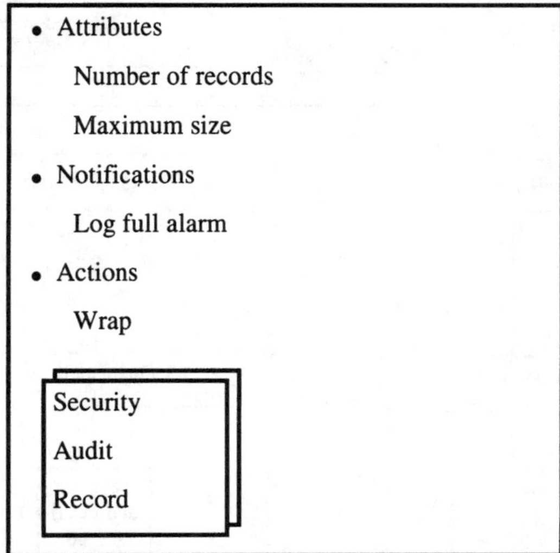

Figure 2.11 A security audit trail MO

MO can be contained inside an instance of a security audit trail MO. The set of rules specifying the allowed containment relationships is called a containment or naming relationship. Figure 2.13 illustrates the containment relationship corresponding to the MIT in Figure 2.12.

Notice that the nodes in Figure 2.13 are MO classes, whereas the nodes in Figure 2.12 are instances of those classes. Although the containment relationship illustrated in Figure 2.13 is a tree, this is not always the case; the containment relationship can be any directed graph.

2.3.7.6.2 Addressing MOs. CMISE allows several ways of addressing MO instances:

- Naming an MO instance by specifying its unique identifier
- Specifying an MO class, thereby referring to all the instances of that MO class within a MIB

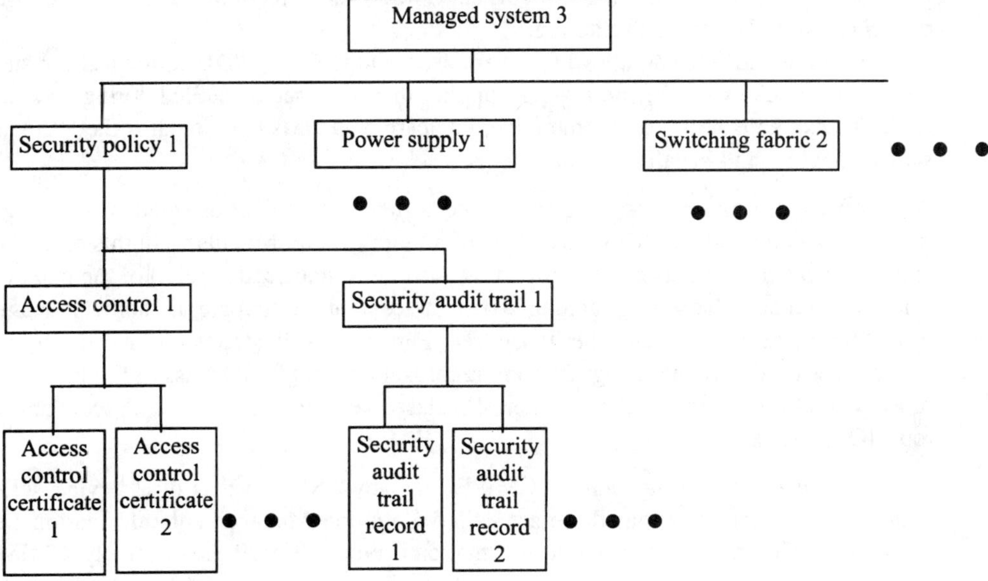

Figure 2.12 Example of a management information tree

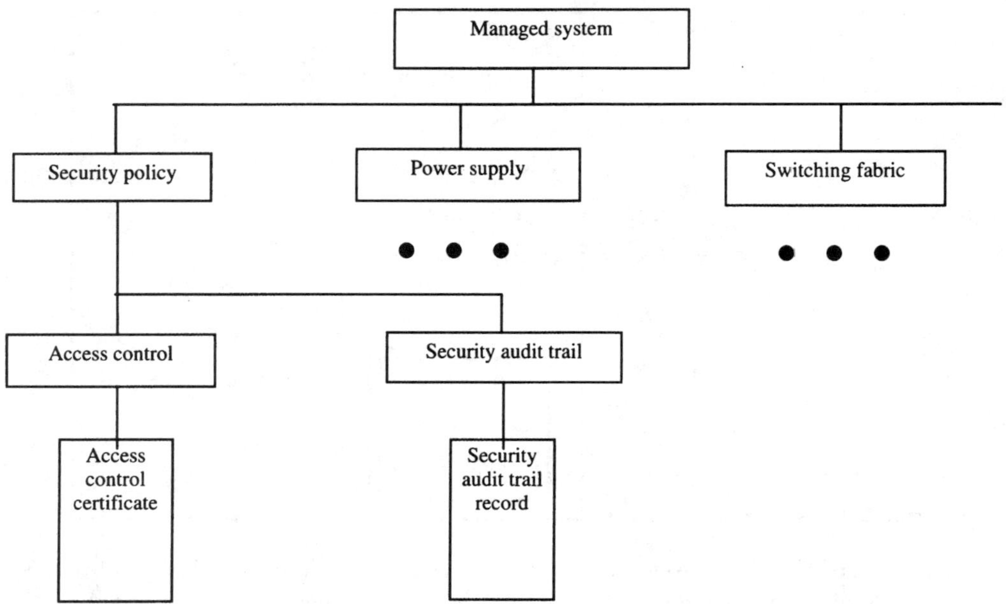

Figure 2.13 Example of a containment relationship

- **Scoping**—consists of specifying the name of an MO instance as the root of a subtree of the management information tree and specifying that
  - all the n-level nodes of that subtree are being addressed
  - all the leaves of that subtree are being addressed
  - all the nodes of that subtree are being addressed

CMISE further allows specifying any subset of the nodes in a scoping subtree through **filtering.** A CMISE filter is any Boolean expression on the attribute values of an MO instance; only MO instances within the scoping subtree whose attribute values conform to that Boolean expression are selected by filtering. CMISE also allows selection of any subset of the instances of a given MO class through filtering.

Scoping and filtering allow, for example, a single CMIP PDU to request the status of all the line cards in a CO from a given supplier that have been installed during a certain period. The example assumes that there is a line card MO class that contains the attributes for status, supplier, and installation date.

2.3.7.6.3 Inheritance. In accordance with basic object-oriented methodology, an MO class can be defined as a special case of a more general MO class. In this case, the specialized MO class inherits all the properties (attributes, notification, etc.) of the parent class. The inheritance relationship among MO classes that participate in some management process is represented as an inheritance tree. Figure 2.14 illustrates an inheritance tree for MO classes used to manage logs of event records. (The "top" MO class in Figure 2.14 is the most general MO class in OSI; all other MO classes are direct or indirect descendants of the top MO class.)

2.3.7.6.4 CMISE Services. CMISE, like any ASE, consists of a service definition and a protocol specification; these are called Common Management Information Service (CMIS) and Common Management Information Protocol (CMIP), respectively. CMISE services consist of request–reply services and notification service. Each service is provided through corresponding CMIP PDUs. Each invocation of a request–reply service results in a

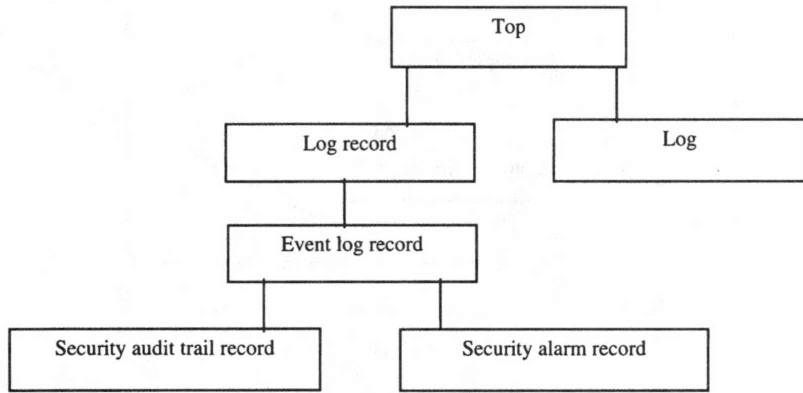

Figure 2.14 Example of an inheritance tree

CMIP request PDU, or a management operation PDU, and zero or more response PDUs. The CMISE requests (management operations), resulting in corresponding CMIP request PDUs, are:

- **Create** a managed object
- **Delete** a managed object
- **Set** (write) the value of one or more attributes in one or more managed objects
- **Get** (read) the value of one or more attributes in one or more managed objects
- **Cancel get** to cancel an outstanding, previously submitted get request in order to stop a flood of multiple responses
- **Action,** which can be just about anything defined by the user

The CMISE response, resulting in CMIP response PDU, is:

- Send a **response** to a management operation; a single get request can result in several response PDUs returning the information that has been requested.

The CMISE notification service, resulting in CMIP notification PDU, is:

- Send a **notification** (for example, a security alarm) when some user-defined trigger event occurs (for example, unauthorized attempted access); the receiver of a notification may return a confirmation, depending on the type of the notification.

2.3.7.6.5 Managers and Agents. Supporting any of the CMIP PDUs carries a price tag. Some savings may accrue if some CMIP PDUs that are not needed are not implemented. To this end, the sender of CMIP request PDUs and CMIP notification confirmation PDUs is called a **manager** or is said to be acting in the **manager role.** Similarly, the sender of CMIP response PDUs and of CMIP notification PDUs is called an **agent** or is said to be acting in the **agent role.** Typically, a managing system need only send manager PDUs and receive only agent PDUs. Similarly, a managed system (for example, a CO) may only need to send agent PDUs and receive only manager PDUs. The manager–agent relationship, with the corresponding PDU flows, is illustrated in Figure 2.15.

The system acting in the agent role must maintain the instances of the MO classes that are relevant to the management process between the two systems. The set of those MO instances is the **Management Information Base (MIB).**

In principle, two interacting systems can each support both the manager and agent roles, as illustrated in Figure 2.16. Thus, CMIP is a fully symmetric protocol, although it allows for

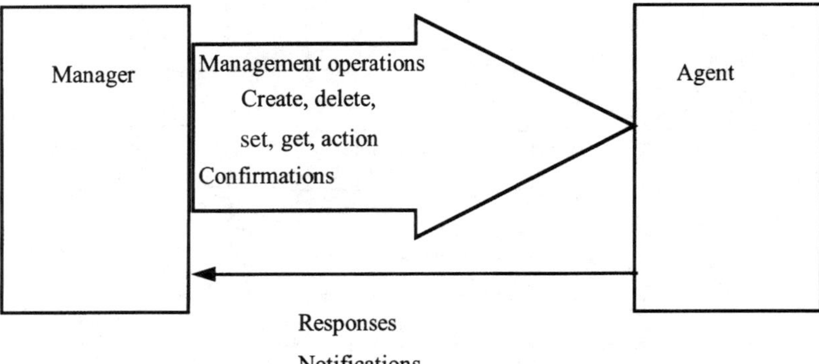

Figure 2.15 Manager and agent systems

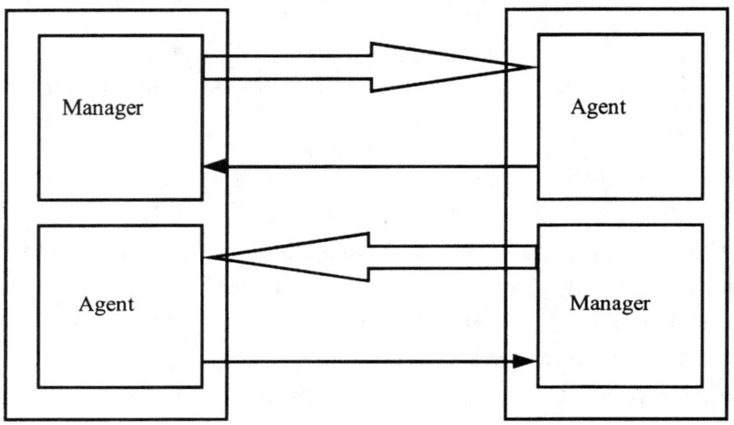

Figure 2.16 Symmetric CMIP interface

asymmetric implementations. Perhaps a more common occurrence is a cascading chain of command, as illustrated in Figure 2.17.

CMISE uses ACSE for association setup and ROSE for all subsequent exchanges. The minimal configuration for systems using CMISE is therefore as illustrated in Figure 2.18.

2.3.7.7 SMASE (System Management Application Service Element). SMASE provides the services that are used to control the managed system (e.g., managing a security audit trail, managing access control information). In doing this, the SMASE uses the services provided by CMISE. SMASE consists of several Management Functions, such as log manage-

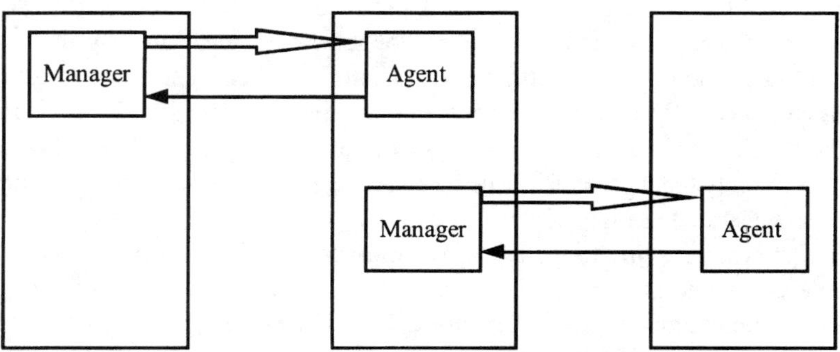

Figure 2.17 Cascading management

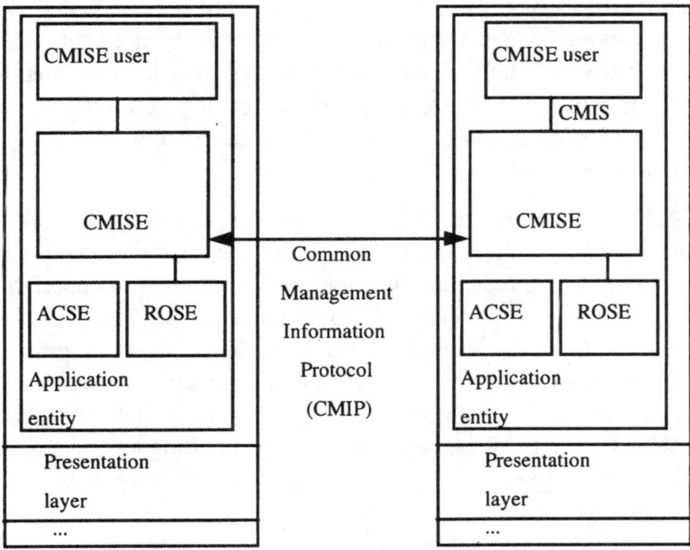

Figure 2.18 Basic setup for systems using CMISE

ment and access control management. Each Management Function consists of one or more Functional Units (FUs). FUs that are not needed over a given interface need not be supported. For instance, an OS that manages the configuration of a digital cross connect system need not support the FU needed for managing access control information. ACSE is used during association setup time to negotiate which FUs will be used during the association. Figure 2.19 illustrates the minimal setup of ASEs for systems using SMASE.

2.3.7.8 Proper Naming. For productive interactions to happen, any two communicating entities must be able to uniquely and unambiguously identify themselves to each other. They must also be able to uniquely and unambiguously identify any component of the TMN

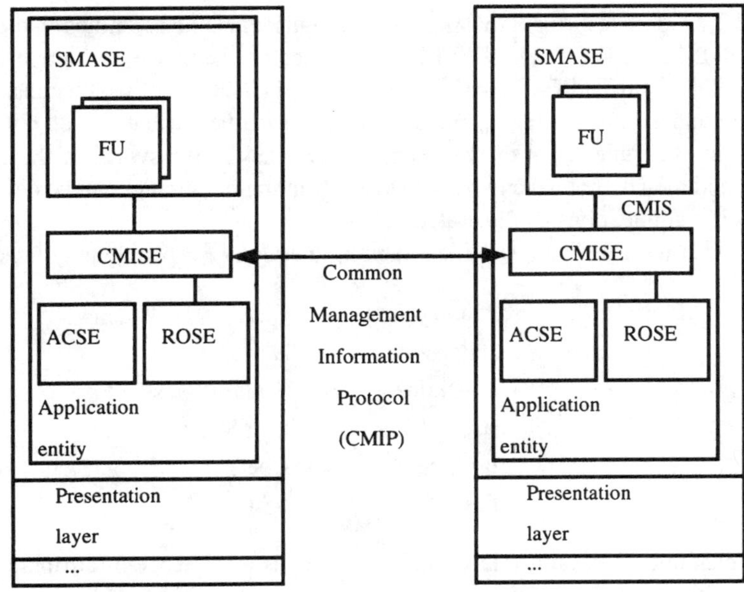

Figure 2.19 Minimum configuration for systems using SMASE

about which they exchange information and any software module(s) that they use in the course of the communication (for example, the set of ASEs used for an association). Within TMN there are two well-defined structures that support unique naming: OBJECT IDENTIFIER and DistinguishedName. Although these naming conventions were not developed specifically for security purposes, they are used extensively in providing security for the TMN. Both are introduced below.

2.3.7.8.1 OBJECT IDENTIFIER. There is little danger that a mother will give the exact same name to any two of her children. Therefore, each child is clearly distinguished, by its name, from its siblings. It is quite common, however, for different mothers to choose the same name for a child. Thus, "John" does not uniquely identify a single person. The goal of an OBJECT IDENTIFIER is to allow different organizations (for example, corporations, people, societies) to independently name entities within their respective domains (just as mothers name their children independently of each other), yet ensure that every entity has a unique, unambiguous name.

OBJECT IDENTIFIER achieves this goal by requiring any naming authority (i.e., anyone that gives names) to ensure that all the names it issues are distinct from each other. To help ensure this requirement, all those names are registered with the issuing naming authority. To make matters simpler (though perhaps less creative), each name must be a positive integer. Among other things, a naming authority can give names to other naming authorities, thereby allowing the emergence of a naming hierarchy.

The definition of OBJECT IDENTIFIER specifies that one of three specific naming authorities must be at the beginning of each hierarchy: the ITU (assigned the "name" 0), ISO (assigned the "name" 1), or both ISO and ITU combined (assigned the "name" 2). Within each of these authorities there are several subordinate, registered, naming authorities. In particular, under ISO (1), member-body is specifically established to assign and register names for ISO members. Under ISO (1), the name of member-body is 2. One of the ISO members, the USA, has the name 840 under member-body. ANSI T1 standards are registered under USA. Thus, any registered ANSI T1 standard has an OBJECT IDENTIFIER that starts with the sequence {1 2 840 }. To improve legibility by humans, such sequence is often written as { iso(1) member-body(2) usa(840) } Only ANSI can issue OBJECT IDENTIFIERs that begin with this specific sequence.

2.3.7.8.2 DISTINGUISHEDNAME. DistinguishedName is defined in the X.500 directory standard. Like OBJECT IDENTIFIER, it is based on the concept of a tree-like naming hierarchy. A DistinguishedName is a SEQUENCE of pairs, each of which consists of an attribute identifier and a value for that attribute. The values of the attributes need not be integers; they can be any printable string of characters as specified by the syntax of the corresponding attribute, and they do not have to be registered. Naming hierarchies can be developed as needed for specific applications as illustrated below:

A DistinguishedName within a naming hierarchy for people may look like:

```
countryName  =  "USA"
stateName    =  "NY"
cityName     =  "White Plains"
streetName   =  "Elm Street"
lastName     =  "Doe"
firstName    =  "John"
```

As the example illustrates, a DistinguishedName is a sequence of attribute value assertions. Each attribute value assertion consists of the name of an attribute (e.g. countryName) and a value for that attribute (e.g., "USA"). The sequence of attribute value assertions cannot be arbitrary; rather, it must correspond to a path from the root of the directory schema to one of its leaves. Figure 2.20 illustrates a possible directory schema for defining DistinguishedName for people. The example above is a valid DistinguishedName for the schema in Figure 2.20.

2.3.7.8.3 Security Audit Trail—An Example. Before we charge ahead with any more TMN components and modules, we can take a short break and contemplate what we have covered so far. This is best done with a simple example: A security auditing application program uncovers an unusual pattern in Eve's access of some systems. It determines that more data are needed. It calls upon the local ACSE to set up an association with a remote system that contains a security audit trail (i.e., a log of events that might be relevant to security analysis). It instructs the ACSE that the association should involve the security audit trail FU of SMASE (this FU supports the information model for the management of security audit trails), CMISE, ROSE, and STASE-ROSE in addition to ACSE. After ACSE has set up the association, the application program tells SMASE to access the security audit trail in the remote system and retrieve all the security audit trail records that show unauthorized access requests by Eve during the last three years. Since the SMASE FU uses the information model for security audit trail, it can use CMISE to formulate a get request that uses scoping and filtering to specify which security audit trail records are needed. CMISE produces a CMIP PDU which it passes to ROSE. ROSE assigns to this message an invocation number, produces a ROSE PDU, and passes it to STASE-ROSE for encryption and integrity protection. STASE-ROSE computes the message authentication code (MAC) for the ROSE PDU, encrypts the ROSE

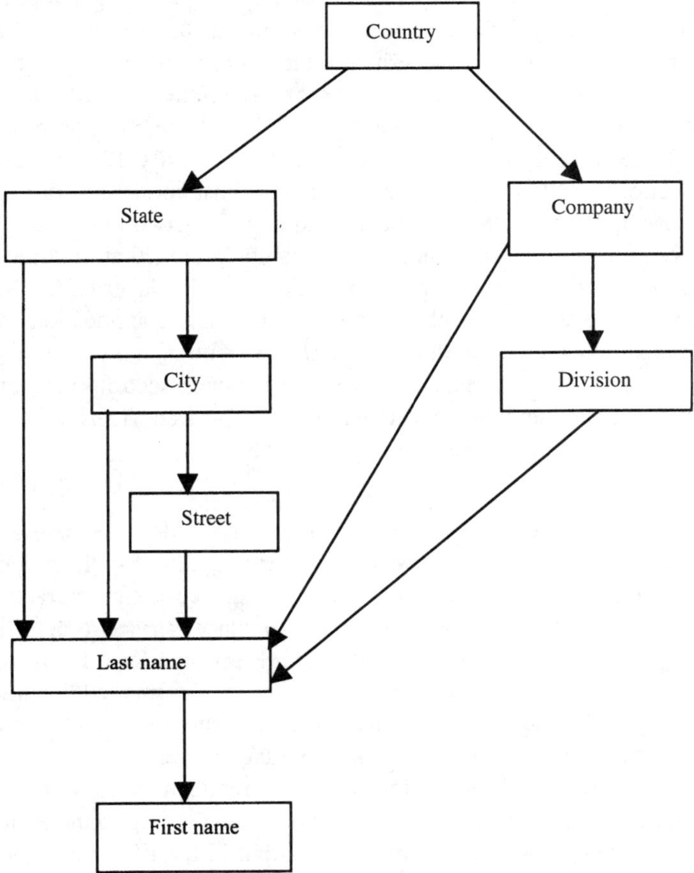

Figure 2.20 Example of a directory schema for a people directory

PDU, and passes the encrypted ROSE PDU and the corresponding MAC, inside a STASE-ROSE PDU to the presentation layer. From there it proceeds down the stack, over to the remote system, up the stack, and from the presentation layer to STASE-ROSE. STASE-ROSE decrypts the encrypted ROSE PDU, verifies that the MAC is valid, and passes the clear text ROSE PDU to ROSE. ROSE takes note of the invocation number and passes the CMIP PDU to its client CMISE. With the help of the security audit trail FU of SMASE, CMISE retrieves the relevant records from the security audit trail. As each record is retrieved, CMISE produces a CMIP response PDU and passes it to ROSE, which appends the invocation number of the initial request. It then passes a ROSE PDU to STASE-ROSE for integrity and confidentiality protection. When that message, decrypted and verified, is received by ROSE in the requesting system, ROSE uses the invocation number to match the response with the proper request and passes it to CMISE. CMISE provides the information to SMASE which converts it to a form usable by the requesting application. This example illustrates some of the bookkeeping and other grunt work done by the OSI protocol stack to transparently provide the application with the requested information.

At the transport layer, the different protocols can provide their own security measures. For TCP/IP, security can be provided through the Secure Sockets Layer version 3 (SSL3), which is being standardized as Transport Layer Security version 1 (TLS1) by the IETF; it resides on top of TCP. SSL3 can provide authentication, data integrity, data confidentiality, or non-repudiation.

At the network layer, the X.25 Closed User Group (CUG) is used to provide controlled network access. This is done by allowing only members of the CUG to communicate with other members. For IP, there are a number of ways to provide network layer security. These include the use of firewalls and IP Security (known as IPSec). Firewalls are placed at the entrance to a network to provide security for the entire network from external sources. This is done by controlling what information can pass through the firewall. The screening criteria include the IP address, the type of tranport protocol used, the transport protocol port used, time, and date. Firewalls only work to provide islands of security. IPSec is another option used to provide security for IP. IPSec secures the host-to-host communications (where the host can be a user or a gateway). IPSec can be used to provide networkwide security.

Since numerous security mechanisms can be provided at multiple layers, the question becomes, Which mechanism should be supplied at which layer [ISO7498-2]? The answer to this question depends on a number of factors, such as the applications on the network, how much of the network and its applications must be secured, and where this security is most efficiently applied. One important consideration is which security mechanisms and protocols are supported in commercially available products. These considerations will be applied in this book to derive a framework for securing the TMN.

2.3.7.9 EDI. Although CMISE provides all the desirable features for network management interactions, it may be too expensive for an application that requires only modest capabilities. For instance, when a small or medium-size customer interacts with a network service provider's TMN over the X interface to place service orders, a simple document exchange protocol will suffice. For this reason Electronic Data Interchange (EDI) has been incorporated into TMN. EDI supports the exchange of formatted documents such as service requests. It provides the means for specifying the structure of each document, and it allows the nesting of one or more documents within a larger one.

When an EDI-based application is ready to send a message to a remote entity, it passes the message to an EDI translator, which formats it according to the applicable EDI standard and the agreed upon document format. The output of the EDI translator can be delivered to the remote entity via electronic mail, or it can be carried directly over the transport layer. Most EDI-based applications use TCP/IP for transport.

Using electronic mail for delivery of EDI messages is too slow for many network management applications. In order to send an EDI message directly over TCP/IP, the EDI translator needs the services of an entity that sets up and terminates the TCP connections as required. Those services are provided by the Interactive Agent (IA). The IA is also used for flow control. It allows the receiving entity to signal the sender that an overload condition, or system malfunction, requires a temporary halt to the flow of messages. The IA can also perform security services for an EDI message.

2.3.7.10 CORBA. The Common Object Request Broker Architecture (CORBA) [OMG-COS2.2, OMG-ORB2.2] has been developed and continues to be refined by the Object Management Group, a broad-based computer industry consortium. The purpose of CORBA is to support distributed processing, software portability, and location transparency. As such, it is expected to become quite ubiquitous. If indeed CORBA becomes universally available, along with user-friendly, low-cost, effective toolkits, then it may well be used in network management. TMN has therefore been expanded to incorporate CORBA in addition to CMISE and EDI.

CORBA can be introduced with a simple example: a user (a person or a program) needs to sort a large file and calls a sort procedure. In most cases the procedure will be resident on the same computer and will proceed with sorting the file. In some cases the right procedure may not be available on the local computer, or the file may be too large to sort on the user's platform. The user therefore needs to find a computer that can and will undertake the sorting task, establish a connection with that server, pass on the file to be sorted along with all relevant instructions and parameters, and wait for the result. CORBA takes care of all this tedium transparently to the user. The user may think that the file was sorted locally, even if CORBA got another system to do the job. And the remote system may have used yet another server for some preprocessing of the cumbersome file.

CORBA manages this accomplishment with two Application Programming Interfaces (APIs) and an interface protocol. The API on the client (user) side is a stub. In our example the stub looks to the user as a local sort routine, to be called as needed. However, instead of containing code that sorts files, it contains code that allows it to search for a system that can actually sort the user's file. It then uses the Inter ORB (Object Request Broker) Protocol (IOP) to pass the job on to the remote server. The server's API is a skeleton. The skeleton calls upon the (purportedly) local sort routine with the necessary parameters, waits for the result, and returns it to the client ORB, which returns it to the user. The key to the IOP is the Interface Definition Language (IDL), which is used to specify what information the server needs from the client to do its job (in our example: the file to be sorted, the field used for sorting, and perhaps the structure of the records) and what will be returned from the server to the client (for example, a sorted file or error messages). IDL is object oriented, like GDMO. It can be used to model shared knowledge between systems. However, since CORBA was designed for distributed processing, and not for network management, it does not have all the expressive power of CMIP and GDMO. CORBA, for example, does not have constructs for scoping and filtering.

A major benefit of CORBA, completely lacking in the OSI world, is software portability. It is made possible by publishing a standard API. Thus, any application that runs on CORBA can (in principle) run on any platform that supports CORBA. Where CORBA is a bit fuzzier in ensuring interoperability. In principle, each supplier can create a product line based on the CORBA architecture and supporting the CORBA APIs, but with a proprietary IOP. To remedy this shortcoming, the Object Management Group (OMG) has defined the Generic IOP (GIOP). It is generic in the sense that it is independent of any underlying transport protocol. When the transport is TCP/IP (it usually is for CORBA), GIOP rides on top of the Internet IOP (IIOP). IIOP provides the mapping of GIOP to TCP, quite similar to the corresponding functionality of the IA for EDI.

When security transformations are needed to protect CORBA interactions, they are performed on GIOP messages, since those messages are interpreted unambiguously by the communicating entities. A protocol is needed to carry protected GIOP messages, analogous to the way STASE-ROSE carries protected ROSE messages. This protocol, Secure IOP (SECIOP), is sandwiched between GIOP and IIOP.

Transport layer security can be provided for CORBA by using Secure Socket Layer version 3 (SSL3) between IIOP and TCP, regardless of Secure IOP (SECIOP) [OMG-SSL].

2.3.7.11 SNMP—Keeping It Simple. When the Internet started to materialize, it was clear that it needed a management protocol to watch over its disparate components. At the time CMIP was still in its infancy, with few information models or actual products. Besides, CMIP provides a lot more than what is needed to manage a simple router or a modem bank. This unnecessary richness comes at a price that far exceeds the processing budget of a small router, let alone a modem. The Simple Network Management Protocol (SNMP) was created to fill the gap.

The simplest alphabet consists of just two characters, 0 and 1. It is ideal for representing the state of a switch as being off or on. It is also powerful enough to express anything that can be expressed, though it would be woefully inadequate for writing this book. The simplest network management protocol, the equivalent of the two-character alphabet, consists of just two management operations: get and set the value of a simple information item (for example, an integer). Of course, it must also support responses to get requests. SNMP goes a bit further. It also allows a manager to get the value of the next information item (this is possible because all the information items are strictly ordered), and it allows an agent to send a trap or a notification to a manager. SNMP can accomplish so much with so little because of the simple, yet clever, arrangement of management information.

2.3.7.11.1 A SIMPLE STRUCTURE OF MANAGEMENT INFORMATION. The simplest information element in an SNMP Management Information Base (MIB) is an **OBJECT TYPE.** It has simple (yes, the word is used a lot here) **syntax** which can be one of the following:

- INTEGER
- OCTET STRING
- OBJECT IDENTIFIER
- NULL

It has an **access** mode that can be one of the following:

- Read-only
- Write-only
- Read-write
- Not-accessible (a value that cannot be written or read by the manager may seem superfluous; we shall shortly see, however, how important it actually is)

It has a **status** that can be:

- Mandatory (must be supported)
- Optional (may be supported)
- Obsolete (need not be supported any longer)
- Deprecated (i.e., mandatory but about to become obsolete)

It has an **ObjectName,** represented as an OBJECT IDENTIFIER (OID), so it can be referenced in a management operation. If there is only one instance of a given OBJECT TYPE in an agent system's MIB, then it might be addressed just by invoking that OBJECT TYPE's ObjectName. For instance, if there is only one instance of OBJECT TYPE counter in an agent system, then a manager could, for example, request the value of the counter by specifying the OBJECT TYPE's OID. If a MIB has several instances of the same OBJECT TYPE, for example, if it has several instances of OBJECT TYPE counter, then more specificity is needed, as we shall see shortly.

An OBJECT TYPE can (but doesn't have to) also include:

- DescrPart: a textual description of the semantics of the OBJECT TYPE
- ReferPart: a textual reference to an object in another MIB
- IndexPart: can be used if this OBJECT TYPE is part of a more complex structure, as we shall see shortly
- DefValPart: defines a possible default value that can be used when the agent creates an instance of this OBJECT TYPE

A simple, flat database consists of a file of records, in which all the records share the same format, that is, the same sequence of fields. In an SNMP MIB, the file is viewed as a table and each record is a row in that table. For the mathematically inclined, a table can be viewed as a two-dimensional array; a row can be viewed as a vector or a one-dimensional array; a simple information element (i.e., a cell in the table) can be viewed as a scalar.

A row (record) object is defined as a SEQUENCE of simple OBJECT TYPEs. Thus, a row is defined as:

SEQUENCE {<list of ObjectName referring to simple (scalar) objects>}

In defining a row, the IndexPart is a list of ObjectName components of the row (i.e., a list of one or more fields within the row) whose combination of values is unique within the table. The IndexPart is used to uniquely identify and select a row within a table.

A table object is a SEQUENCE OF rows of the same type. Thus, the syntax of a table object type is:

SEQUENCE OF <row object>

Now it becomes clear how we can address an individual, scalar information element within a MIB: There is at most one instance of a table of a given type in an MIB. There is a set of fields within each row whose combination of values is unique. Those fields make up the "index" of the table. They fully distinguish each row in a table. Therefore, an information element can be addressed by specifying the ObjectName of the table in which it resides, the values of the index fields of the appropriate row, and the ObjectName of the type of the desired information element.

We can now appreciate the significance of the "not accessible" access mode. A manager is not allowed to set the value of an index field, for this would change the identity of the row. A manager is not allowed even to read the value of an index field because the manager must supply that value in order to access the row.

Actually, the addressing of scalar objects is more streamlined than described above. It is based on some straightforward naming conventions. The OID of a row object type is typically chosen to be the OID of the table where it resides followed by the integer value 1. (Remember that an OID is a sequence of integers.) The OID of a scalar object type is typically chosen to be the OID of the row object type where it resides, followed by an integer that specifies the location of that scalar object type within the row. Thus, the OID of a scalar object type also designates a column in the table. It is therefore called a **column object.** A scalar object instance can be referenced with that object type's OID, followed by the values of the index fields of the corresponding row. The order of the values of the index fields is the order in which the fields appear in the row. This object instance specification can be converted into an OID by converting each index field value into a sequence of one or more integers. The conversion is quite intuitive:

- An (nonnegative) integer needs no conversion.
- An octet string of fixed length n is converted into n integers by reading each octet as an integer.
- An octet string of variable length with n octets is converted into $n + 1$ integers, the first one has the value n, and the others are obtained by reading each octet in the string as an integer.
- An OID with n subidentifiers is converted into $n + 1$ integers, the first one has the value n, and the others are the n subidentifiers of the OID.
- An IP address is a ready-made sequence of integers.

Thus, every scalar object instance in the MIB has its own OID. This provides a natural lexicographical ordering of the object instances. Indeed, the OIDs define a tree, with the object instances as the leaves of the tree. Every path from the root to a leaf corresponds to that object instance's OID. If the children of each node in the tree are ordered in ascending order, then a depth; first traversal of the tree provides the desired lexicographical ordering.

This lexicographical ordering is not just a clever arrangement; it is essential for SNMP. Indeed, the agent can create and delete objects at will, so the manager does not really know what is in the MIB. With the get next request, the manager provides the OID of a known object and requests the value of the next object in the lexicographical order. The response includes both the OID and the value of that next object. The manager can now use that new OID and reiterate the process. A well-informed manager, who knows the OIDs of several object instances in an MIB, can even request multiple items in a single message. No object can hide from the scrutinizing manager, though it may take a while to find out that nothing has changed since the last exhaustive search.

2.3.7.11.2 SNMP PDUs. With the structure of the MIB established, we can now examine the SNMP PDUs. There are three types of PDUs from the manager to the agent:

- GetRequest
- GetNextRequest
- SetRequest

And two types of PDUs from the agent to the manager:

- GetResponse
- Trap

Each PDU starts with a preamble stating which **version** of SNMP is used. At present, version 1 is most prevalent; it is the only version discussed so far in this book. The much-improved version 2 is not making much progress. The even better version 3 is over the horizon.

Next it presents the user login, called **community** since several users can share the same user login. This monumental affront to security has deterred many practitioners from allowing use of the SetRequest capability.

An implicit **tag** specifies the type of PDU. After this, the PDUs start to gain some individuality.

The Trap PDU contains:

- **enterprise**—the type of object generating the message
- **agent-addr**—address of the agent generating the message
- **generic-trap**—representing one of the following values: coldStart, warmStart, linkDown, linkUp, authenticationFailure, egpNeighborLoss, or enterpriseSpecific
- **specific-trap**—an integer
- **time stamp**—local clock value measured since the last reset
- **variable-bindings**—the real payload of the message. It consists of a list of variables (for example, the number of packets received on a port in the last hour) and their values (for example, 1,500), and it is represented as a sequence of object instance identifiers and the values of those object instances.

The other four PDU types—GetRequest, GetNextRequest, SetRequest, and GetResponse—continue to share the same additional fields:

- **request-id**—it is used for correlating a response with a request.
- **error-status**—in a response PDU, it can represent one of the following values: noError, tooBig, noSuchName, badValue, readOnly, or genError; in a request PDU it has the value 0.
- **error-index**—in a response PDU, it is an integer; in a request PDU, it has the value 0.
- **variable-bindings**—This is a sequence of object instance identifiers and their values.

In the variable-bindings section of a GetRequest PDU, the OID of the object whose value is sought must be specified, but its value must be NULL.

Since variable-bindings can contain several OID-value pairs, a single message can be used to get, set, or provide the values of several scalar objects.

2.3.7.11.3 VERSION 2—BIGGER AND BETTER. As the use of SNMP grew, so did the realization of its limitations. SNMPv2 was created to remedy the most glaring weaknesses of SNMP. Its innovations address three areas:

- Structure of management information (SMI)
- Protocol operations
- Security

The first two bullets are discussed here. The last one is reserved for Part 3 of this book.

2.3.7.11.3.1 Structure of Management Information. The changes to SNMP's SMI introduced by SNMPv2 include some minor enhancements as well as substantive improvements. In this high-level overview, we consider only the latter.

SNMPv2 has two types of tables: it keeps the table structure of SNMP, and it adds a new type of table where the manager can **create** and **delete** rows. A table supports row creation and deletion if its rows include a field of type RowStatus with access of read-create. The simplest way for a manager to create a new row in such a table is to

1. Issue SetRequest where the value of the index field(s) is currently unused in the table.

2. Set the value of RowStatus to createAndGo. (This is one of the several possible values for RowStatus.)
3. Set the values of other fields within the row for which the manager has write privileges.

SNMPv2 borrows a familiar concept from object-oriented design: inheritance. It allows the addition of extensions (i.e., new fields) to existing row object types. Thus, a newly defined **extended row** is essentially inherited from the initial row type, with new fields added at the end. The new row type has the same index fields as its parent.

SNMPv2 unifies traps and responses under NOTIFICATION. Part of the definition of a NOTIFICATION-TYPE is a list of the scalar objects whose values are to be provided in an instance of that NOTIFICATION-TYPE.

To improve interoperability between different SNMP systems, SNMPv2 introduces the concept of information modules. An **information module** is a collection of object types that are typically used together in support of a management function. An information module comes with a conformance statement that specifies the minimum set of capabilities that an implementation must support for conformance to that module. It also further provides for explicit statement of any variations from the generic module definition in a specific implementation. A variation may be, for instance, a more constrained limitation on the set of values that an INTEGER object can have.

2.3.7.11.3.2 Protocol Operations. SNMPv2 builds on the PDUs of SNMP by enhancing the scope of existing PDUs and defining a couple of new PDUs.

In SNMP the **GetRequest** and **GetNextRequest** are atomic operations: either the agent responds with the values of all the objects in the request, or the agent responds with an error code and no values. In SNMPv2 these two requests are **"best effort"** operations. The agent responds with values for all the objects in the request that it can provide; for others it may return error codes.

Probably the major contribution of SNMPv2 is the **GetBulkRequest** PDU. Its structure is similar to the GetNextRequest PDU with two additional INTEGER valued fields: nonrepeaters and max-repetitions. Its semantics, however, is much more powerful then the GetNextRequest PDU. It is best explained with an example. Consider a GetBulkRequest message with N object instances listed in its variable-binding list. Let nonrepeaters have the value $K < N$ and let max-repetitions have the value L. The interpretation of GetBulkRequest is:

- For each of the first K object instances in the variable-binding list, provide the value of the next object instance in the lexicographical order; so far this is the same as for GetNextRequest.
- For each of the remaining N-K object instances in the variable-binding list, provide the values of its L lexicographical successors.

Thus, with a single GetBulkRequest message the manager can request the values of numerous object instances without fully enumerating them.

In order to support manager-to-manager communications SNMPv2 defines the **InformRequest** PDU. It has the same syntax as an SNMP trap PDU but a somewhat different behavior:

- InformRequest is sent from a manager to another manager, not from an agent to a manager.
- The first variable in the variable-binding list is a time stamp.
- The second variable in the variable-binding list is an event type.
- If the receiving manager cannot properly process the InformRequest PDU, it returns a response PDU indicating an error.

- If the receiving manager does successfully process the InformRequest PDU, it returns a response PDU with the same information as in the InformRequest PDU it received.

2.3.7.11.4 SNMPv3—A VERSION FOR ALL SEASONS. SNMPv1 looks too timid for the forthcoming tasks of managing emerging mega-internets. SNMPv2 is not making much headway. These circumstances are ripe for creating v3.

2.3.7.11.4.1 SNMPv3 Architecture. At this point, with foresight enhanced by hindsight, the v3 architecture graciously accommodates v1 and v2, as well as future versions. This architecture is illustrated in Figure 2.21.

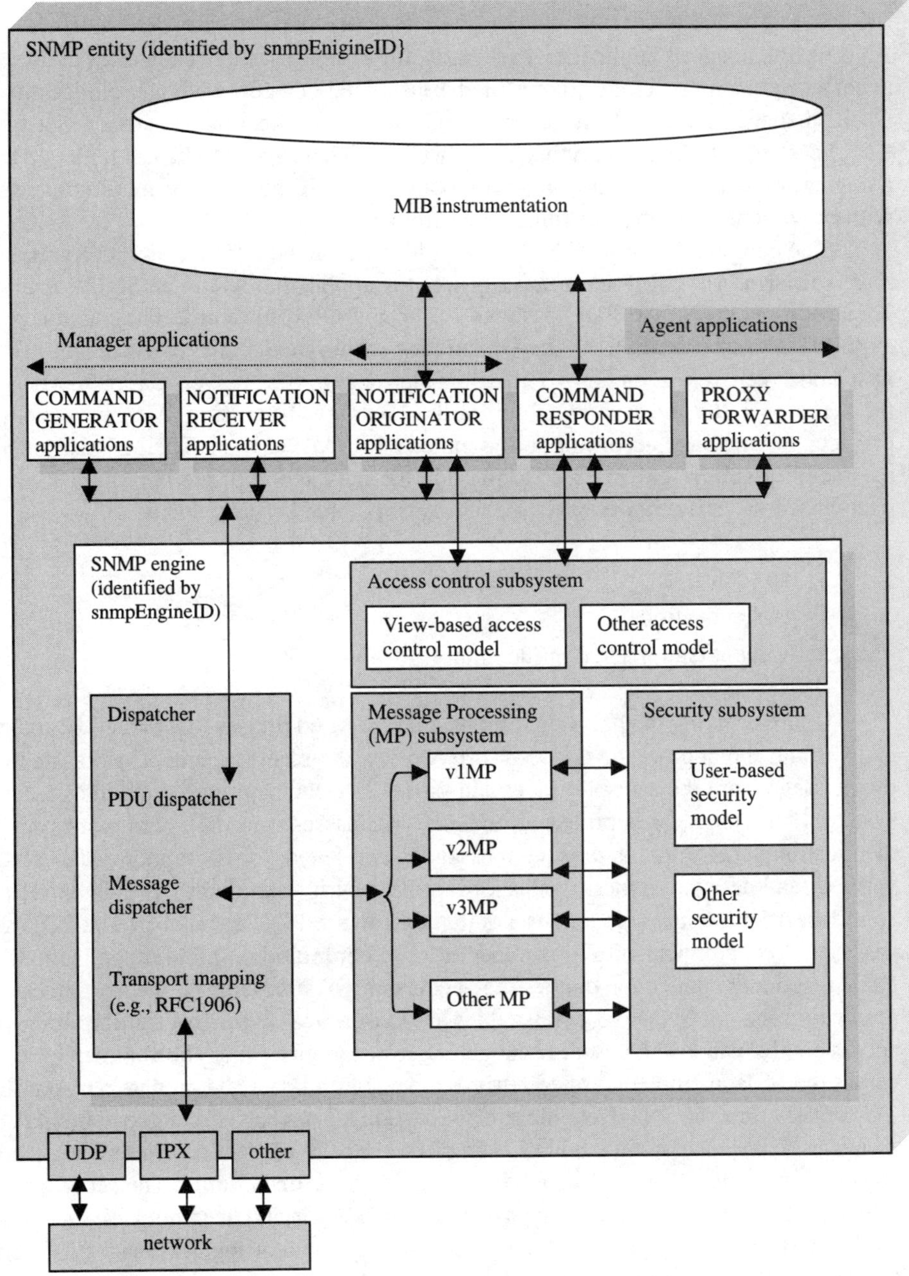

Figure 2.21 SNMPv3 architecture

Each **SNMP entity** has one **SNMP engine;** both are identified with an **snmpEngineID** which is unique within a management domain. Each SNMP entity has one or more **SNMP applications.** Each SNMP application belongs to one of the following categories:

- Command originator
- Notification receiver
- Notification originator
- Command responder
- Proxy forwarder

An SNMP manager's applications belong to any of the first three categories above; an SNMP agent's applications belong to any of the last three categories above. Notice that unlike a CMIP manager, an SNMP manager can originate a notification to another manager. (In the CMIP case the notification originator is always called an agent.) The last bullet above denotes an agent application that forwards an incoming SNMP message to its ultimate destination, with or without translation to some other protocol.

The components of the SNMPv3 architecture can be introduced with a simple management scenario. An SNMP **command generator** application within an SNMP manager entity prepares a get command PDU destined for an SNMP agent entity. The command generator application passes the PDU to the **Dispatcher** within the local SNMP engine. (There is one Dispatcher per SNMP engine.) The dispatcher determines the version of SNMP and passes the message to the corresponding unit of the **Message Processing** (MP) subsystem. The MP prepares the message according to the indicated SNMP version. The MP also interacts with the **security subsystem** in order to provide the message with the appropriate security. The **user-based security model** of SNMPv3 supports three security levels:

1. No security
2. Authentication
3. Authentication and confidentiality

Upon receiving the properly formatted and secured message from the MP, the dispatcher maps it into the appropriate transport protocol. At the receiving side, the dispatcher receives the message from the transport layer and passes it to the appropriate module in the MP subsystem. If the message is protected, then the MP passes it to the security subsystem for authentication verification, and, if the message is encrypted, for decryption. After receiving the verified (and decrypted) message back from the security subsystem, the MP extracts the relevant data from the message and passes that data to the PDU dispatcher. The PDU dispatcher passes the get command with its parameters to the **command responder** application. The command responder application queries the **access control subsystem** if the get request should be granted or denied. SNMPv3 defines the **view-based access control model;** other models of access control can also be used. If the get request is granted, then the command responder accesses the MIB instrumentation to retrieve the requested data and prepare a response.

Every time the SNMP engine stirs into action it is on behalf of some **principal.** The principal can be one or more individuals acting in some roles and/or one or more applications. The principal has a human-readable, model-independent **securityName.** The securityName maps into a security model-dependent **security ID.** The security ID need not be human readable.

An SNMP entity can represent several physical and/or logical devices. It is convenient to group the information elements within an SNMP entity according to the devices to which they correspond. To that end, information elements within an SNMP entity can be grouped into one or more **contexts.** Typically, each device within an SNMP entity would have its own context. Contexts can overlap (i.e., an instance of a managed object type can belong to several contexts).

Each object instance has a unique identifier within each context in which it is included. Each context has at least one **contextName** that is unique within the SNMP entity that contains that context. Each context is uniquely identified through its **contextEngineID** (which is the corresponding snmpEngineID) and its contextName.

A **scopedPDU** is an SNMP PDU with a unique context identifier (contextEngineID and contextName). The information elements within the PDU need only be uniquely identified within the scopedPDU's context.

2.3.7.11.4.2 SNMPv3 Protocol. An SNMPv3 message is a SEQUENCE of the following fields:

- msgVersion is an INTEGER that identifies which version of SNMP is used for the message.
- HeaderData is itself a SEQUENCE consisting of:
  - msgID, an INTEGER that allows correlating a response with a request
  - msgMaxSize, an INTEGER that specifies the maximum acceptable size for a response
  - msgFlags, an OCTET STRING, individual bits within that string specify if authentication and/or privacy protection are provided and if a Report PDU must be returned to the sender under those conditions that cause generation of a Report PDU
  - msgSecurityModel, an INTEGER that identifies which security model is being used
- msgSecurityParameters is an OCTET STRING that contains information that is generated and used only by the security subsystems.
- ScopedPduData which is a CHOICE between:
  - ScopedPDU, which is a SEQUENCE consisting of a contextEngineID, contextName, and the actual PDU; and
  - encryptedPDU which is an encrypted ScopedPDU value, represented as an OCTET STRING.

Any security-related information is encapsulated within msgSecurityParameters, which is dependent on the security model. Part 3, which addresses the security of TMN transactions provides a full description of msgSecurityParameters for SNMPv3's User-based Security Model.

2.3.8 Security-Related Components of the TMN Stack

Now that we have presented the major components of the OSI stack as it applies to the TMN, let us look briefly at the security aspects of interactions among these components. The relation among the different modules in TMN is summarized in Figure 2.22. Figure 2.22 also illustrates the flexibility of the OSI reference model, which allows various underlying lower layer protocol stacks to be used within the OSI 7-layer reference model. Figure 2.22 also includes two security modules that are implemented within individual user applications rather than in the communications protocol stack: TR 40 [TR40] and Application-Based Security [T1.254]. These modules are discussed in Chapter 8.

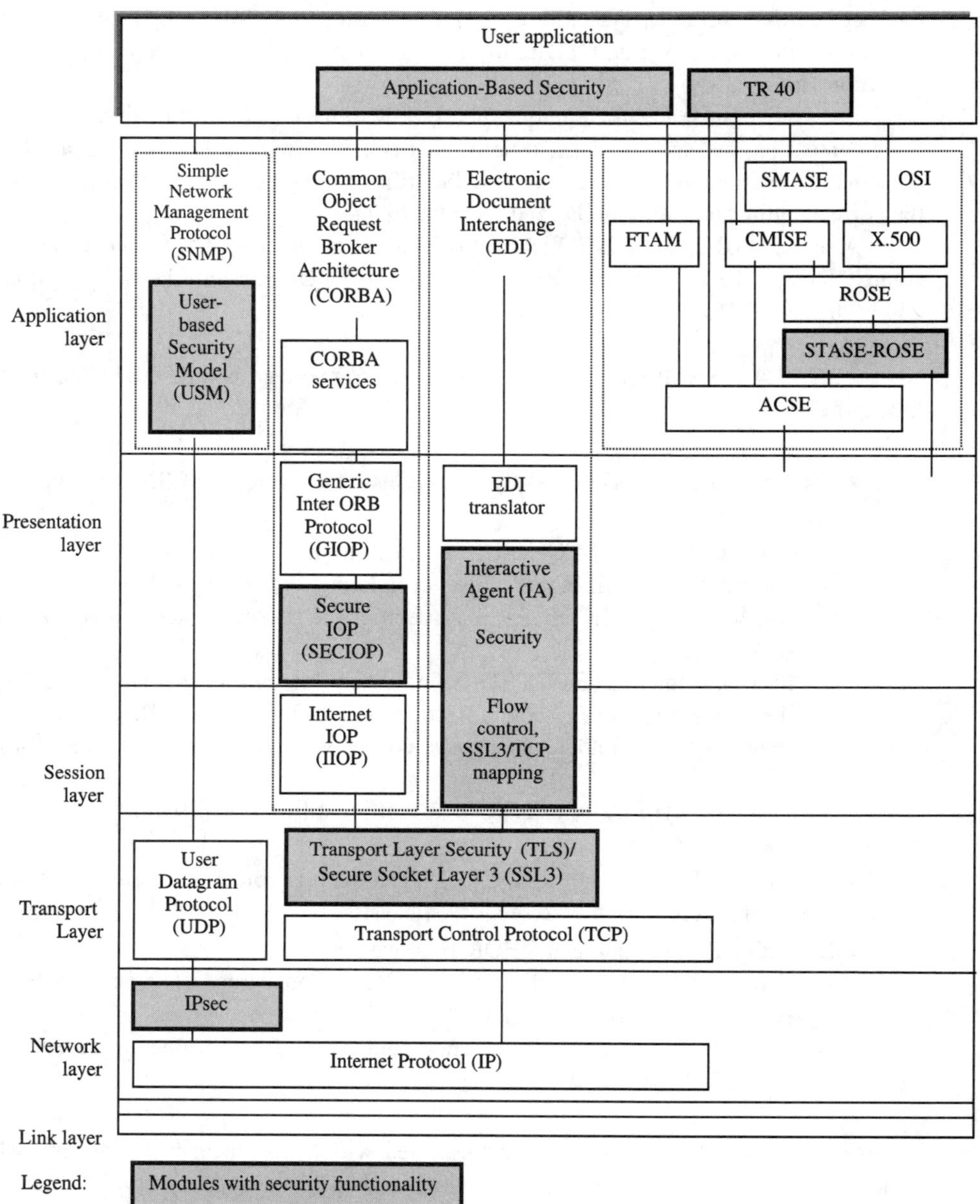

Figure 2.22 Main interactions of security modules

PART II
SECURITY OVERVIEW

Every system faces some potential threats. The more important the system, the larger the risks associated with such threats. When a system is connected to a communication network, the potential threats are greatly magnified. When thousands of systems, dispersed over a continent, are interconnected by several communications networks, they present a very large target to a would-be intruder. Besides being such an awesome target, the TMN has several unique vulnerabilities that require special attention.

This section starts with a roundup of the usual suspects and a brief discussion of the threats (for example, eavesdropping, corrupting information) they present. The significance of such threats is then expounded by correlating them to potential risks with direct business impact. Most threats and risks are quite generic; they apply to most systems and networks. Many of them, however, are greatly amplified in the context of the TMN. Unfortunately, the extensive discussion of TMN-specific threats and risks provided here is illustrative rather than exhaustive. Fortunately, several security services are available to counteract the potential threats (or this book would not have been written). Correlating security services to TMN security threats and risks helps security planners to evaluate and prioritize the security needs of their TMNs.

A vast arsenal of security mechanisms is available to support the salutary security services. This part of the book provides an introduction to such mechanisms and shows how they support security services. For some security mechanisms (for example, encryption), we can choose from a broad selection of algorithms. Rather than attempt to catalog all the existing security algorithms, this book focuses on those algorithms that are currently used or contemplated for TMN security. It offers an overview of their workings and provides an assessment of their strengths and weaknesses in the context of the TMN.

This part of the book is intended as a tutorial for nonsecurity experts; security experts can perhaps move to the next part. However, the discussion of TMN-specific threats and risks may be relevant to security experts and novices alike.

3

TMN Attacks and Defenses

An unprotected TMN offers a large target to several types of broad-brush attacks. In addition, a sophisticated intruder can perpetrate highly targeted attacks to achieve specific goals. Fortunately, an array of security services offers several options for defending the TMN against any currently known attacks.

3.1 TMN GENERAL THREATS AND VULNERABILITIES

This section provides a generic overview of security threats and risks, favoring the TMN as the source of specific examples.

3.1.1 Potential Security Attackers

A good marketing strategy usually starts by identifying the potential customers that might be interested in some product or service. Similarly, a good security analysis often starts with identifying the potential attackers. Each class of potential attackers is characterized by

- The methods of intrusion
- The risks that may result from a successful attack
- The size of the class—roughly, how many people are in that category
- The likelihood that a member of that class would undertake an attack

Obviously, a class of potential intruders characterized by the ability to mount sophisticated attacks, with grave potential risks, including numerous individuals of questionable ethics, would require strong, comprehensive security countermeasures. Similarly, a small class of people with neither capabilities nor inclination to commit serious crime would hardly justify the investment needed for powerful security measures. In most cases, as illustrated below, the situation is less clearcut: for most classes of potential attackers some of the criteria listed above may cause concern, while others may promote complacency. It is a matter of local security policy to evaluate the different threats and decide on the appropriate security measures.

Potential security attackers that may launch an attack against the TMN through a number of methods include:

- An external intruder using low-tech methods (for example, guessing passwords)
- A determined external intruder using sophisticated, high-tech methods
- An intruder to a client's organization
- A client serving as a front for a criminal operation
- An internal intruder

Although it is impossible to secure any network or facility absolutely against all attacks, prudent and responsible precautions will allow almost any degree of security desired. The above list is arranged so that the level of precautions, and consequently the effort, needed to guard against an attack rises with each possible attacker. In addition, the probability that the listed party would perpetrate an attack should decrease through the list. Therefore, each organization must choose a level of risk and security at which the organization is comfortable with the tradeoff between the security precautions that have been taken against the remaining risk of attack.

3.1.2 Potential Security Threats

The TMN faces possible security threats such as:

- Unauthorized access
- Eavesdropping
- Masquerade
- Modification of information
- Repudiation
- Replay, reroute, misroute, delete messages
- Network flooding

3.1.2.1 Unauthorized Access. Unauthorized access involves an intruder gaining access and gathering information from a resource to which it is not entitled. Access can be achieved through either physical or electronic methods. This book deals only with electronic methods, and not with threats and protections against physical intrusion.

Access to a restricted resource can occur in one of two ways. Either a resource can be breached by an intruder (for example, through an unprotected dial-up modem access to the target system), or an intruder may have obtained surreptitiously valid credentials to fool the resource into giving it access. In the first case, the security provided by the network is insufficient, whereas in the second case, the authorized user's credentials are not being securely stored. Either method for breaching the security of the resource can have devastating effects on that resource or system. Confidential information can be examined, removed, and modified by an intruder. In addition, applications on that resource can be tampered with to the benefit of the intruder.

3.1.2.2 Eavesdropping. Eavesdropping refers to an intruder being able to interact with the communications channel so that information can be extracted from the channel, possibly without the knowledge of any of the communicating parties. Eavesdropping can compromise the network and allow an intruder to obtain restricted information.

3.1.2.3 Masquerade. Masquerade occurs when an intruder is able to mimic an authorized user, so that the resource believes that the intruder is actually the authorized user. This can be accomplished through an intruder impersonating the source address of the authorized user or the login and password of an authorized user, or even impersonating the signature of the authorized user. The outcome of this deception is to allow an intruder to access and, possibly, change restricted information.

3.1.2.4 Modification of Information. Modification of information occurs when unauthorized information is entered into a resource. Modification can occur as either an entire unauthorized transmission or as part of an authorized transmission (e.g., a "Trojan horse" attack, in which an intruder places additional code in the authorized message and has the resource execute that code). The modification can either be a change in a legitimate transmission or an entirely spurious transmission. This can lead to a large security breach—for example, changing the security practices of the resource to allow the intruder to enter whenever it pleases.

3.1.2.5 Repudiation. Repudiation refers to a party having received services and then denying having either requested them or received them. This can lead to disputes between parties in the billing for services.

3.1.2.6 Replay, Reroute, Misroute, Delete Messages. In a replay attack, an intruder copies a valid message and then reuses the message in order to access a resource. This can be done in order to receive services to which the intruder is not entitled. In a reroute attack, an intruder can change the route a message takes in order to facilitate capturing the message as it travels between parties. Misroute attacks cause a message to reach the wrong destination. A delete attack prevents a message from ever reaching any destination.

3.1.2.7 Network Flooding. Network flooding occurs when an intruder floods the network with false or irrelevant messages. Flooding the network with bogus association requests requires the targeted system to waste resources in processing such messages. If the network flooding is large and sustained, the network will not be able to allocate resources to legitimate requests.

3.1.3 Potential Security Risks

This section focuses on the possible consequences of the security threats discussed in the previous section. These security risks can be grouped as follows:

- Theft of information
- Unauthorized use of resources
- Theft of service
- Denial of service

3.1.3.1 Theft of Information. Theft of information occurs when an intruder has access to restricted information, either at a node or in the communications channel between nodes. Theft of information can arise as a consequence of

- Unauthorized Access
- Masquerade
- Eavesdropping

3.1.3.2 Unauthorized Use of Resources. Unauthorized use of resources allows an intruder the use of resources to which the intruder is not entitled. It is a possible security consequence of

- Unauthorized access
- Masquerade
- Modification of information
- Replay
- Reroute, misroute messages

3.1.3.3 Theft of Service. Theft of service involves usage of services without authorization or without proper billing. It can be a consequence of a combination of security threats:

- Unauthorized access
- Masquerade
- Modification of information
- Repudiation
- Replay attack
- Reroute, misroute, delete messages

3.1.3.4 Denial of Service. Denial of service is preventing a resource from performing as expected. It generally arises from a combination of five possible security threats:

- Unauthorized access
- Masquerade
- Corruption of resource management information
- Reroute, misroute, delete messages
- Network flooding

Denial of service can target a single user or all network users.

3.1.3.4.1 SINGLE-USER DENIAL OF SERVICE. An intruder can deny a single user access to a resource in one of two ways. One is for an intruder to actively intercept restricted information to and from a user, not allowing that information to reach its destination. The other is to change the security management information at a resource to deny access to an authorized user.

3.1.3.4.2 NETWORKWIDE DENIAL OF SERVICE. An intruder can attempt to stop all valid traffic flow through a communications channel. This can be done, for example, by flooding the channel with large amounts of bogus messages. Given a flood of irrelevant information, resources will become overloaded and can stop listening to *any* information coming from that channel. In the case of a flood of falsified information, an intruder places many requests on the resource that the resource believes to be true requests. The resource will then try to process these falsified requests and eventually not be able to act on valid requests that are received. The most widely known example of this is the SYN flood attack on Internet Service Providers.

Table 3.1 presents a summary listing of security threats and their consequences discussed above.

TABLE 3.1 Summary of Security Threats and Risks

Security Threats	Security risks: Theft of Information	Unauthorized Use of Resources	Theft of Service	Denial of Service
Unauthorized access	✔	✔	✔	✔
Masquerade	✔	✔	✔	✔
Eavesdropping	✔			
Modification of information		✔	✔	✔
Replay		✔	✔	
Reroute, misroute, delete messages		✔	✔	✔
Repudiation			✔	
Network flooding				✔

3.1.4 Impacts of Security Risks on the TMN

The security risks mentioned above can have large impacts on the TMN in all five Management Functional Areas (MFAs). To show the extent of these security vulnerabilities, a sample set of Management Application Functions (MAFs) for each MFA are examined for possible security risks. (MAFs are subcategories that further describe and specialize functionality within a MFA.)

3.1.4.1 Configuration Management—Provisioning. The Provisioning MFA describes the process by which services are ordered and activated. An intruder can invade the TMN at many points in the process. Here are a number of examples of such points:

- *Theft of information*—An intruder can gain access to restricted information, either about the network itself or about the user requesting the service (for example, possible service offerings by the network and which customers are requesting which services).
- *Theft of service*—An intruder can gain access to services without payment either through tampering with the TMN or through repudiating the request for service (e.g., an intruder can order a service, use that service, and then deny having ever ordered the service).
- *Denial of service*—An intruder can cause denial of service by submitting false requests for service, which might eventually cause the TMN not to be able to honor true requests for service; or an intruder can stop valid information from reaching its proper destination (e.g., intercepting and blocking a customer's request for service).

3.1.4.2 Performance Monitoring. Performance Monitoring describes the functionality used to monitor the real-time performance of the network. This information is very valuable to the TMN, since it shows the current state of traffic. It is used to control and tune the performance of the network. Some security vulnerabilities in Performance Monitoring are:

- *Theft of information*—An intruder can gain access to information regarding the performance of the network, such as which resources are currently in use or the current capacity of the resources.

- *Unauthorized use of resources*—An intruder can modify, delay, block, or submit falsified traffic reports to an Operations System (OS). This can cause the network to react to inaccurate or possibly nonexistent traffic conditions.

3.1.4.3 Fault Management—Trouble Administration. Trouble Administration refers to the process by which error conditions in the network are reported to the OSs. As in the other MAFs, intruders can cause damage to a network by corrupting trouble administration messages. Examples include:

- *Theft of information*—An intruder can gain access to information regarding current faults in the network, such as which elements are currently experiencing outages, at what stage of repair those outages are in, or which resources have been allocated to repair the outage.
- *Denial of service*—An intruder can intercept, modify, block, or falsify reports showing network problems (such as trouble tickets). This can cause the network either to allocate resources to a nonexistent problem or to ignore problems that need correction.

3.1.4.4 Accounting Management

3.1.4.4.1 USAGE MEASUREMENT. Usage Measurement measures a customer's amount and time of network usage. This data is used for billing customers. Attacks on the usage measurement of the network can impact the amount of revenue that can be billed by the provider, or compromise the privacy of customers' data, as shown by:

- *Theft of information*—An intruder can gain access to information regarding usage of the network, such as user's usage, which services are currently being used, and the history of usage.
- *Theft of service*—An intruder can intercept, modify, block, or falsify usage information at the resources to affect the billing records being reported to the proper OS. This can result in improper usage records, which will lead to inaccurate billing.

3.1.4.4.2 TARIFFING/PRICING. Tariffing/Pricing deals with the pricing of services within the network. An attack in the tariffing of a service can have large effects on the revenue of the provider:

- *Theft of information*—An intruder can gain restricted information about the pricing of a service.
- *Theft of service*—An intruder can intercept, modify, or block transmission of tariffing information between OSs, thereby preventing proper billing for services.

The attentive reader can now extend the preceding discussion to the more sensitive case of security management. The ease with which this can be done demonstrates the opportunities that await an enterprising intruder.

3.2 THREATS UNIQUE TO THE TMN

This section describes some threats to the TMN that may not appear in other cases or not with the same urgency. It starts by pointing out some generic TMN vulnerabilities that are not tied to any specific applications. It then lists some examples of application-specific threats.

3.2.1 Generic TMN Vulnerabilities

Although the emerging TMN offers new opportunities for improved network management, it also presents new vulnerabilities, as follows.

- The advent of TMN promotes more automation and centralization of network management functions, resulting in **remote management** of NEs. Such remote management, replacing on-site technicians, offers new opportunities to intruders.
- The focus on standard interfaces is especially propitious for interactions with **external entities,** outside the TMN. Priming the TMN to interact with external entities further facilitates the intruder's task.
- In pre-TMN days, an intruder needed to know which protocols were used in network management communications. In many cases the protocols were proprietary. With the TMN's use of **standard interfaces,** an intruder can get complete and detailed information regarding the communication protocols from several Web sites. Use of standard information models also gives an intruder easy access to the semantics of network management transactions in addition to the syntax of the protocols.
- In pre-TMN days, an intruder often needed to know the complete details of operating systems in targeted systems (in conjunction with the interface protocol). With **open system interfaces** an intruder doesn't even need to know which operating system is used in the target system.
- The use of a common **transfer syntax** in open systems interfaces helps an intruder make sense of bits flowing on the wire. For instance, the bits may read as follows: here comes a string of characters, it has 158 characters, and the string is ". . .". With CMIP and publicly available information models it may also be clear what those characters represent—for example, a customer's name and address.
- Suppose an intruder wants to alter the status of line cards that satisfy certain conditions in a large CO. Using SNMP, the intruder will have to issue tens of thousands of messages to the CO to find out which of the line cards satisfy those conditions and later make the changes in the few selected cases. With CMIP this can be accomplished, using **scoping and filtering,** with a single message. Even in the absence of any security measures, someone is bound to notice tens of thousands of unexpected messages flowing into a CO. Without stringent security measures, the chances are that no one will notice a single spurious message.

The bulleted list above illustrate how an intruder can manipulate the beneficial features of the TMN.

3.2.2 TMN Domain-Specific Vulnerabilities

Recent legal and regulatory developments in the United States have opened, even mandated, new electronic interfaces between Incumbent Local Exchange Carriers (ILECs) and Competitive Local Exchange Carriers (CLECs) for such purposes as service ordering and trouble administration. The resulting Inter-Carrier Electronic Commerce (ICEC) presents some new security risks, which are the focus of this section. This illustrative analysis should help the reader develop similar risk analyses for other TMN areas.

The ICEC presents some unique security challenges beyond those traditionally found in telecommunications networks. Indeed, with ICEC an ILEC offers electronic connections to its own competitors, who are competing with each other over publicly known, standard interfaces. Some of the threats are generic (i.e., application independent), whereas others are application specific. Both categories are examined here.

3.2.2.1 Generic ICEC Threats. ICEC interfaces exacerbate some of the common threats to communication interfaces and introduce additional, unique threats. Some of these unique threats are:

- ICEC promotes new applications with external entities, outside the ILEC's TMN, thereby increasing external exposure of TMN elements.
- As competition increases, employees under pressure are more likely to resort to unethical methods (for example, illegally changing, or "slamming," a customer's preferred interexchange carrier) in order to achieve their goals.
- ICEC interfaces may be carried over extended networks, and operated by different providers, thereby offering potential intruders a broad selection of points of attack.
- An ILEC with excellent security in place still cannot be sure that all of its connected CLECs have equally good security; a vulnerability in a CLEC's system may allow an intruder to attack connected systems belonging to others.
- An unscrupulous CLEC might surreptitiously gain access to other CLECs' confidential information maintained by the ILEC. The aggrieved CLECs may then seek compensation from the ILEC for any perceived losses.
- Although many CLECs are established, reputable companies, nothing prevents a criminal organization from setting up its own CLEC as a front or for the purpose of committing theft, fraud, or other mischief. In this case even bona fide messages coming from this CLEC, not from an intruder, may cause harmful consequences.
- The ICEC interfaces are based on publicly known and available standards.
- Communications stacks and protocol analyzers are commercially available and can be used by intruders.

3.2.2.2 Application-Specific Threats. It is neither practical nor necessary to compile an exhaustive list of the threats that are specific to applications. The following cases illustrate what might happen.

3.2.2.2.1 SERVICE ORDERING. Service ordering presents several opportunities for mischief, as illustrated below.

False Inquiry Without proper access control mechanisms in place, a CLEC may inquire about the services of its competitors' customers and obtain information it is not entitled to.

Order Repudiation A CLEC employee may have erroneously placed a service request (perhaps in anticipation of a contract that did not materialize). To avoid any charges for the request, that employee and the CLEC may deny having placed the service request. Meanwhile, the ILEC may have spent considerable resources in fulfilling the request (for example, starting new construction for a large request). The ILEC needs the equivalent of a legally binding signed contract in order to recover its costs.

False Order Cancellation A CLEC employee, under pressure to show better results than competing CLECs, may cancel a competitor's service request in order to degrade that CLEC's service as perceived by their customers.

Order Modification A CLEC employee, under pressure to show better results than competing CLECs, may modify a service request from a competing CLEC by, for example, changing a request for 100 lines to a request for 1,000 lines in order to cause a dispute between the targeted CLEC and the ILEC.

Delay Notification of Service Availability A CLEC employee, under pressure to show better results than competing CLECs, may delay or delete a notification from the ILEC to the targeted competing CLEC that the requested service is now available.

Customer Private Information By capturing service inquiry and service request messages, an intruder can gain access to customers' private information, such as name, address, phone number, and social security number. The aggrieved parties might hold the ILEC (partly) responsible for compromising their private data.

3.2.2.2.2 Trouble Administration. At first glance, trouble administration does not seem to raise any security concerns. However, upon closer scrutiny even such an innocuous looking application offers an intruder some opportunities for mischief.

False Trouble Reports By masquerading as a valid CLEC, an intruder may submit fictitious trouble reports. The result (desired by the intruder) may be lots of ILEC technicians chasing nonexisting problems. This could increase the ILEC's operating costs while reducing the quality of service it provides to its customers, CLECs, and otherwise.

False Claim of Trouble Report Submission A CLEC employee may have erroneously prevented or delayed the submission of an important trouble report. The result may be a long-lasting failure, with consequent losses to the CLEC. The employee, eager to protect his or her reputation, may falsely claim to have submitted the trouble report. The employee may bolster such a claim by doctoring the CLEC's log. The CLEC's management may innocently believe their employee and seek compensation from the ILEC.

Delay/Delete Notifications re Trouble Report Status/Disposition This attack is similar to the deletion/delay of a notification of the availability of requested service. A CLEC employee, under pressure to show better results than competing CLECs, may delay or delete a notification from the ILEC to the targeted CLEC that the reported trouble is now fixed.

3.2.2.2.3 Testing. *Request/Perform Unnecessary Intrusive Testing* A CLEC employee, under pressure to show better results than competing CLECs, may request/perform unnecessary intrusive testing that disrupts service to others.

Flood Testing Capabilities This attack is similar to the submission of false trouble reports. By masquerading as a valid CLEC, an intruder may submit unwarranted test requests. The result (desired by the intruder) may be lots of ILEC technicians performing unnecessary tests. This could increase the ILEC's operating costs while reducing the quality of service it provides to its customers, CLECs, and others.

Falsify Testing Results A CLEC employee who has accidentally caused a failure may falsify test results in order to shift the blame to an innocent party.

3.2.2.2.4 Performance Management. A CLEC employee, under pressure to show better results than competing CLECs, may falsify performance data sent from the ILEC to targeted CLEC(s), or change the requests for performance data from CLEC(s) to the ILEC. In either case, the CLEC(s) may fail to undertake appropriate actions (ordering new services, rerouting traffic), resulting in unnecessary expenses or reduced quality of service.

3.2.2.2.5 Alarm Notification. A CLEC employee, under pressure to show better results than competing CLECs, may send a false alarm notification to another CLEC, causing the targeted CLEC to undertake unnecessary actions. This may increase the targeted CLEC's operating costs while reducing the quality of their service. The aggrieved CLEC may innocently believe that the alarm notification came from the ILEC and seek compensation.

3.3 SECURITY SERVICES

Various security services can protect the TMN from the attacks described in the preceding section. These services have varying coverage; they protect against different threats from intruders with different levels of technical sophistication and tenacity. Some of those services are interdependent, and some have overlapping functionality. When the cost of security services is an issue, as it usually is, it is the challenge of the TMN security architect to select the right mix of security services.

3.3.1 Connection Access Control

Connection access control is the first line of defense. It prevents connection establishment by an unauthorized party. It prevents all messages, including association setup request messages, from unauthorized sources from reaching their destination. It therefore saves the targeted systems from having to deal with such unauthorized messages. It often saves the network from even carrying such messages.

Connection access control compares the source address and the destination address against a list of authorized source/destination pairs. It is typically provided through Closed User Groups (CUGs) over X.25 transport networks and through Fire Walls (FWs) over TCP/IP transport networks. It can be circumvented by falsifying the source address and replacing it with that of an authorized party. This is the essence of IP address spoofing attacks in TCP/IP networks.

Although connection access control may be bypassed by some intruders, this service provides good protection against a potentially large number of would-be intruders with little technical sophistication or motivation. It has the advantage of simplicity: it is provided through only one, or a few systems, in order to protect a large number of systems.

3.3.2 Peer Entity Authentication

Peer entity authentication is the application-to-application equivalent of secure login. It allows each party to strongly authenticate the other through cryptographically secured exchanges. It provides a (usually successful) second line of defense against intruders who have successfully bypassed connection access control. Peer entity authentication is performed during association setup. Quite frequently, it also serves as a vehicle for negotiating security measures that will be enforced/available during the course of the connection.

Peer entity authentication assures the responder of an association setup request that the request comes from the party identified as such in the request message. It further assures the initiator of the association setup request that the response came from the intended receiver of the request, and not from an intruder.

3.3.3 Data Origin Authentication

After a successful association setup by an authorized party, an intruder can "hijack" the association by inserting its own messages toward either of the corresponding parties. Data origin authentication protects against this type of intrusion by ensuring that messages received in the course of the association are indeed from the purported sender. Data origin authentication can be applied to all the messages being exchanged in the course of the association, or only to some selected messages.

3.3.4 Integrity

Data origin authentication assures the receiver of a message that the message was sent by the correct party. However, it does not provide any assurance that the message has not been modified while in transit. Indeed, a determined intruder may physically tap into the transmission line carrying the messages, read every message, and selectively modify the tenor of some or all of the messages. Although such an intruder may not be able to freely insert false messages into the association, the ability to modify any of the regular messages offers tremendous opportunities for mischief. Integrity service detects such willful alterations, and it automatically provides data origin authentication. Detection of mischievous alterations usually prompts the user to undertake recovery measures. Recovery, however, is outside the scope of the integrity service. Integrity service comes in several forms, as follows.

3.3.4.1 Selective Field Integrity. In some cases only a few, critical fields within a large message require integrity protection, and so selective field protection may be appropriate. In most cases, however, it is just as easy to protect the integrity of the entire message. However, depending on the facilities that are available, whole message protection may not be practical.

3.3.4.2 Whole Message Integrity. Whole message integrity allows the receiver to detect whether any alterations have been made to the message. In general, it does not allow the receiver to automatically recover the correct message from the corrupted message.

3.3.4.3 Session Integrity. An intruder who cannot modify messages without the intrusion being detected can still cause considerable mischief by manipulating valid messages. An intruder can delay messages, delete messages, change the order of messages, replay a valid message to the receiver at a later time(s), reroute a valid message to unintended destination(s), and misroute a valid message to the wrong destination (a combination of message deletion and message rerouting). Session integrity service allows the detection of some or all of these types of intrusion.

3.3.5 Confidentiality

The security services discussed so far provide protection against active intrusion. An intruder can opt for only passive intrusion: read messages in transit without affecting them and thereby acquire some confidential information. Confidentiality protects against this intrusion. Although confidentiality protects against unauthorized disclosure of information, it does not automatically protect against message alteration. In general, when a message is protected for confidentiality, an intruder cannot change its content in any controlled way. For example, an intruder cannot change the customer's name from "Alice" to "Bob." However, an intruder may cause some arbitrary changes to a few fields, and the receiving party may not be able to detect the fact that the message has been altered. Like integrity, confidentiality comes in different forms.

3.3.5.1 Selective Field Confidentiality. Selective field confidentiality, as the name implies, protects only a few, selected fields within a message. Selective field confidentiality is used in the following cases:

- Only a few fields (for example, customer name, address, phone number) in a large message need confidentiality protection; encrypting only the selected fields would reduce processing overhead and in some cases transmission overhead.

- While some fields need confidentiality protection, it is important that other fields be available to users who are not authorized to read the confidential information.
- The message contains several confidential fields, and different users are allowed to read different fields. In this case the various fields may be encrypted with different keys and perhaps with different algorithms (if the different protected fields need different levels of security).
- Selective field confidentiality may be used out of necessity if whole message confidentiality is not available.

3.3.5.2 Whole Message Confidentiality. Whole message confidentiality protects the privacy of an entire message. It is used in the following cases:

- It is important that the whole message remain confidential.
- Only a portion of the message requires confidentiality protection, but it is easier to encrypt the whole message than a few selected fields.
- Only a few fields need confidentiality protection, but selective field confidentiality is not available (typically because confidentiality is provided below the presentation layer).

3.3.5.3 Traffic Flow Confidentiality. In some rare cases, it is important that an intruder not be aware that messages are flowing between some locations, let alone be aware of the length and number of such messages. This is quite relevant in military applications but is not required for network management applications. This service had been considered for the management of Lawfully Authorized Electronic Surveillance (LAES). LAES allows law enforcement agencies, with proper lawful authorization, to listen to and gather information about communications involving suspected criminals. However, this security service was eventually dropped because it was not considered crucial and it is rather expensive to provide. Therefore traffic flow confidentiality will not be considered any further here.

3.3.6 Non-Repudiation

Non-repudiation prevents an entity from falsely denying it has sent or received a message. It is used mostly for legal protection.

3.3.6.1 Non-Repudiation of Origin. Non-repudiation of origin provides the receiver of a message with irrefutable proof, which can be verified by a neutral third party (for example, a judge), that the message was produced by the purported originator and that not a single bit has been changed in the message since it was produced.

Non-repudiation of origin automatically provides data integrity. Although integrity provides acceptable proof to the receiver that the message comes from the purported originator without any subsequent alterations, non-repudiation of origin provides such proof to any third party. In particular, it proves that the message could not have been fabricated by the receiver.

Non-repudiation of origin may be important, for example, for electronic service ordering. Thus, if after issuing a request for 1,000 circuits a CLEC denies having issued the request or claims the request was only for 10 circuits, the ILEC will have the electronic equivalent of a signed contract to prove the authenticity and veracity of the original service request.

3.3.6.2 Non-Repudiation of Receipt. Non-repudiation of receipt provides the sender of a message with irrefutable proof, which can be accepted by a neutral third party (for example, a judge), that the message was received by the intended recipient without any alterations. In general, it also proves that the message was received at a specific date and time.

Non-repudiation of delivery may be quite important to the CLECs. When a CLEC submits a service request or a trouble report that is not acted upon by the ILEC, it will prevent the ILEC from claiming it has never received the request or report.

Non-repudiation of delivery may also be valuable to the CLEC; it will prevent an ILEC from falsely claiming that it has not received a notification that a requested service is available or that a network problem has been fixed.

3.3.7 Access Control

At the network level, access control ensures that only authorized parties can request the establishment of a session with TMN elements. At the application layer level, access control ensures that only authorized entities can successfully establish associations with TMN elements.

At the application level the primary purpose of access control is to ensure that only authorized entities are granted access to TMN resources. For instance, access control is used to guarantee that an ILEC cannot access any CLEC data or resources that the ILEC is not authorized to access. Perhaps equally important, access control is used to ensure that an ILEC cannot access any data, kept by the CLEC, pertaining to another ILEC.

Access control also ensures that a valid user has only limited access to certain resources. In some cases, a CLEC has limited access even to data related to itself; for instance, a CLEC may be able to read any information regarding the status of a service request it had submitted, but it may not be allowed to change the "promised availability date" as set by the ILEC. Access control can further accept or deny requests based on the time of day, day of the week, or physical location of the origin of the request message.

A TMN needs to allow valid entities to designate proxies for some applications and to confer certain access privileges to such proxies. For instance, in some cases a CLEC may allow an ILEC to designate one or more proxies that will act on its behalf. In such cases, the ILEC may extend some or all of its access privileges to its proxies.

3.3.8 Security Alarm

A basic security requirement for each system is that it should be able to issue a security alarm whenever an attempted security breach is suspected. The CLEC should be able to determine what information shall be contained in the security alarm notification and to what destination the notification shall be routed. The destination may depend on the type and severity of the alarm, time of day, and day of the week. Such notifications should be acknowledged by the recipient. If proper acknowledgment is not received within a predetermined (and manageable) time window, the alerting system should attempt to send the alarm to an alternate destination(s).

Once a system has been corrupted, it may not be able to send any security alarms. Worse yet: it may send only misleading notifications. It is therefore important that each system be able to issue a security alarm notification when it suspects (attempted) intrusion into a system with which it interacts. In particular, all the ILEC systems interfacing with the ILEC gateway(s) should be configured to spot unwarranted exchanges and issue security alarm notifications when such exchanges are detected.

A system may be configured to enter a "panic mode" when it detects a serious intrusion and it receives no acknowledgments for the security alarm notifications it sends out. In this mode it may fully or partially suspend all activities except broadcasting security alarm notifications to all the systems with which it communicates. This defense requires that TMN systems be able to recognize such distress messages and relay them to the proper destination.

TABLE 3.2 Security Services against Security Threats

Security Services	Security Threats						
	Unauthorized Access	Eavesdropping	Masquerade	Modification of Information	Repudiation	Replay Reroute, Delete	Network Flooding
Access control	✔			✔			✔
Peer entity authentication	✔		✔				
Data origin authentication	✔		✔				
Data integrity[a]			✔	✔		✔	
Data confidentiality		✔	✔				
Non-repudiation					✔		

[a]Data Integrity cannot stop the security breach but can only detect the security breach.

3.3.9 Security Audit Trail

Some intrusions are massive and immediately visible: they may disable part or all of the network, or they may delete all billing records. Other attacks are more insidious: the intruder may acquire some confidential data without making any changes; the intruder may make small configuration changes that increase its own available bandwidth without proper billing and without having a major impact on overall network performance; or the intruder may effect modest reductions in its billing records. When such attacks succeed, their effect is not immediately noticeable. However, by analyzing transaction data collected over long periods (for example, a few months), from several sources, the cumulative effects of the intrusions may become apparent. If adequate records are kept, it may be possible both to detect the intrusion and to identify the intruder. A security audit trail is an event log that captures such security-relevant information.

A security audit trail records all security alarm notifications it has issued and the events that triggered such alarms. In general, it may also record innocuous events, such as association setup attempts, and access to certain types of data, which may reveal, in the aggregate, some unexpected patterns.

A security audit trail is especially valuable when a persistent low-level intrusion is suspected. The security administrator may then configure the security audit trail to capture specific events, based on the type of suspected intrusion.

Table 3.2 lists the various security threats described in Section 3.1.2 and shows the security services that can attempt to thwart the security threats. Table 3.2 does not include security alarm and security audit trail services since those service are used to detect rather then prevent attacks. The network security threats listed in this table can be mapped to a set of possible consequences that arise from those threats.

Table 3.3 combines Tables 3.1 and 3.2 by correlating the possible security services with the security risks from the security threats.

TABLE 3.3 Security Services Correlated with Security Risks

	Security Risks			
Security Services	Theft of Information	Unauthorized Access to Resources	Theft of Service	Denial of Service
Access Control	✔	✔	✔	✔
Peer entity authentication		✔	✔	
Data origin authentication		✔	✔	
Data integrity[a]	✔	✔	✔	✔
Data confidentiality	✔	✔		✔
Non-repudiation			✔	

[a]Data Integrity cannot stop the security breach but can only detect the security breach.

4

Basic Security Mechanisms

Security mechanisms consist of various algorithms and protocols that support security services. Although some of the security mechanisms discussed in this chapter support single security services (for example, digital signatures are used essentially for non-repudiation), others are used in the provisioning of several security services (for example, hashing is commonly used for peer entity authentication, data origin authentication, data integrity, and in non-repudiation). This chapter describes the generic security mechanisms, for example, hashing and digital signatures, as well as some specific algorithms used for such mechanisms. The focus is only on some of the algorithms most commonly used in network management. Other algorithms may be more suitable for different applications, for example, electronic commerce over the World Wide Web. While this chapter provides some aspects of those algorithms that are most relevant to their use in network management, it omits the mathematical details of their exact definition, performance analysis, and cryptographic robustness.

4.1 HASHING

A hashing operation takes as input a (potentially long) plain text message M and produces a fixed-length, short (for example, 16 octets in the case of MD5 [RFC1319]) message digest m. The hashing function is considered a secure, one-way function if it satisfies four related requirements:

- Any change to the initial bit string (for example, adding, deleting, or changing a bit) changes the message digest completely and not just a small portion thereof. (This is part of the **secure** property.)
- Even when the hashing function, the message, and its digest are known, it is practically impossible to construct another message that would produce the same digest. (This requirement is called **second preimage resistance** property.)
- It is computationally infeasible to find any pair of messages that have the same hash value. (This requirement is known as **collision free.**)
- It is practically impossible to derive the original message from its digest, even if the hashing function is known. (This is the **one-way** property.)

Several hashing algorithms have gained some popularity in the last few years: MD2, MD4, MD5, SHA (Secure Hashing Algorithm) [FIPS180-1], and MDC-2. MD2 may be too slow; MD4 [RFC1320] has some security weaknesses; MD5 [RFC1319] would be preferable. SHA, which is similar to MD5, is used with the NIST Digital Signature Algorithm (DSA) as part of the Digital Signature Standard (DSS). MDC-2 is based on the NIST Digital Encryption Standard (DES).

No known hashing functions have been proven to be secure. For instance, at present there is no proof that no one could ever invent a simple way to produce different strings that result in the same SHA digest. Some hashing functions are known to be insecure. For instance, for some hashing functions it is fairly easy to produce two distinct messages that result in the same digest. Other hashing functions are not known to be insecure. Obviously, only hashing functions in the second category are used. If a hashing function moves from the second category to the first one, its usage should be stopped.

MD5 produces a 128-bit digest; it is not collision free. Indeed, pairs of strings that result in the same MD5 message digest have recently been produced. Although these developments cast some doubt on the usefulness of MD5 for some applications, they are not threatening to the use of MD5 in network management. Indeed, MD5 may be used to ensure the integrity of network management messages. While two strings that result in the same MD5 message digest can be found, it is most unlikely that either one of them, let alone both strings, will resemble valid network management messages. Furthermore, at present MD5 is still believed to be second preimage resistant. As such, it is still useful for network management applications. SHA is a more recent hashing algorithm, producing a 160-bit digest. MD5 and SHA are currently the leading candidates for hashing in network management security.

Hashing is rarely used directly to provide security services; rather, it is generally used as an ingredient in keyed hashing and digital signatures which are described further below. The structure of MD5 is representative of most secure hashing functions. Since this structure has important security implications, it is briefly presented here. MD5 starts by padding an arbitrary length initial message into a multiple of 512 bits as follows: it appends a single 1 bit to the message, followed by enough 0 bits to produce a message that is 64 bits short of a multiple of 512 bits. It then appends a 64-bit representation of the initial message's length. The core of MD5 consists of a nonlinear function that operates on two bit strings—one of 512 bits, and the other of 128 bits. The result of the core MD5 function is a string of 128 bits. The core MD5 function operates sequentially on the 512-bit blocks of the padded message. For the first 512-bit block, it uses the following 128-bit string (in hexadecimal notation) as its second input:

Initial value = 0123456789abcdeffedcba9876543210

For each of the subsequent blocks, it uses the output from the previous round as its 128-bit input. The output from the last round is the message digest produced by MD5. The operation of MD5 on a single block is illustrated in Figure 4.1. Figure 4.2 illustrates the operation of MD5 on a whole message.

Hashing by itself does not provide security (except for the detection of transmission errors). Rather, hashing is used by other security mechanisms, such as digital signatures and keyed hashing (as discussed below).

4.1.1 Keyed Hashing

Keyed hashing consists of appending a secret key, known to the two communicating entities, to a clear text message hashing the result, and then sending the original message, either encrypted or not but without the previously appended key, along with the message digest computed with the hashing function. Keyed hashing can be used to provide **strong peer**

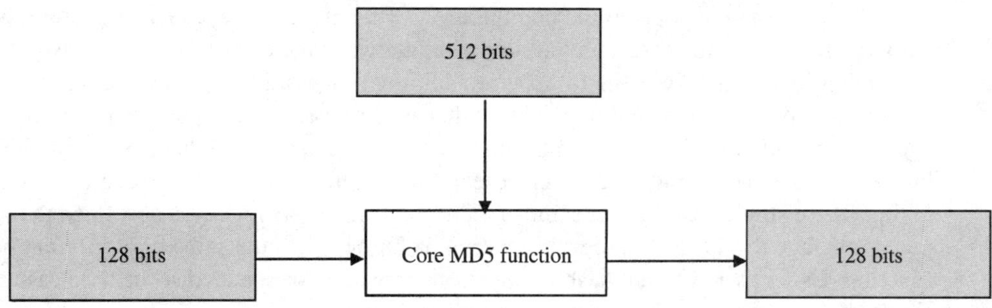

Figure 4.1 Structure of MD5 for one 512-bit block

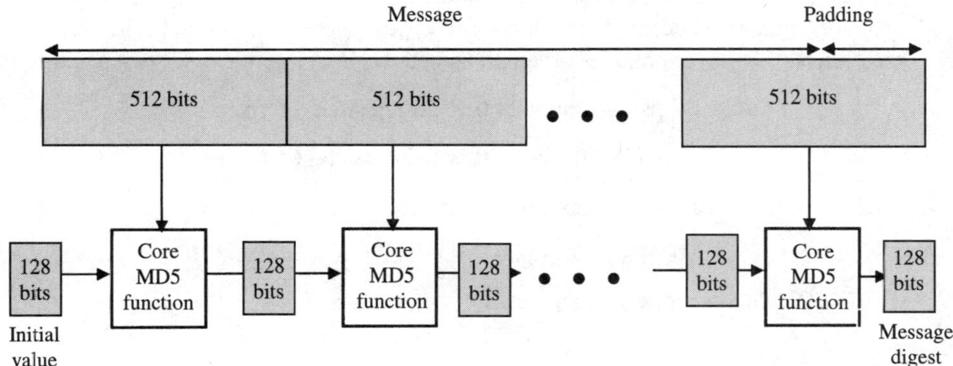

Figure 4.2 MD5 operating on a whole message

entity authentication. Each party can compose a string consisting of its own identifier, the intended recipient's identifier, and a unique time stamp. It then computes the keyed hash of that string by appending the secret key to it and hashing the result. The original string is then transmitted, in the clear, along with the message digest. The receiving party can authenticate the sender by appending the shared secret key to the clear text string and compute the same hash as the sender. If the resultant message digest is the same as the one received from the sender, then the identity of the sender has been authenticated. This type of authentication is resistant to replay since every authentication will include a different time stamp. It is further resistant to misrouting or rerouting since the identity of the intended receiver is included within the string protected by the hashing.

Keyed hashing can also be used to provide **strong data origin authentication.** The sender can simply append an authenticator, as described above, to each message to be protected. Notice that in order to avoid replay each authenticator needs to use a different time stamp.

Keyed hashing can also be used to provide **integrity** protection. The sender appends the secret key to the clear text message, computes the message digest of the resulting string, and sends the initial message along with the message digest. If the message has been altered in transit, then the keyed hash computed by the receiver on the received message will not match the received message digest. In order to protect the message against replay, a unique (nonrepeating) string must be included in the original clear text message. It can be a unique time stamp, a sequence number, a nonrepeating random number, or any combination of those. Inclusion of a time stamp will further allow detection of delays. Inclusion of a time stamp or a sequence number will allow detection of reordering of messages within an association. Inclusion of a sequence number allows detection of message deletion. In order to prevent rerouting or misrouting attacks to other parties that share the same key, the identity of the intended receiver should be included in the original message.

Simple usage of some of the popular hashing functions in keyed hashing presents some vulnerabilities. In some cases, an intruder who intercepts the clear text message with the corresponding digest may be able to append additional material to the clear text message and compute a corresponding message digest that will be accepted by the receiver. To protect against such vulnerabilities, double hashing is used instead of single hashing. In double hashing, a secret key is appended to the clear text message and the message digest is computed as in the case of single hashing; the same or a different shared secret key is appended to the computed message digest; and the resulting string is hashed with the same or a different hashing function. Using two different hashing functions provides some added security in case a security flaw is uncovered in one of the hashing functions.

HMAC is a commonly used keyed double-hashing procedure [RFC2104]. It has been initially defined with MD5 hashing for authentication over IP networks. HMAC uses a key *K* of length equal to 64 octets. If the shared secret key is less then 64 octets, then it is padded with zero octets (0 × 00) to obtain *K* of length 64 octets. HMAC further uses two padding constants:

$$\text{ipad} = \text{the octet } 0 \times 36 \text{ repeated 64 times}$$

$$\text{opad} = \text{the octet } 0 \times 5\text{C repeated 64 times}$$

HMAC-MD5 for a message "text" is defined as:

$$\text{MD5(K XOR opad, MD5(K XOR ipad, "text"))}$$

where XOR denotes bitwise exclusive-OR.

4.1.2 S-key

S-key was developed by Bellcore in order to provide peer entity authentication with the simplicity of password authentication, yet with resistance to replay attack.

A major weakness of simple password authentication is that an intruder who intercepts a valid password can subsequently replay that password, thereby successfully masquerading as the original, valid party. With S-key, the original password of a valid user is successively hashed *N* times, for example, 1,000 times. Only the last hash is conveyed, securely, to the system that needs to authenticate that user, and that value is stored locally (by the receiver). The user keeps the original password and a counter *I* of how many times the password has been used. For each session establishment, the user hashes the password *N*-*I*-1 times and uses the result as the authentication value. The receiver hashes this authenticator once and compares the result to the locally stored value for that user. If the hashed value of the authenticator (i.e., the original password hashed *N*-*I* times) matches the locally stored value, then the user has been properly authenticated and the newly received authenticator (i.e., the password hashed *N*-*I*-1 times) replaces the locally stored value for that user. The same original password can thus be used *N*-1 times.

Including an S-key based authenticator with the messages of an association provides data origin authentication for those messages (though without data integrity protection), with protection against replay, deletion, and reordering without the use of time stamps or sequence numbers.

4.2 ENCRYPTION

Encryption is the process by which information is encoded so that only an authorized user has access to that information. A user needs the "key" that will unlock the encoded information and change it back to its original form. There are two forms of encryption:

- Symmetric encryption
- Asymmetric encryption

Symmetric encryption is a procedure by which the same key is used for both encryption and decryption. Asymmetric encryption, on the other hand, uses a pair of keys: one for encryption and the other for decryption.

4.2.1 Symmetric Key Encryption

Symmetric key encryption is the oldest and most commonly used type of encryption to provide confidentiality service. With symmetric key encryption, two communicating entities share a secret key used to encrypt and decrypt messages.

A brute force attack against symmetric key encryption consists of trying all possible keys until the decrypted message looks like a valid clear text message. With a good encryption algorithm, no attack is significantly more efficient then the brute force attack. The strength of such algorithms (i.e., the amount of computation it takes, on the average, to uncover the secret key) increases exponentially with the size of the key.

The prime candidate for symmetric key encryption in the TMN is the Data Encryption Standard (DES) [FIPS46-1, FIPS46-2, FIPS74, FIPS81]. Although DES has been in use for over 20 years, there are no publicly known successful attacks against DES that are significantly more effective than brute force attack. DES uses a 64-bit key, of which 8 are parity bits; thus, a DES key has 56 random bits. Therefore, on the average, the DES algorithm has to be run 2^{55} times in order to decrypt a DES encrypted message and retrieve the encryption key. With sufficient processing power (for example, thousands of PCs), a DES encrypted message can be decrypted in a matter of days or even hours. Therefore a DES key should be changed at least once a day. Furthermore, DES encryption should not be used for encrypting highly sensitive information, for example, lists of encryption keys.

DES is based on the Data Encryption Algorithm (DEA). DEA operates on two inputs:

- A 64-bit DES key that must be secret
- A 64-bit block of plain text that needs confidentiality protection

It produces a 64-bit ciphertext. The reverse operation uses the ciphertext and secret key to produce the original plain text.

Encrypting the same plain text block with the same key will result in the same ciphertext. Simply observing identical ciphertext blocks may provide an intruder with some useful information. To protect against this attack, the plain text block is first exclusive-ORed with an arbitrary 64-bit Initialization Vector (IV). The receiving party must know the IV, which need not be secret. Typically, it is transmitted, unencrypted, along with the ciphertext. This mode of using DES, which is called the Electronic Code Book (ECB) mode, is illustrated in Figure 4.3.

An obvious disadvantage of the ECB mode is that for a large message an IV needs to be sent with every 64-bit block of plain text. This overhead is dealt with in the Cipher Block Chaining (CBC) mode which uses the ciphertext from one block as the IV for the next block. Only the first block needs an explicit IV, as illustrated, without showing the DES key, in Figure 4.4. The single IV is transmitted along with the ciphertext blocks. Besides ECB and CBC, there are two more DES modes of operation. However, the CBC mode is most appropriate for encrypting long messages; therefore, it is the one used in TMN applications. Figure 4.5 illustrates decryption of DES in the CBC mode, without showing the encryption key.

4.2.1.1 Padding. The DES algorithm takes as input clear text of length that is an integer multiple of 8 octets (and an encryption key). If the clear text has a length that is not an integer multiple of 8 octets, then more **padding** octets must be added to comply with the DES specifications.

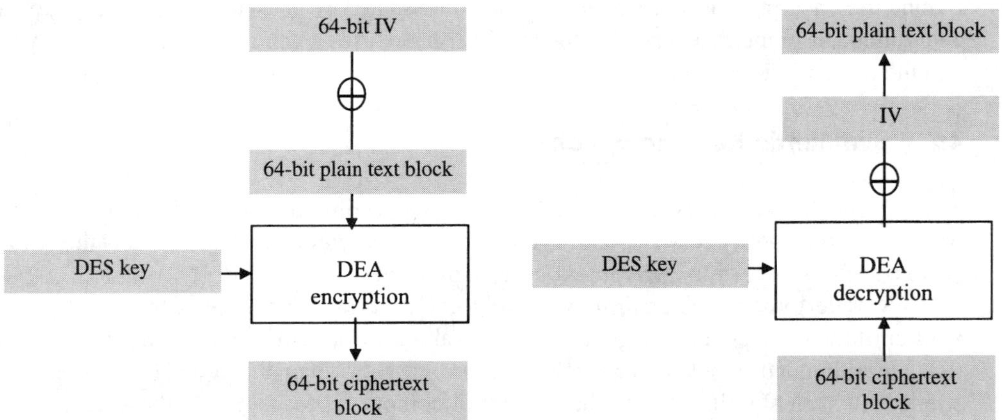

Figure 4.3 DES encryption and decryption in the ECB mode

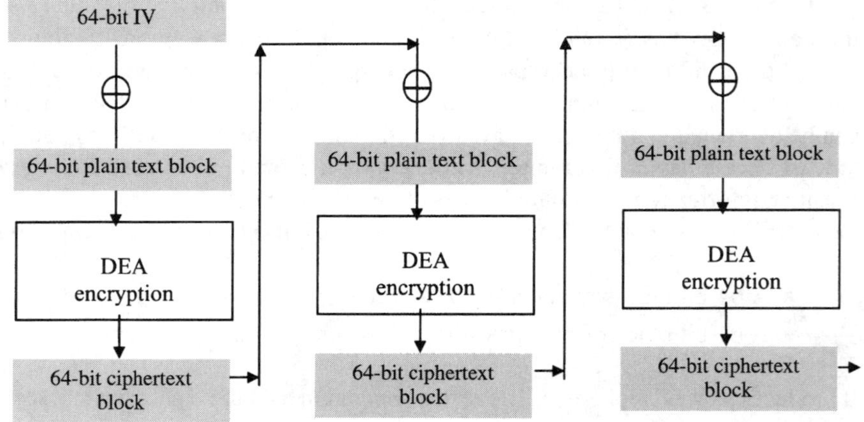

Figure 4.4 DES encryption in the CBC mode

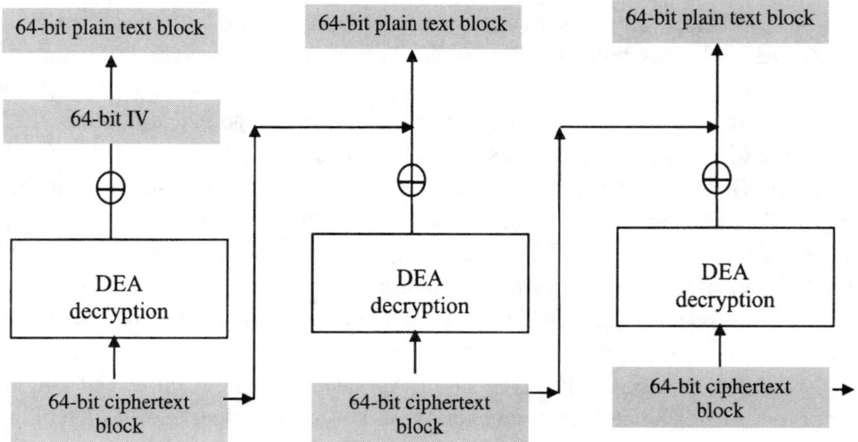

Figure 4.5 DES decryption in the CBC mode

The convention adopted for the initial TMN applications in the United States is that the last octet of text shall denote how many of the last 8 octets are relevant data octets; the remaining octets, if any (i.e., after the last relevant data octet and before the last octet of the text), are arbitrary. The OIW (Open Systems Environment Implementers Workshop) Implementation Agreements specify that rather than being arbitrary, those octets should be the same as the last octet. If the clear text has a length that is an integer multiple of 8 octets, then 8 padding octets are added, the last octet having the value 0.

If the padding overhead is an issue, **zero overhead "padding"** can be used, as specified for some TMN applications.

If the clear text is a multiple of 8 octets, it shall be encrypted in CBC mode without any padding. Otherwise, if the length of the clear text is $n.8 + k$ octets ($n > 0, 0 < k < 8$) then the first $n.8$ octets are encrypted in DES CBC, the last 8 octets of ciphertext are transmitted, as usual, then they are encrypted again, the first k octets of this last encryption operation are XORed with the remaining k octets of the original clear text, and are transmitted. If the clear text is less than 8 octets (i.e., $n = 0$), then a confounder may be used (8 random octets in front of the clear text). If a confounder is not used but an initialization vector (IV) is used (the two provide the same function), then the IV (which must be known or transmitted in clear text) shall be encrypted, and the first k octets of the resulting ciphertext shall be XORed with the k octets of clear text. If neither an IV nor a confounder is used, then one encrypts the key itself and XORs the first k octets of the encrypted key with the clear text. (This solution may be more vulnerable to replay, known clear text attack, and electronic codebook attack.)

4.2.1.2 IV Selection. In principle, the IV should be an arbitrary string of 64 bits. This specification has led to a subtle vulnerability in some of the early TMN implementations in the United States. It is described here mainly in order to illustrate potential vulnerabilities that may be associated with even fairly simple security procedures.

The authentication procedure for those applications [TR40] specified the use of a DES-encrypted time stamp, in the CBC mode. It further specified the use of arbitrary IV, to be transmitted in clear text. Finally, it specified that the DES encryption key be changed every four days. An intruder can make copies of the authenticators transmitted in the course of one or more days on a given association. For each copy the intruder notes the time when the authenticator has been transmitted and the ID of the encryption key. On any subsequent day, while the same encryption key is used, the intruder can replay a captured authenticator at the exact time of day when it was initially transmitted, with an appropriately designed IV as follows: For the initial authenticator, the intruder knows the IV and the first 8 octets of GeneralizedTime: YYYYMMDD (i.e., 4 octets for the year, 2 for the month, and 2 for the day). Therefore, the intruder also knows IV XOR YYYYMMDD (the exclusive OR of the IV and the first 8 octets of the GeneralizedTime that went into the DES encryption in the first round of the CBC mode). When the intruder replays the authenticator on date YYYYMMDD′, the IV will be changed to

$$\text{IV}' = (\text{IV XOR YYYYMMDD}) \text{ XOR YYYYMMDD}'$$

This way

$$\text{IV}' \text{ XOR YYYYMMDD}' = \text{IV XOR YYYYMMDD}$$

Thus, the input into the first round of DES encryption is the same in both cases. All communications after the first day the association is established, as well as for the duration of the usage of the key, are vulnerable to this attack. After this vulnerability has been discovered, the specifications are changed to require that the DES key shall be changed at least once every 24 hours.

4.2.1.3 Error Propagation. DES in the CBC mode has the advantage that a transmission error in the ciphertext stream has only a limited impact on the decrypted message. Suppose one or more bits within the *n*th 64-bit ciphertext block have been corrupted. From Figure 4.4 we see that the corresponding *n*th block clear text will be completely garbled. Furthermore, the $n + 1$th 64-bit block of clear text will have the wrong bit wherever the *n*th ciphertext block has a wrong bit. No other clear text bits are affected.

Whereas the consequences of changing a few bits are contained, the effects of deleting or adding bits are more severe. If a single bit is deleted or added in a ciphertext block, the corresponding clear text block as well as all subsequent clear text blocks will be garbled.

Although limited error propagation is desirable, it also allows an intruder to corrupt a message without detection. Indeed, by changing a single bit in a ciphertext block, the corresponding clear text block, as well as 1 bit in the next clear text block, will be in error. Since the rest of the message is correct, the receiver may not realize that a small portion has been corrupted. The intruder cannot be sure what the decrypted message will be. However, the mere ability of ensuring that a specific field within a message is incorrect may be sufficient to cause the harm desired by the intruder. Therefore, integrity must be provided independently of confidentiality. There are three ways to provide both integrity and confidentiality:

1. Encrypt the message, then compute the MAC of the encrypted message.
2. Compute the MAC of the clear text message, then encrypt the message and transmit the encrypted message with the MAC.
3. Compute the MAC of the clear text message, then encrypt the message along with the MAC.

Option 1 has the disadvantage that the MAC is computed for an encrypted message. If the encryption and the MAC are performed by different entities, then the entity providing the MAC may not be sure of what is being authenticated. Option 3 provides a bit of additional security by encrypting the MAC in addition to the message.

4.2.1.4 Triple DES. A DES key can be broken by a brute force attack by running the DES algorithm no more than 2^{56} times. This is adequate for most TMN applications but not good enough for highly sensitive material, such as the transmission of lists of encryption keys. Better security can be bought by using DES twice, with two different keys, but the resulting security is not so much superior. Double DES is vulnerable to the following attack:

A block of plain text *P* is encrypted using keys *k*1 and *k*2, resulting in ciphertext *C:*

$$E_{k1}(E_{k2}(P)) = C$$

Decrypting both sides with k1 gives:

$$D_{k1}(E_{k1}(E_{k2}(P))) = D_{k1}(C)$$

Therefore:

$$E_{k2}(P) = D_{k1}(C)$$

Suppose the intruder knows *P.* This is not terribly unreasonable, for many messages start with a time stamp, the receiver's identity, or the recipient's identity. All the intruder has to do is to encrypt *P* with all possible keys *k*2 and decrypt *C* with all possible keys *k*1 and then look for a match between the two lists. The intruder will have to run the DES algorithm 2**57 times, although the attack requires 8 × 2**56 octets of memory.

When a high level of confidentiality assurance is required, the DES algorithm should be used three times, with three different keys. The resulting **triple DES** (3DES) encryption is resistant to any currently conceivable attacks. 3DES is generally used in the Encrypt-Decrypt-Encrypt (EDE) mode rather than in three consecutive encryptions. This does not impact security in any way, but it has the property that if the three keys are the same, then 3DES in the EDE mode reduces to a single DES.

4.2.1.5 Digital Seals. As discussed earlier, keyed hashing can be used to provide data integrity. Another method is encrypted hashing or digital seals. A digital seal of a message is computed by hashing the message, without appending to it any secret key, and then encrypting the resulting message digest with a symmetric key encryption algorithm such as DES. Digital seals and keyed hashing provide comparable levels of security with nearly identical use of resources. One advantage of keyed hashing is that it does not require the use of strong encryption which may be subject to export controls by the U.S. government. (DES with a 56-bit key is subject to such controls.) Perhaps for this reason keyed hashing is gaining popularity over digital seals.

4.2.2 Asymmetric Encryption

In asymmetric encryption one key is used for encryption, and a close relative is used for decryption. A user starts by creating one member of the pair of keys—the private key. This key remains a closely guarded secret. From the private key the user derives the matching public key. The user can (and usually does) tell the world the value of the public key. Knowing the public key, it is not possible to deduce the private key. When either key is used for encryption, the other is used for decryption.

The implications of public key encryption are amazing. Anyone who knows Alice's key can send her a confidential message by encrypting it with Alice's public key. Although anyone can send such a message, only Alice, who knows the corresponding private key, can decrypt the message. Therefore, Alice need not exchange secret information with anyone in order to receive confidential messages. Eliminating the cumbersome step of initial exchange of some secret information has moved cryptography from the privileged few (super-spies, major financiers) to the masses (Web surfers, digital cross connects). Section 4.3 describes even more fantastic applications of public key encryption.

The main contenders for the popular public key encryption are:

- RSA (Rivest-Shamir-Adelman) [RSA78]
- El Gamal (which can be used only for digital signatures, not for general-purpose encryption) [FIPS186]
- Elliptic curve cryptosystems [KOB87, MIL85]
- Rabin [RAB79]

RSA is the oldest and most commonly used public key encryption algorithm. An RSA private key consists of two large prime numbers (more accurately, it is a simple function of two large primes), whereas the corresponding public key is the product of the two prime numbers. The RSA procedure is illustrated in Figure 4.6. The strength of the RSA algorithm rests on the difficulty of factoring large numbers. Ongoing advances in mathematics and computers challenge this premise. In 1984 the largest number that had been factored was 71 decimal digits long; by 1994, the record moved to a 129-digit number. Therefore, public keys must be chosen large enough to withstand increasingly sophisticated attacks over the time of their expected usage. At present, keys with 512 bits are considered fairly secure against an attacker with modest resources, but they can be broken with the use of massive processing capabili-

RSA procedure	example
Alice generates a private and public key pair	
Alice selects 2 prime numbers p and q.	$p = 7, q = 17$
Alice's modulus $n = pxq$	$n = 7 \text{ x } 17 = 119$
Alice generates e which is prime with respect to $(p-1)(q-1)$	$e = 5$, prime with respect to $(7-1)(17-1) = 96$
Alice generates d such that $Dxe = 1 \text{ mod}((p-1)(q-1))$	$d = ((p-1)(q-1)(e-1)+1)/e = 77$ $dxe = 385 = 4 \times 96 + 1$
Alice's public key is (n,e)	(119, 5)
Alice's public key is (n,d)	(119, 77)
Bob sends a message to Alice, encrypted with Alice's public key	
Bob has message, the binary representation of that message is read as an integer m $(m < n)$	$m = 19$ (<119)
Bob creates an encrypted message $s = m^e\text{mod}(n)$	$19^5/119 = 20807$ with remainder $= 66$
Bob sens s to Alice	$s = 66$
Alice decrypts Bob's message with her private key	
$M = s^d\text{mod}(n)$	$66^{77}/119 = 1237\ldots$ with remainder $m = 19$
Works thanks to the Euler–Fermat theorem: For any two prime numbers p and q, if m is prime with respect to pxq then: $m^{(p-1)(q-1)} = 1 \text{ mod}(pxq)$	

Figure 4.6 Example of the RSA procedure

ties (thousands of PCs) in a few months. Keys with 768 bits are considered adequate for most applications. Critical applications, such as certifying public keys, require keys of at least 1024 bits. Keys that must be secure several years from now must be even larger.

4.3 DIGITAL SIGNATURES

Important business transactions are usually sealed with mutually binding, signed contracts. If any party fails to live up to its contractual obligations, the aggrieved party can produce the signed contract in a court of law as it seeks redress. Even lesser transactions culminate in signed checks with legal standing. As electronic commerce expands, an electronic equivalent of ink on paper signature is required. Consider, for example, a telecom company that buries fiber in the ground in order to fulfill an electronic request for service. It would like to have the equivalent of a signed paper contract that would assure eventual payments for its investments. Digital signatures, based on public key cryptography, provide the solution.

A digital signature of a message (of any length) is typically a fixed-length string of bits that can be derived from the original message and a secret private key. Anyone who knows the corresponding public key can check if a claimed signature is indeed the correct signature corresponding to the original message and obtained with the claimed secret key.

The remainder of this section provides an example of a simple digital signature scheme. (Other digital signature schemes are possible; the approach described here is provided as an illustration, not as a generic description.) It is illustrated in Figure 4.7.

1. A (potentially long) plain text message M is being transformed by a secure hashing function, which is publicly known, into a message digest m. For any given digital signature standard, all the users apply the same hashing function.
2. The originator of the message encrypts the message digest m using its own private key. The result of this encryption is the signature s of message M.
3. Upon reception of the signed message (i.e., M and s), the recipient computes the message digest m of M as in step 1 above. The recipient further decrypts the signature s using the sender's publicly known key. If the result of this operation is identical to the representative m obtained before, then the sender knows that the message

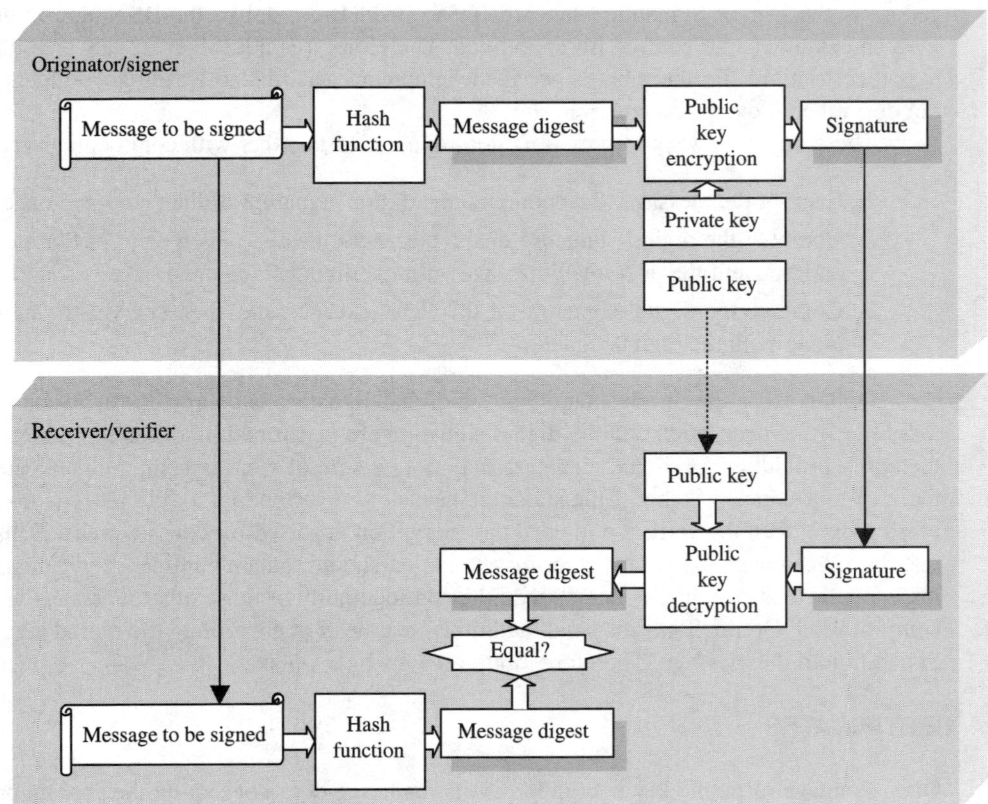

Figure 4.7 Signing and verifying

was indeed originated by the putative sender, and furthermore, that the message has not been altered in any way. The digital signature will therefore detect any accidental transmission errors as well as malicious tampering.

In addition to providing data origin authentication and data integrity, a digital signature also provides non-repudiation of origin. Indeed, if the sender of a signed message later denies having sent that message, any impartial third party (for example, a judge) can perform the validation procedure described above (since the public key of the sender is indeed publicly known) to be convinced that only the claimed sender could have signed that message.

In addition to providing non-repudiation of origin, digital signatures can also be used to provide non-repudiation of delivery. For instance, the receiver of a message can append his or her own name and a time stamp to the message, sign the message, and return the signature, along with the time stamp that was used, to the message originator. The message originator can verify the validity of the signature, which proves that the producer of that signature was in full possession of the message at the noted time. This proof of delivery requires the cooperation of the recipient. If the recipient fails to cooperate, the sender may decide to suspend business transactions with the uncooperative partner. For most TMN applications, this would be sufficient incentive to cooperate. In some cases, for example, when an inventor sends a patent application to an attorney who then claims to be the inventor, this failure to cooperate can be catastrophic for the message originator. In such cases, different mechanisms are required for non-repudiation of delivery. Since the need for such mechanisms, which can be quite cumbersome, is not anticipated for TMN applications, they are not covered in this book.

The most commonly used digital signature algorithm is based on the RSA public key encryption algorithm [RSA78, PKCS1, PKCS7]. The main challenger to this established procedure is the Digital Signature Standard (DSS) [FIPS186]. Unlike the RSA algorithm, the DSS algorithm cannot be used for encryption. Therefore, it is not subject to any government export restrictions. Furthermore, since its development was funded by the government, it is not encumbered by any patents.

There are three ways of providing non-repudiation together with confidentiality:

1. Encrypt the message, then compute the digital signature of the encrypted message.
2. Compute the digital signature of the clear text message, then encrypt the message and transmit the encrypted message with the digital signature.
3. Compute the digital signature of the clear text message, then encrypt the message along with the digital signature.

Option 1 has the disadvantage that the digital signature is computed for an encrypted message. If the encryption and the digital signature are performed by different entities, then the entity providing the digital signature may not be sure of what is being authenticated. A more serious concern is that if the signature needs to be verified by a third party (for example, a judge), then the verifier will need the encryption key used for confidentiality. This key may have been used to encrypt other messages that should remain confidential. If the key is made public (for example, in court), then the confidentiality of those other messages may be compromised. Option 3 has the small additional overhead of encrypting the digital signature in addition to the message. Therefore, option 2 wins by a whisker.

4.4 CERTIFICATES

One advantage of public key encryption is that its usage does not require the sharing or distribution of any secret information. Only the public keys need to be distributed. Such distribution can be done by broadcasting or via a directory, either electronic or paper-based. This distribution still must be done securely. If Alice obtains Bob's public key from a directory, she must be sure that the entry for Bob's public key in the directory has not been changed by an intruder. Such assurance is provided with a public key certificate.

A public key certificate is a record that binds Bob's public key to Bob's unique identifier. The record is signed by a trusted Certification Authority (CA). In order to verify the validity of the certificate, Alice must know the CA's public key. There are no automated mechanisms for the secure distribution of the CA's public key. Alice must obtain it directly, in person, from the CA; through the intermediary of a trusted courier; or from multiple, independent postings of the key (for example, in newspapers and television broadcasts). The security of all of Alice's communications depends on the authenticity of the CA's public key that Alice uses.

ITU-T Rec. X.509 version 3 [X.509] defines the following fields for a certificate:

- Version, the default is version 1
- Serial number
- Algorithm used for signing the certificate
- Name of the issuing CA
- Validity period of the certificate (from, until)
- Name of the subject for whom the certificate is issued
- Public key of the subject and algorithm corresponding to the public key
- Unique identifier of the CA, optional, which can be present only if version 2 or 3 is used
- Unique identifier of the subject, optional, which can be present only if version 2 or 3 is used
- Extensions—which allow for inclusion of additional information
- Digital signature of the preceding fields, properly encoded, by the CA

If Alice has the public key of CA1, which she trusts, and Bob's certificate is signed by CA2, then Alice needs a certificate for CA2's public key signed by CA1. This is the simplest example of a certification path. In principle, a certification path may consist of several certificates. Typically, every entity within a TMN would have the public key of that TMN's CA. All other public keys that it uses must be certified directly or indirectly (via a certification path) by the local CA.

4.5 ACCESS CONTROL MECHANISMS

Access control can be awfully complicated. Hundreds of technicians and applications may have various access privileges to a managed system. A large system, for example, a CO, may have hundreds of thousands of MO instances, each one having numerous attributes. The access privileges of any given user may be restricted to just some operations on some attributes of some MO instances. Furthermore, such privileges may be restricted to only a few hours during some days of the week. Keeping such detailed information for every potential user may be far more complex and voluminous than what's being protected. Fortunately, several mechanisms can dramatically simplify this picture.

The core of any access control mechanism is an access control decision function [X.741]. This function inspects any incoming request to access system resources and decides whether to grant or deny the request. The request is issued by an **initiator** who requests some **operations** on a resource. A CMIP request can specify any of following operations:

- Get (the value of an attribute)
- Set (the value of an attribute)
- Create (an instance of a MO)
- Delete (one or more MO instances)
- Action (on a MO instance, with some specified parameters; for example, "test the circuit with 5V")

The specified operation, along with the information element it addresses (MO or attribute), form the **target** of the request.

The access control decision function reaches its verdict with the help of **access control information (ACI).** There are four types of ACI:

- A set of **rules** kept by the access control decision function
- **Target ACI,** which is typically kept in the system servicing the request (for example, this document is top secret)
- **Initiator ACI,** which is usually supplied by the initiator along with the request (for example, the initiator's authenticated identity, proof of top secret clearance)
- **Request ACI** is bound to an individual request (typically including contextual information, such as the location from which the request originated)

4.5.1 Rules

There are two types of rules: **grant** and **deny;** for example,

- A request from the security administrator to read the security audit trail shall be granted.
- A request to delete a security audit trail record shall be denied.

A rule can be **global,** applying to the entire MIB, for example, the chief technician can read anything. An **item rule** is limited to some parts of the MIB. The limitation can be any of the following:

- All instances of MOs of a certain class
- An enumerated list of MO instances
- MO instances defined by scoping and filtering
- Any combination of the above

A rule can apply to some specific initiators, or it can apply to all initiators. The latter type is a **default rule.**

Several rules may be applicable to a single request. Arbitration among potentially conflicting rules is resolved with the following precedence conventions:

- Global rules have precedence over item rules.
- Deny rules have precedence over grant rules.
- Initiator specific rules have precedence over default rules.

A rule can be restricted to only some days of the week or some hours of the day. For example, a rule that allows a preventive maintenance system to perform disruptive testing may be applicable only during the wee hours of the night, when the effects of disruption will be minimized.

4.5.2 Initiator ACI

Initiator-provided ACI can be partitioned into three categories:

- Authenticated identity initiator
- Anonymous authenticated initiator
- Anonymous unauthenticated initiator

4.5.2.1 Authenticated Identity Initiator. This is the most straightforward type of initiator ACI: the access control decision is based on the authenticated identity of the initiator. In some cases, the identity of the initiator is authenticated at association setup time and is assumed to be valid throughout the association. A safer approach is to provide data origin authentication for every request message. This scenario requires that the target system knows the initiator. More specifically, the initiator is listed in one or more of the access control rules in the target system. The main disadvantage of this approach is that if there are many potential initiators, then lots of information must be encoded in access control rules.

4.5.2.2. Anonymous Authenticated Initiator. *Note:* the title of this section is not an oxymoron. In some cases the initiator is not known to the target system but has acquired certain access privileges from an authorized source. For instance, Alice is a freelance troubleshooter extraordinaire. She got permission from Bob, the chief maintenance technician, to read any of the information in a limping CO. Since Alice's expertise is in high demand, she does the troubleshooting remotely, from her home, rather than visit the patient. The access privileges she got from Bob are embodied in an **Access Control Certificate (ACC).** The ACC typically contains the following information:

- Identity of the issuer (Bob)
- Date of issue
- Validity period (from, until)
- Identity of user (Alice)
- Authorized targets (read anything in the CO)
- Digital signature of the issuer (Bob) of the preceding fields

Since the CO has Bob's public key, it can verify the authenticity of the ACC when Alice presents it. The CO also knows that Bob is empowered to issue such certificates. The CO still needs to be sure that the ACC is indeed presented by Alice who is identified in the ACC. Alice can prove her identity to the CO with a public key certificate that can be verified by the CO. This will also provide the CO with Alice's public key. Therefore, Alice can authenticate all her messages to the CO, including the message that contains the ACC and her public key certificate, by signing them with her private key.

After Alice fixes the CO, the company is so impressed that it makes her an offer she cannot refuse. When Alice joins the company, she gets a more permanent ACC. Instead of specifying the accessible targets, it asserts that Alice is a member of the elite fixers group. Every CO in the company has an access control rule that associates the elite fixers group (as initiators) with all relevant targets. Thus, Alice can tend to the woes of ailing COs from her home, even though none of her charges knows her name. (Her identity does not appear in any of the access control rules.) The introduction of groups of users with identical access privileges has several advantages:

- The company can have a high rate of personnel turnover without changing the ACI in its systems! It only needs to issue fresh ACCs to new recruits and revoke ACCs of attritioned employees.
- The company can change the privileges of a group by changing ACI in the systems but without issuing new ACCs.
- Each ACC is fairly compact since it specifies only group membership rather than extensive lists of detailed privileges.

4.5.2.3 Anonymous Unauthenticated Initiator. *Note:* the title of this section is not redundant. This case is unlikely to be used in TMN security since it does not support accountability. It is included here for completeness. It is illustrated through an example that has little to do with network management.

Carol has tele-telepathic powers: she can read minds over the network. Her powers are demonstrated in hourly shows, narrow cast over the network to pay-per-view customers. Since the show is highly interactive (Carol promises to read and divulge what every tele-spectator thinks), the audience is limited to just 100 tele-spectators for each show. The show promoters do not care much about the identity of the participants, as long as they pay their dues. To this end, the promoters sell electronic tickets. Each electronic ticket, or **token,** consists of

- A serial number (1–100)
- Validity period (date and time of the show)
- Digital signature of the preceding fields, signed with Carol's private key

The buyer of the ticket can check its authenticity by verifying the signature with Carol's public key. A ticket is accepted only for the specified date and time. When a ticket is accepted, its serial number is recorded in a list of used tickets. Before a ticket is accepted, the promoters verify that its serial number is not on the list of used tickets.

The decision to grant or deny access to the target (Carol's show) is correctly accomplished without any information regarding the identity of the initiator.

The holder of a ticket can give or sell the ticket since the ticket does not contain the identity of the buyer. An enterprising crook can sell the same ticket many times, but only one will be accepted. Naive, second-hand buyers may thus be duped. Savvy buyers just shun scalpers.

4.5.3 Request ACI

Request ACI serves two purposes:

- Assure that the request is properly structured.
- Support access rules that use contextual information.

A well-structured request provides the necessary credentials (for example, ACCs) while ensuring that those credentials are not being used by an impostor. In order to bind such credentials to the request, the whole message, including, for example, any ACCs and the initiator's public key certificate, can be protected with a Message Authentication Code (MAC). To prevent replay of the request, it may contain a unique string such as a time stamp.

Access to a protected resource may depend on contextual information such as time of day, day of the week, freshness of the request (how long ago it was issued), or location from where the request originates. The request ACI that supports such access control rules includes a time stamp and the network address of the initiator.

4.5.4 Target ACI

There are three types of target ACI, corresponding to three types of commonly used access control mechanisms:

- Access control lists
- Capabilities
- Labels

4.5.4.1 Access Control Lists. An access control list (ACL) is a list of initiators that are authorized to access a resource; for each user in that list it further specifies which privileges (operations) are associated with the initiator. An initiator in an ACL can be an individually identified person, system, or application. It can also be a role group identifier. In the first case, the initiator ACI must include the authenticated identity of the initiator. In the second case, the initiator ACI must include a proof that the initiator is a member of the specified group or can act in the specified role. Typically, the proof consists of an ACC issued to the initiator along with proof of identity.

4.5.4.2 Capabilities. A capabilities scheme is based on ACCs that specify explicit access privileges rather than on group membership. The target ACI identifies, for each target, who is authorized to issue an ACC that grants access to that target. Typically, this information will contain or consist of that authority's public key.

4.5.4.3 Security Labels. Security labels provide the simplest and least granular mechanism for access control. It supports the classification of documents as confidential, secret, and top secret, while providing users with corresponding clearance levels. A security label (typically an integer) is associated with every target and with every initiator. If the initiator's security label matches or has a higher value than the target's security label, then access is granted. In a security labels scheme, the target ACI is typically a single security label. If finer granularity is desired, a single-target ACI may consist of several security labels (for example, corresponding to "medical confidential" and "personnel secret"). The various labels then belong to different, independent security classification hierarchies. The initiator ACI in a security labels scheme must include proof that the initiator has the proper clearance (i.e., the proper security label).

4.6 DIFFIE–HELLMAN KEY EXCHANGE

The Diffie–Hellman key exchange algorithm allows two parties to establish a secret key (for example, for symmetric key encryption or for keyed hashing) over an insecure channel. First, the two parties, say Alice and Bob, must agree on a large prime number n and a number g which is a generator mod n (i.e., for each b from 1 to $n - 1$ there exists some a where $g^{a} = b(\text{mod } n)$); n and g need not be secret. Alice and Bob can agree on some commonly used n and g. The Diffie–Hellman key exchange algorithm then consists of the following steps:

1. Alice chooses a large integer x and sends Bob $X = g^{x}(\text{mod } n)$.
2. Bob chooses a large integer y and sends Alice $Y = g^{y}(\text{mod } n)$.

3. Alice computes $k = Y^x \pmod{n}$.
4. Bob computes $k' = X^y \pmod{n}$.

$k = k' = g^{x.y} \pmod{n}$ is the shared secret key established through the exchange. An eavesdropper who knows *g, n, X,* and *Y* cannot compute *k* without first finding out *x* or *y.* This requires a discrete logarithm calculation, which is about as difficult as factorization.

4.6.1 Ephemeral Diffie–Hellman Key Exchange

An intruder cannot derive the secret key generated through a Diffie–Hellman key exchange. However, an intruder can mount a spoofing attack against Alice and Bob. The intruder can intercept the messages that Alice and Bob are exchanging, and replace them with bogus messages produced by executing the Diffie–Hellman key exchange with both Alice and Bob. At the end of the exchange the intruder will share one secret with Alice, who believes she shares that secret with Bob, and another key with Bob who is similarly duped.

To prevent such a man-in-the-middle attack, the Diffie–Hellman key exchange messages must be authenticated. Most commonly, this authentication is provided by signing the messages using the DSS algorithm. If necessary, Alice and Bob can also exchange their public key certificates. There is no danger that an intruder could fake those certificates or signed messages.

4.6.2 Certified Diffie–Hellman Parameters

If Alice and Bob belong to a community of users that have all agreed to use the same values for *g* and *n,* the authenticated Diffie–Hellman key exchange can be simplified. Every member of that community can choose once and for all their large, random integer *x,* compute their own Diffie–Hellman parameter $X = g^x \pmod{n}$, and obtain a certificate for their own *X.* When Alice and Bob want to establish a secret key, they simply exchange their certified Diffie–Hellman parameters and proceed to compute the shared secret. Every time Alice and Bob use this procedure, they will derive the same secret. However, for every pair of users the result will be a different secret. If Alice and Bob want to exchange confidential information, they can derive a symmetric encryption key from the shared secret resulting from their Diffie–Hellman parameters. A simplistic derivation of the symmetric key from the shared secret may simplify the task of an eavesdropper. For example, the symmetric key may be a DES key where the 56 random bits are the first 56 bits of the shared secret. In this case Alice and Bob will use the same DES key for as long as their Diffie–Hellman certificates are valid. This can be months or years. Such extended use of a single DES key would give an eavesdropper plenty of time to crack the key. To foil this attack, Alice and Bob need to generate a new DES key for each session or every day if a session can stay on for several days. A more secure way of deriving a symmetric session key may be, for example, to append the current date to the shared secret, hash the resulting string, and select the first 56 bits of the message digest for the DES key.

4.7 AUTHENTICATION PROTOCOLS

An authentication protocol is akin to the spiders' mating dance: as the two insects approach each other, each knows what it wants, but it is not sure if the other is interested in romance, or dinner, or both. The only way to find out is to get close enough, though not too close before it's too late. While getting closer, knowing the right moves can determine success or failure.

Authentication protocols are security protocols, which are altogether different from communication protocols. A security protocol specifies the type of information (semantics) that needs to be exchanged to accomplish certain security features, rather than its syntax.

A good authentication protocol allows two parties to correctly ascertain each other's identity while communicating over an untrusted network. The untrusted network allows intruders to see, delete, delay, modify, replay, reorder, or reroute any messages carried over the network. Authentication protocols are designed to prevent such enterprising intruders from successfully masquerading as either party.

Authentication can be accomplished directly between the two principals, or it can be mediated through a trusted third party. In the latter case, each of the principals still needs to achieve full authentication with that third party with no further help, so direct authentication capability is still a basic requirement. Using a third party increases the total number of messages that need to be exchanged. It is not expected to be used often for TMN security, and therefore it is not discussed any further in this book.

Direct authentication protocols can be divided into two categories:

- Challenge–response authentication
- Stateful authentication

4.7.1 Challenge–Response Authentication

A challenge–response protocol can be illustrated with a simple (and simplistic) example:

- Alice wants to establish secure communications with Bob. She sends a message to Bob requesting association establishment. Alice generates a unique string of bits that she has never used before for authentication purposes. (It can be a time stamp, a random number, a sequence number, or any combination of these.) This is Alice's challenge. Alice sends it to Bob as part of her association request message.
- Bob encrypts Alice's challenge with his private key. This is Bob's response. Bob sends it to Alice. In the same message Bob can also include his own challenge: a unique string that he generates for authentication purpose.
- Alice decrypts Bob's response using Bob's public key and verifies that the result matches her challenge (the string she sent Bob). Alice then encrypts Bob's challenge with her private key and sends the response to Bob.
- Bob decrypts Alice's response with Alice's public key and verifies that the result matches his challenge.

Three messages are exchanged in the preceding example. That is the bare minimum needed for challenge–response authentication: most challenge–response protocols require more exchanges. The additional exchanges improve the security and can be used to establish shared secret information, such as a symmetric key, that can be used in the course of the newly established association.

In the preceding example Alice and Bob require each other's public key. They can each send their public key certificate along with their response, or they may retrieve each other's public key certificate from a directory.

A major attraction of challenge–response authentication, which makes it ideally suited for Web interactions, is that neither of the communicating entities needs to maintain any information about the other. The communicating entities need not even know about each other prior to connection setup. This allows a Web server to be accessed securely by any of hundreds of millions of potential users. The server does not and cannot possibly know the identities of all the potential users that might want to reach it. Furthermore, it is not practical for the server to maintain state information about every user that has ever accessed it (and may not do so again). The only crucial information that each party needs about the other is its public key. Each party may send its own public key certificate to the other; or any party may retrieve the other party's public key certificate from a directory.

4.7.2 Stateful Authentication

Although challenge–response authentication is quite popular, it cannot be readily used in an OSI environment that allows only two messages for successful association setup: the ACSE AARQ and AARE. Fortunately, the same level of security can be achieved with only two messages. The price to pay for saving a message is the need to maintain some state information, as illustrated in the following example.

- Alice wants to establish secure communications with Bob. She sends a message to Bob requesting association establishment. Alice generates a unique string of bits that she has never used before for authentication purposes. (It can be a time stamp, a random number, a sequence number, or any combination of these.) Alice appends this string to Bob's identifier, signs the result with her private key, and includes both the plain text string and the signature in her message.
- Upon receiving Alice's association request, Bob verifies that the signature matches the plain text string. For his response message Bob generates a unique string of bits that he has never used before for authentication purposes. (It can be a time stamp, a random number, a sequence number, or any combination of these.) Bob appends this string to Alice's identifier, signs the result with his private key, and includes both the plain text string and signature in his message.
- Upon receiving Bob's association response, Alice verifies that the signature matches the plain text string.

There is a serious weakness in this protocol: an intruder can make a copy of Alice's message and send it later to Bob claiming to be Alice. The signature will match the plain text string. To thwart this threat, Bob must keep a file with all the unique strings he has received from Alice. If Bob and Alice communicate frequently, the file will get pretty large. If Bob communicates with lots of people, he will have lots of large files. With every incoming authentication protocol message, Bob will have to verify that new string is not already present in the corresponding file.

Fortunately, there is a simple solution to Bob's predicament: within every string Alice generates she includes the time when the string expires, for example, 4:00 P.M. today. In this case the random portions of Alice's strings need to be unique only within 24-hour periods. Bob can discard all the strings he received from Alice before 4:00 P.M. yesterday.

The solution used in many TMN applications is simpler yet: Alice includes a time stamp indicating when her message was prepared. Since this time stamp is unique, she does not have to include any random string. Furthermore, by comparing the value of Alice's time stamp with his local system clock, Bob can detect any unusual delays. If the delay is excessive, Bob may suspect that the message has been tampered and reject it.

Even this simple approach requires maintaining state information. Indeed, an intruder can capture a valid message sent by Alice to Bob, and then send to Bob another message using the authenticator copied from Alice's message. If the intruder is sufficiently fast, the apparent delay will not seem excessive and Bob will be duped into accepting the fake message. To protect against this attack, Bob must keep the last time stamp he received from Alice. Any new message must have a later time stamp than the one Bob uses for checking Alice's messages. If the new authenticator is accepted, then Bob will update the value of the most recent time stamp he received from Alice.

Things get a bit more complicated if we allow for network-caused reordering. If different messages can travel from Alice to Bob via different routes, with different delays, then a message sent at 8:08 A.M. may arrive before a message sent at 8:07 A.M. With the simplistic scheme outlined in the previous paragraph, Bob will reject the late-arriving message sent at 8:07 A.M. In order to allow for variable network delays up to some limit, say five minutes, Bob can keep all the time stamps received from Alice whose value is not more than five minutes older than the time stamp with the most recent value (which is not necessarily the most recently received time stamp). When a new message arrives, Bob will check that its time stamp is different from any of the time stamps in the list he maintains and that its value shows it was generated no more then five minutes before the message with the latest time stamp.

Things get even more complicated if we allow for clock failures. Suppose Alice's clock was running too fast so that Alice had to reset it. Alice's next message may have a time stamp considerably older than her previous message and may therefore be rejected. Or her next message may have a time stamp that is identical to a previous time stamp. Things can get even worse: Alice's clock may have stopped, and all her time stamps will have the same value. Or her clock may have been reset several years back to some default value, again causing all her messages to be rejected. In order to allow for secure authentication in the face of such adversity (which is not so outlandish, for network elements are known to experience catastrophic clock failures), Alice can maintain a monotonically increasing virtual time, ensuring that each new time stamp has a more recent value than its predecessor. The following paragraphs illustrate a possible construction of a monotonically increasing time for security purposes [T1.259, Q.813].

In this construction we distinguish among four types of time:

1. GMT is the "astronomically correct" time.
2. System Clock (SC) is time shown by the system clock.
3. Virtual Time (VT) is the only time used by security mechanisms (and possibly by other system components).
4. External Time (ET) is the time that appears in an incoming PDU.

VT is read every time an outgoing PDU containing a time stamp is generated and every time an incoming PDU containing a time stamp is received (as well as for other purposes outside the scope of this discussion). Every time the VT is read it is first updated; then the updated value is provided in response to the read request. The updated value is also stored in nonvolatile memory. The procedure for updating the VT is defined by following pseudocode:

$$\text{If VT} < \text{SC then VT} = \text{SC},$$

$$\text{else VT} = \text{VT} + 1 \text{ tick}$$

where 1 tick is the smallest amount by which the (virtual) clock can be incremented; it should be small enough that the maximum possible rate of (virtual) ticking should be higher then the peak rate at which the VT is read. Typically, 1 tick may be 10 ms.

If the SC stops or is reset to some default value, VT continues to "tick" at a "virtual" rate, corresponding to the frequency of its usage. VT catches up with the real time after the SC is updated to GMT.

In order to allow for substantial clock (SC) drift (for example, 30 minutes) between consecutive resets of the SC to GMT, systems may accept PDUs with ET that differs by up to twice the allowed drift (for example, 1 hour) from the VT. The exact value of this permissibility parameter can be adjusted to match the characteristics of the clocks of the communicating systems. Of course, with drifting clocks delay detection will be reduced to a broader granularity.

In order to accommodate catastrophic SC failures (extended stoppage and/or loss of current time), a second, more generous (for example, four hours) permissibility parameter may be introduced. If a PDU arrives with ET value between the two permissibility parameters, it will still be accepted, but the event might be logged in a security audit trail. If the incoming PDU has an ET which is outside the limit allowed by second permissibility parameter, then a security alert might be issued in addition to logging the event in a security audit trail. The decision on whether to continue, release, or abort the association in this case is a matter of local security policy. All the security audit trail logs can be done with VT. This enssures that the relative order of events is strictly maintained.

Every time the SC is reset to GMT, the event may be logged in the security audit trail. The log may contain the values of the SC and VT before and right after the update. This information may be useful for correlating VT and GMT for a security audit trail analysis.

Besides requiring fewer messages than challenge–response authentication, time-based stateful authentication allows the detection of message reordering and delay. If it includes a sequence number, it allows the detection of message deletion.

Stateful authentication requires that each communicating entity maintain some information about its potential interlocutors. In the preceding example it keeps the last (few) time stamp(s) it receives from each of its correspondents. In the context of network management, this is not an overwhelming burden. A typical TMN element can be accessed only by a few dozen other systems, so it need not keep vast amounts of state information.

4.8 MAPPING SECURITY SERVICES TO SECURITY MECHANISMS

Table 4.1 lists the various security services discussed in 3.3 and identifies which of the security mechanisms presented earlier can support those services.

TABLE 4.1 Security Mechanisms to Provide Security Services

	Security Mechanisms					
Security Service	Access Control Mechanisms	Keyed Hashing	Symmetric Encryption	Asymmetric Encryption	Digital Signatures	Digital Certificates
Access control	✔					
Peer entity authentication		✔	✔	✔	✔	✔
Data origin authentication		✔			✔	
Data integrity		✔			✔	
Data confidentiality			✔	✔		
Non-repudiation					✔	

Table 4.1 can be combined with Table 3.2 to provide a direct correlation between security threats and security mechanisms to protect against those threats. Table 4.2 provides this correlation. It shows which security mechanisms might be drafted to provide protection against security threats. The provisioning of some security service may automatically provide some other security services. For example, non-repudiation automatically guarantees integrity, which in turn guarantees data origin authentication, which, in most though not all cases, provides the necessary and sufficient credentials for access control. Figure 4.8 illus-

TABLE 4.2 Security Mechanisms to Protect against Security Threats

	Security Threats						
Security Services	Unauthorized Access	Eavesdropping	Masquerade	Modification of Information	Repudiation	Replay Reroute, Delete	Network Flooding
Access control	✔			✔			✔
Keyed hashing	✔		✔	✔		✔	
Symmetric encryption	✔	✔	✔				
Asymmetric encryption	✔	✔	✔				
Digital signatures	✔		✔	✔	✔		
Digital certificates	✔		✔				

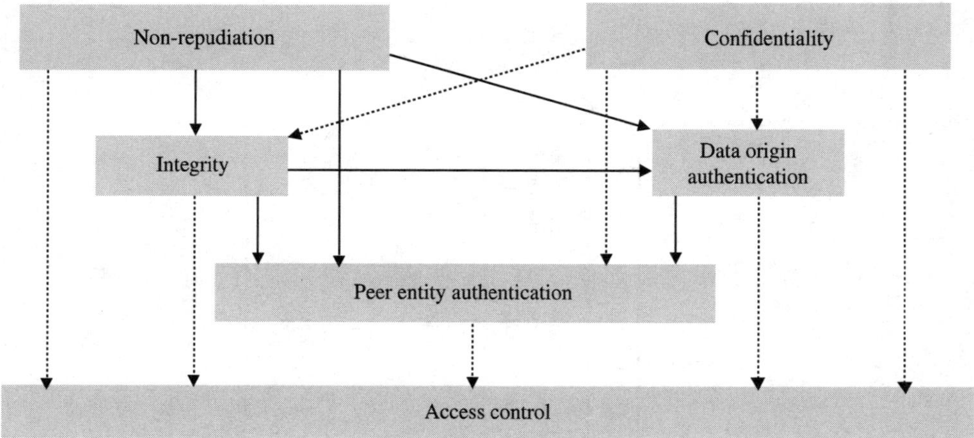

Figure 4.8 Partial order of security services

trates the partial ordering (directed acyclic graph, for the mathematically inclined) of security services. A solid arrow in Figure 4.8 denotes that one security service inherently provides another, while a dashed arrow denotes that one security service supports and may be sufficient for another, though not always.

5

Support Mechanisms

The security mechanisms outlined in the previous chapter are intended to prevent security intrusions. Unfortunately, it is impossible to guarantee that no intrusion would ever succeed. Since there is always a possibility that an attack may be successful, it is necessary to be able to detect such intrusions and recover from them. A security audit log is generally used to detect insidious intrusions that may not have an immediate, obvious impact. A security alarm is usually the first response to an intrusion and triggers a recovery procedure. Both are discussed below. The security mechanisms described previously require the secure distribution of either symmetric or public keys. Support mechanisms for such key distribution, especially use of the TMN directory for the secure distribution of public keys, are also discussed in this chapter.

5.1 SECURITY ALARMS

A managed object generates security alarms when a security violation (such as an expired password or key, or an attempt to access unauthorized resources) is detected. The agent then sends the security alarm to the manager using the CMIS M-EVENT-REPORT service. This information can later be used in understanding the nature of the possible security breach and providing information on how to close that breach. The information presented in a security alarm includes:

- Identity of the managed object affected by the security event
- Time of the security event
- Event type (integrity violation, operational violation, physical violation, security service or mechanism violation, or time domain violation)
- Cause of the security alarm (the possible values of this parameter depend on the specific event type)
- Severity of the security alarm—indeterminate, critical, major, minor, or warning
- Security alarm detector—who detected the security event
- Service user—who requested the service that led to the generation of the security alarm

- Service provider—the intended service-provider of the service that led to the generation of the security alarm

5.2 SECURITY AUDIT LOG

The security audit log supports the collection of security-related events, especially those produced through security alarms. This information can be used, when needed, to detect and trace a security breach to its source. The security audit log records various information, including:

- The identity of the managed object that was involved in the security-related event
- The event type (either a service report or a usage report)
- The time at which the event occurred
- If the event type was a service report, then the cause for the service report

A *service report* consists of a report of events related to the provision, denial, or recovery of service. A *usage report* is a report of statistical information related to security.

The data collected in the security audit log can be used to analyze various security violations and identify changes needed in the security policies. A security audit record (a collection of which is contained in the security audit log) is generated as a result of a security-related event. The security events are based on predetermined conditions that are defined in the network's security policies. The security audit log also provides the ability to collate security information from many sources. This increases the probability of discovering an intruder that attacks the network through different venues. Therefore, even though the security audit log cannot stop a security breach, it can help prevent future breaches.

5.3 KEY DISTRIBUTION

All of the security mechanisms we have discussed require secure distribution (assuring data origin authentication, data integrity, and confidentiality of private keys [IEEE802.10C]) and management of secret information. This section starts with the description of a simplistic process for exchanging and managing keys between Manager and Agent. This process is used in Electronic Bonding (EB)—an electronic interface between Local Exchange Carriers and Interexchange Carriers for SML applications such as trouble administration and service ordering [TR40]. This section then continues with the description of a more effective and scalable approach based on public key encryption.

5.3.1 Key Lists

The procedure established for key distribution and management in EB, though quite effective for key distribution within a small community of users (any single user may need to share secret keys with up to a dozen other users), becomes cumbersome if used for a large community (for example, all the entities of a TMN). Therefore, this section addresses only the case of EB between a Service Provider's TMN and a Customer over the X interface. In this scheme, data origin authentication and peer entity authentication are provided by including authentication information in the CMIP access control field. (This field is available in the CMIP m-Get, m-Set, m-Create, m-Delete, and m-Action PDUs.) The authentication information must include a key identifier that identifies the key used in the current PDU. The syntax for this authentication information is provided in ANSI T1.228-1995 [T1.228].

The Service Provider provides the Customer with a list of keys. The default size of the list is 1,000 keys; however, the Customer and the Service Provider may agree on a different size list. For the remainder of this document, we assume the default value of 1,000. The keys are numbered, and thus identified, from 1 to 1,000. The list must be in a mutually agreed upon electronically readable form.

The Customer changes the keys at random time intervals, not to exceed one day. The change of keys is accomplished by a change in the key identifier field, which must be part of the access control structure, if this scheme is used. The key identifier is used as an index into the list of keys. Discarded keys are not reusable. If a discarded key identifier is reused or an invalid key is used, the behavior is at the discretion of the recipient.

A backup list (of the same number of keys as in the initial list) of keys will be kept on hand by both the Customer and the Service Provider for use in case it is suspected that the entire key list has been compromised.

After a major portion of the keys (for example, 700 of the 1,000 keys) have been used, or after a year has passed, whichever comes first, the Service Provider will provide the Customer with a new list of keys. The numbering of the keys in each subsequent list will be a continuation of the numbering of the previous list. Every key will have a unique key Identifier (e.g., 1–1,000, 1,001–2,000, 2,001–3,000, etc.).

Every new list of keys will be triple-DES encrypted using three unused keys from the current list. Those three keys will not be used again. The encrypted list of keys, along with the key identifiers for the three keys used to encrypt it, can be sent over public channels.

5.3.2 Public Key Distribution

The public keys of various users can be placed in a central repository available to all users. Thus, any user wishing to conduct secure communications can do so by obtaining the receiver's public key and encrypting the message with it. Because of the complexity of the encryption algorithms, one common use for public key encryption is to securely transmit a symmetric key to be used in future communications. The sender encrypts the symmetric key with the public key of the intended receiver in order to ensure its confidentiality. It can further be signed by the sender with its private key in order to provide data origin authentication. Therefore, public key encryption solves the problem of secure sharing of secret keys.

Public key encryption allows secure communications, without the need to share secret information. In addition, the information needed to securely communicate (i.e., public keys of TMN entities) can be held in a central repository. Other useful information can also be held in the repository. These include the Access Control Lists (ACLs) and Access Control Certificates (ACCs) that were described for use in access control. A repository known as the X.500 Directory has been defined for this purpose.

5.4 DIRECTORY

The X.500 Directory [X.500] is based on the use of public key encryption. It is designed as a central repository for any information that needs to be shared among many users. In particular, each user has an entry in the Directory. One attribute of this entry is the user's public key certificate [X.509]. Therefore, whenever a TMN user needs the public key of another user, it requests the public key certificate of the intended receiver from the X.500 Directory. The Directory can also contain the Access Control Certificates (ACC) of TMN entities. ACCs can be requested from the Directory to check a user's access privileges.

5.4.1 Automatic Registration

Before any information can be retrieved from the directory, it must be entered into the directory. Populating the directory with the public key certificates of TMN entities and updating such information as new TMN entities (for example, new add-drop multiplexers) are added presents an administrative burden. Such a burden can be minimized if each TMN entity can register itself automatically with the directory. One requirement for the TMN Directory is that it support populating the Directory Information Base (DIB) by automatic registration. In order to prevent registration of false information, the Directory must be able to ascertain the data origin authentication and integrity of all the messages that cause automatic registration. The basic premise of automatic registration is that prior to such registration the Directory does not know about the newly registering entity and therefore cannot possibly have any secret information shared with that entity. In order to have proper security, authentication must be established by using a public key system, or have recourse to a third party such as a Kerberos™ [MILL87] system. Of the two options, a public key system is the only one that also supports non-repudiation. This scheme is easy to manage since for *n* entities it includes exactly *n* private and *n* public keys. Furthermore, each entity needs to know only its own private and public key, as well as the local Certification Authority's (CA) public key. It also results in less traffic, and it is extensively developed in X.509. In addition, as shown below, it can be implemented in steps that increase flexibility when the Directory architecture becomes more distributed. As a result, this book focuses on the use of a public key system.

The simplest scenario for Directory security is that of a single security domain within a TMN. In this case, the automatic registration and secure use of the Directory proceed as follows [T1.252]:

1. When an entity (including the Directory) is first installed, it receives the following four items from the Certification Authority (CA) of its security domain:

   - A unique identifier (such as an AE title)
   - The CA's public key
   - Its own private and public key pair (no other entity has a copy of the private key)
   - A certificate (as defined in X.509), signed by the CA, linking that entity's identifier to its public key

   These items can only be updated by the CA. An entity can read its own private key; no other entity can read that key.

   In order to support this step, all TMN entities must be programmable to accept these four items and keep them securely. One way of achieving this is with a Programmable Read Only Memory (PROM) or an Electronically Programmable Read Only Memory (EPROM). Other procedures may be possible. The exact procedure for inputting the information into TMN entities is for further study.

   How the public key is input into the TMN entities depends on local security policy. It is critical that this procedure be done securely since the whole security infrastructure depends on it.

2. When an entity first registers with its Directory, it presents the certificate it received from the CA. (The Directory can request this certificate from the new entity upon noticing traffic from this entity for the first time.) Since the Directory has also received the CA's public key, it is able to validate the certificate. Based on this certificate, it creates a new entry for the registering entity, which includes the entity's identifier and its certificate. Any additional information, or modification of existing information, that this entity wants to enter subsequently in the Directory will be included in messages signed by that entity. This procedure ensures the integrity of the DIB.
3. Whenever a (registered) entity communicates (securely) with the Directory, it will authenticate itself to the directory by signing the message (in this case the signed

message must include a nonrepeating string such as a time stamp) or by appending a security token which is an encrypted nonrepeating string. The Directory will authenticate itself to that entity using the same procedure, but it will also include its own certificate. Since the requesting entity, like any other entity in this scheme, knows the CA's public key, it can verify the validity of the certificate.

4. If an entity requests the public key of another entity, and it is entitled to receive it, then the Directory will send the certificate it has received at registration time to the requesting entity. Since the requesting entity, like any other entity in this scheme, knows the CA's public key, it can verify the validity of the certificate.
5. If an entity requests any other information to which it is entitled from the Directory, and if it is deemed necessary to protect the integrity of that information, then the Directory will sign this information with its private key. In this case, the Directory must also include its own certificate in order to provide the requesting entity with the Directory's public key.
6. If an entity requests any information to which it is entitled from the Directory, and if it is deemed necessary to protect the confidentiality of that information, then the Directory will encrypt this information with the requesting entity's public key (which the Directory has in its DIB). If the transferred information is very long (for example, several thousands octets, which is rather unlikely for Directory services), then public key encryption may not be practical. In this case, the Directory may encrypt the requested information using symmetric encryption (such as DES) with a secret key and send that secret key encrypted with the recipient's public key.
7. Procedures 4 and 5 above may be combined to provide both integrity and privacy protection.

The scenario discussed above can be generalized by allowing more types of access control mechanisms or allowing several security domains that may belong to different TMNs. These topics are discussed below.

5.4.2 Directory Access Control

The simplest scenario, discussed above, assumes that the only Access Control Information (ACI) required by the Directory is the authenticated identity of the Initiator. This is sufficient if, for example, the following security policy for access control is established:

- Each certified entity can read any entry in the Directory.
- The CA can delete any entry in the Directory.
- Each certified entity can make changes to its own entry in the Directory and no other changes.

It also supports more fine-grained access control mechanisms, such as **Access Control Lists** (ACL), which specify what the corresponding access rights are for each entity (or alternatively, which entities can access each target).

ISO/IEC 10164-9 | CCITT Rec. X.741 (Information Technology—Open System Interconnection—System Management—Part 9: Objects and Attributes for Access Control) and the Network Management Forum (NMF) Forum 016 (Application Services; Security of Management) describe additional types of access control mechanisms, including security labels and Access Control Certificates. Both are discussed next.

In an access control scheme based on **security labels,** each entity is assigned to one or more security groups. Each security group has well-defined access privileges. The Initiator of a request must demonstrate to the Directory that it belongs to a security group that has the access privileges required for the requested operation. To this end, the Initiator must have a certificate, issued and signed by the CA, stating that the Initiator belongs to the relevant security group. If an entity belongs to several security groups, it may have several such certificates; each one can attest to membership in one or more security groups. Each certificate contains the entity's identity (such as an application-entity title) and a list of one or more security group identifiers. It may also contain the date of issue of the certificate, time interval (from, until) during which the certificate is valid, and other information.

In order to prevent unauthorized replay of a security label certificate, it may be bound to a specific request message. To achieve such binding, the security label certificate may be contained in a message signed by the Initiator. That message must include a nonrepeating string such as a time stamp. Alternatively, the security label certificate may be sent to the Directory outside any message signed by the Initiator, in which case the message from the Initiator that actually requests the operation must be signed by the Initiator. Again, the signed message must include a nonrepeating string such as a time stamp. In the second case, if the Directory keeps a copy of the security label certificate, then it may not be necessary to submit this certificate again for subsequent requests to the Directory. In either case, the Directory must have received (previously or in the same message) the Initiator's public key certificate (containing the Initiator's identifier and public key).

An **Access Control Certificate** (ACC) is issued and signed by the CA. It contains the identifier of the entity to whom it was issued and a list of the access privileges to which that entity is entitled. It may also contain the date of issue of the certificate, time interval (from, until) during which the certificate is valid, and other information. (Notice that whereas an ACL is securely kept by the Directory, an ACC is kept by the Initiator.) The binding of an ACC to a request message is identical to the procedure discussed above for security label certificates.

5.4.3 Multiple Security Domains

If a TMN contains several security domains and entities from different domains need to communicate, then certification paths (as defined in X.509) must be established. (In a simple certification path, an entity's CA will provide that entity with a certificate that some other entity is a valid CA and that CA's public key.)

The certification path mechanism can also be used for secure communication between a Directory in one TMN and entities in other jurisdictions. X.509 defines the certificate object class, as well as all the other objects and attributes needed to support certification paths.

5.5 PROTOCOLS FOR SECURITY

Security mechanisms such as encryption and digital signatures provide neat solutions to security threats, but they can wreak havoc with interoperability. In order to make sense of any secure message, the receiver must know, for instance, if a digital signature is placed before or after the actual message; if the digital signature is represented as one or more integers or as an octet string; what algorithms were used to sign and encrypt the message; and which keys were used. Special protocols have been devised to untangle this knot.

Before we peer into such protocols, a bit of semantic discussion is needed to clarify some basic phrases. Indeed, the engineering profession is so taken with software reuse that the same set of words is being recycled with much gusto. To avoid undue confusion, here are some clarifications:

Communication protocols (for example, IP, TCP, X.25, and CMIP) deal primarily with the syntax and procedures needed to exchange information.

Security protocols (for example, Diffie–Hellman key exchange, challenge–response authentication) deal only with the semantics of information exchanged to accomplish some security functions.

Secure communication protocols (for example, SSL3, STASE-ROSE) focus on the syntax and semantics of messages exchanged in support of secure communications.

Part 1 of this book emphasized communication protocols. Part II of this book, so far, has surveyed security protocols. The rest of this book will focus on secure communication protocols.

5.5.1 Anatomy of Secure Communication Protocols

In Part 3 of this book we will visit several security communications protocols. They all say basically the same things, but they say them differently. (Otherwise this book would have been mercifully shorter.) It is helpful to lay out those commonalities.

A typical security communication protocol has two phases:

1. Handshaking
2. Secure transfer

Each of these contains a couple of (potentially overlapping) phases.

During the handshaking phase, the protocol supports authentication and negotiation of a security context. A security context specifies which security mechanisms and algorithms will be supported in the course of the association. It can also specify any number of related security parameters, such as which encryption keys will be used. In general, if explicit security negotiation is omitted, then a default context is automatically adopted.

The authentication is one of three types:

1. **Initiator only authentication** is used when any of several callers (not necessarily authorized) can attempt to establish an association with a target system over a secure communications network.
2. **Target only authentication** is used when an anonymous initiator is calling a publicly available target (for example, a public Web site) over a potentially nonsecure network.
3. **Two-way peer entity authentication** is called for in all TMN applications.

The authentication mechanism can be either stateful (requiring the authenticating party to maintain some information about the authenticated party) or stateless (which may require several message exchanges).

The secure transfer phase allows the communicating parties to exchange information with the desired level of protection. In addition, it may also support dynamic update of security context (for example, use different algorithms for different messages, update the encryption key). Figure 5.1 illustrates the life cycle of a secure communications protocol.

The preceding discussion of secure communications protocols applies to connection-oriented communications. Connectionless communications (for example, datagram-based communications) cannot support true security context negotiation, and they can support only sender authentication.

A secure communications protocol, like any communications protocol, specifies the messages exchanged between two protocol state machines. A typical protocol state machine does not have the wisdom to decide which data need confidentiality protection, let alone which encryption algorithm to use for such service. It certainly does not know how to file a claim in court when it detects some malicious wrongdoing. Therefore, to be of any use it must

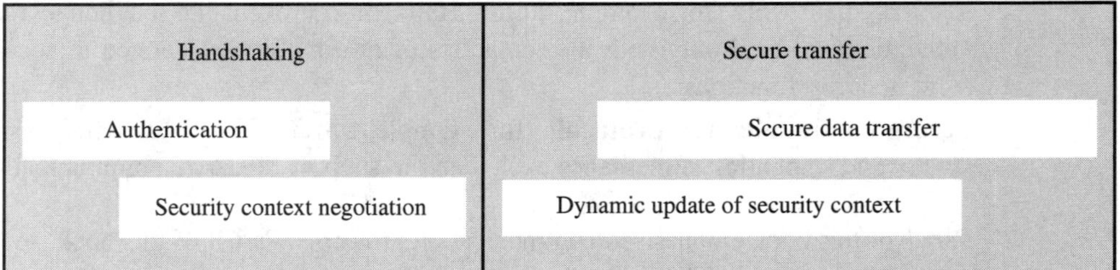

Figure 5.1 Life cycle of a secure communications protocol

have some local **administrative interfaces,** independent of the protocol interface with the remote peer entity. One local interface tells the protocol state machine which security parameters to use when another local interface allows the protocol state machine to alert the proper authorities (typically another software module within the same system) of any suspicious transactions. Yet another local interface may disclose, when asked, what security parameters have been negotiated and which ones have been used so far.

5.5.2 Security in Layers

A secure communications protocol may be implemented within an application or in any layer of the seven-layer OSI stack except for the session layer. So which will it be? No solid answer is possible, but here are a few good pointers.

Implementing a secure communications protocol within an application is the least appealing to application programmers. They are busy enough with network management applications. Moreover, this choice minimizes opportunities for software reuse. It should be contemplated only as a solution of last resort: when the desired security services are not supported within the communications protocol stack.

As we go down the protocol stack, we rapidly lose flexibility in providing security. When an N + 1 level PDU (see Figure 2.5) is received in layer N as a Service Data Unit (SDU), it is an opaque string of octets. Thus, once below the presentation layer, there is no concept of selective field protection. Worse yet: the transport layer can chop large messages into fixed-size blocks and combine tiny messages into larger blocks. As protection is provided for transport blocks, the concept of security per application message disappears. In particular, non-repudiation may not be provided below the presentation layer.

If security resides in layer N, then it protects layer N + 1 PDUs. It cannot protect layer N headers. Indeed, layer N headers are needed for routing to the correct destination, and they contain the necessary data (for example, security parameters) to allow the remote layer N protocol state machine to properly process the layer N PDU (for example, decrypt its payload, which is a layer N + 1 PDU). After such processing, the layer N + 1 PDU becomes available to the layer N + 1 protocol state machine to work its magic. Full encryption, including all header information, is possible only for point-to-point connections; where no routing is necessary and the same protection is used for the whole stream of data, this corresponds to physical layer security. Thus, lower placement of the secure communications protocol can buy some additional traffic flow confidentiality, especially if the security is in the physical layer. However, this kind of protection is not deemed relevant for TMN applications.

Ideally, a secure communications protocol should be independent of any technologies used within the layer where it resides. This is wishful thinking. Closed User Groups are tied to X.25, SSL3 is currently implemented only over TCP, and STASE-ROSE protects only ROSE-based protocols (for example, X.500 and CMIP). So we are stuck with the following tradeoff: if a secure communications protocol is implemented at a higher layer, then it has the advantage of being independent of any underlying lower-layer technology. On the down side, it protects only a few applications. For instance, STASE-ROSE, which resides in the application layer, is equally applicable if used over TCP, X.25, or any other type of connection-oriented transport. However, it does not protect FTAM, CORBA, or EDI exchanges. On the other hand, if a security protocol is implemented at a lower layer, then it can protect any higher protocol. For instance, SSL3, which resides in the transport layer, can protect FTAM, CORBA, EDI, CMIP, and X.500 exchanges. However, it is typically tied to a specific technology; thus, SSL3 is available over TCP but not over X.25.

Data link layer security protects data flow over individual links. It is well suited for security over a Local Area Network (LAN) where all exchanges are over a single (shared) link. However, TMN requires protection across extensive data communications networks. Therefore, the data link layer is not an acceptable home for TMN security.

5.6 GSS-API—SECURITY IN A BOX

To a cryptographer, the design and analysis of encryption algorithms is its own reward. To a security administrator or a system engineer, the proliferation of ciphers is truly an embarrassment of riches. Having to implement various security algorithms in various layers of various protocols on various systems, replacing obsolete algorithms with (hopefully) more robust solutions, and ensuring they all work together can produce an abundance of frustration. The antidote to this headache is a neatly packaged library of security algorithms. Such a library could be used by any protocol, at any layer, in any system. Updates to security algorithms can be done within that library. Thus, implementers of communication protocols can focus on communications rather than worry about the optimum encoding of cryptographic algorithms. Furthermore, reusing the same library would improve the prospects of interoperability.

A reusable library is a quintessential portable software module. It therefore needs a well-defined API—preferably a standardized API so that a system designer can interchangeably get the library from any of several providers that build to that standard. The IETF has obliged by publishing such a standard: the Generic Security Service Application Programming Interface (GSS-API) [RFC2078].

Any application, or protocol implementation, can call upon GSS to perform any required security functions, such as generate or verify authentication credentials. In order for a GSS in one system to verify credentials generated by a GSS in another system, both must adhere to identical conventions regarding the structure of the credential. Although such a credential appears as an opaque string to the calling module, it usually has a well-defined structure to the GSS. Therefore, in order to ensure interoperability between separate GSS implementations, the GSS-API defines a rudimentary inter-GSS protocol in addition to the API. The resulting architecture can be called an **out-of-stack secure communications protocol.** To appreciate this architecture, let's first examine a more conventional in-stack secure communications protocol as represented by Figure 5.2. It exchanges SDUs with the communications protocol state machines just above and below in the stack, and it exchanges PDUs with its remote peer.

In contrast, the virtual GSS PDUs are carried as opaque GSS tokens inside the actual PDUs of the module that use GSS services, as illustrated in Figure 5.3. GSS exchanges messages with its client using the GSS-API. It does not exchange any SDUs with any other communications protocol state machine.

GSS-API supports all the security functions outlined in Section 5.5.1. In fact, it has such a rich functionality that only some of its capabilities are briefly reviewed below.

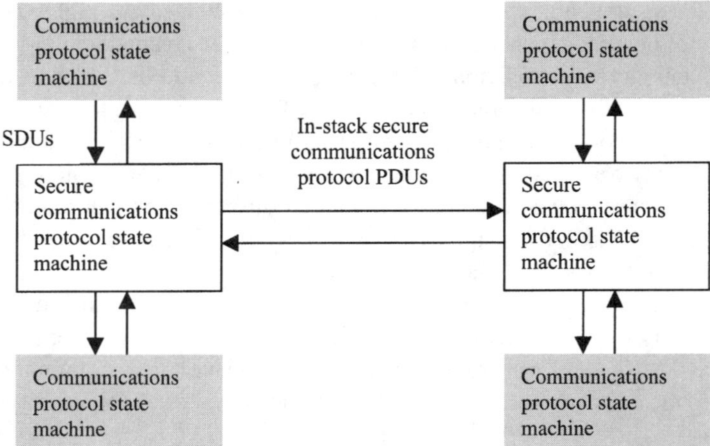

Figure 5.2 In-stack secure communications protocol

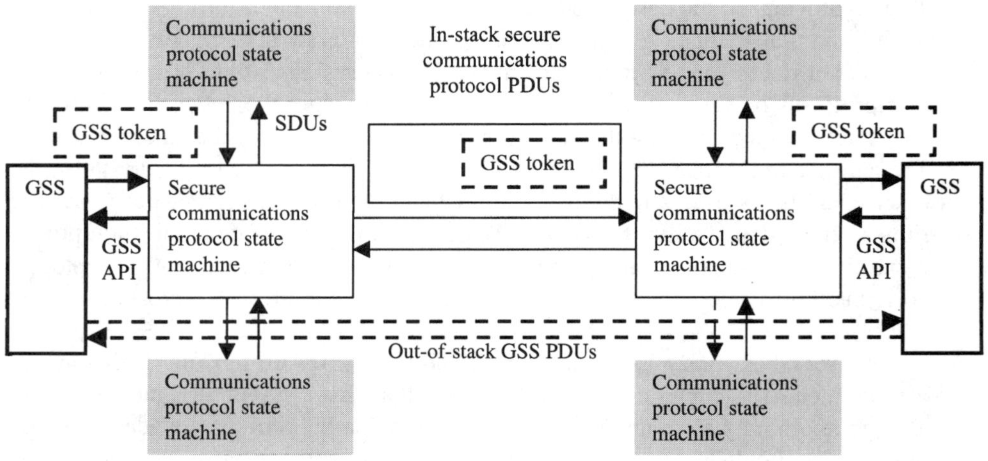

Figure 5.3 GSS out-of-stack secure communications protocol

5.6.1 GSS Handshaking

Figure 5.4 illustrates the calls to GSS, resulting responses, and corresponding PDUs carrying GSS tokens that are exchanged during a simple GSS-supported handshaking phase.

1. The initiator in system A issues a GSS_Init_sec_context() call to its local GSS in order to initiate a security context. Parameters to this call may include the target's name (in system B), security mechanism type, any credentials to be used, an indication of anonymous calls, an indication of delegated privileges, the lifetime of security tokens, and several additional characteristics.
2. The response to the initial call includes an output_token to be forwarded to the target system (B). It may also include additional parameters, and it indicates whether the initial call was successful; if it was not, then why; and whether further interaction is expected. In this example, it returns a major_status code

   GSS_S_CONTINUE_NEEDED indicating that it expects to receive a response from the target's GSS. (For simplicity, the status codes are not shown in Figure 5.4.)

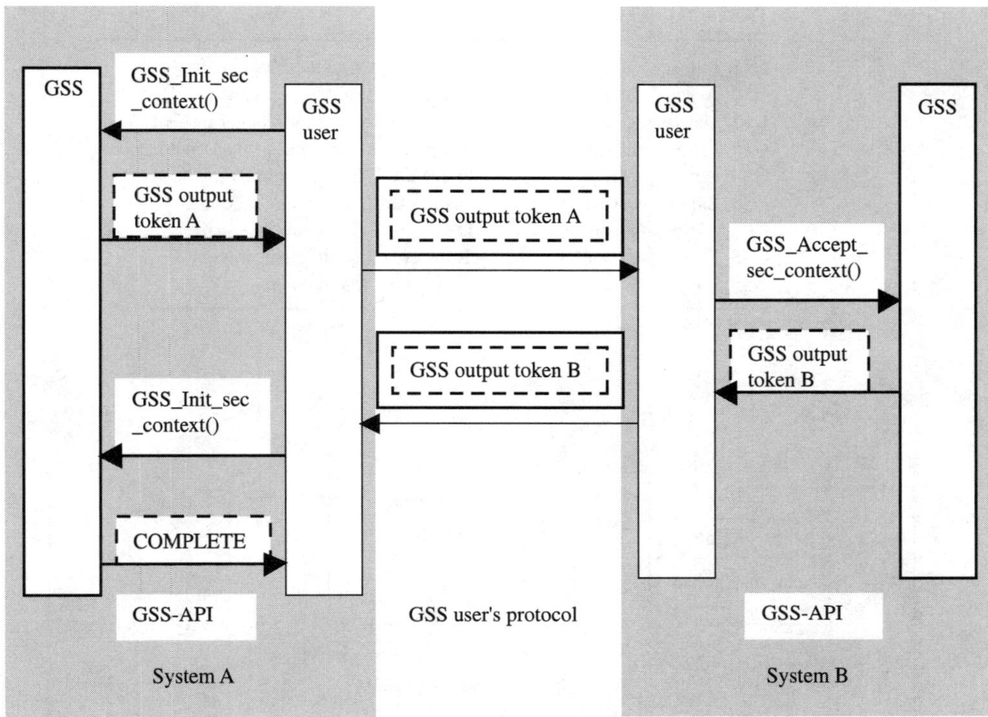

Figure 5.4 GSS-API handshaking sequence

The output_token typically consists of an authenticator along with security context information. The structure of the output_token depends on the security context.

3. The initial GSS user sends a PDU to its remote peer entity, containing the output_token it got from GSS.
4. The target peer entity in system B issues a GSS_Accept_sec_context() call to its local GSS, including the just received output_token as one of the parameters.
5. B's GSS processes the received output_token and, if satisfied, returns a new output_token to be forwarded to A. In our example, it also returns a major_status code GSS_S_COMPLETE, indicating that it does not expect to receive any further calls for setting up the security context.
6. B's GSS user forwards the output_token it just received to its peer entity in A.
7. A's GSS user issues a GSS_Init_sec_context() call to its local GSS. This is a continuation of the previous GSS_Init_sec_context() call since GSS had indicated that a continuation is needed. The call parameters include the output_token just received from B.
8. A's GSS verifies the output_token. In our example, it returns a major_status code GSS_S_COMPLETE indicating that the handshaking phase has completed successfully.

The complete set of handshaking-phase GSS-API calls, under the banner of **context-level calls,** is provided in Table 5.1.

5.6.2 GSS Secure Transfer

Figure 5.5 illustrates GSS secure data transfer.

TABLE 5.1 Context-Level Calls

GSS_Init_sec_context	Initiate outbound security context
GSS_Accept_sec_context	Accept inbound security context
GSS_Delete_sec_context	Flush context when no longer needed
GSS_Process_context_token	Process received control token on context
GSS_Context_time	Indicate validity time remaining on context
GSS_Inquire_context	Display information about context
GSS_Wrap_size_limit	Determine GSS_Wrap token size limit
GSS_Export_sec_context	Transfer context to other process
GSS_Import_sec_context	Import transferred context

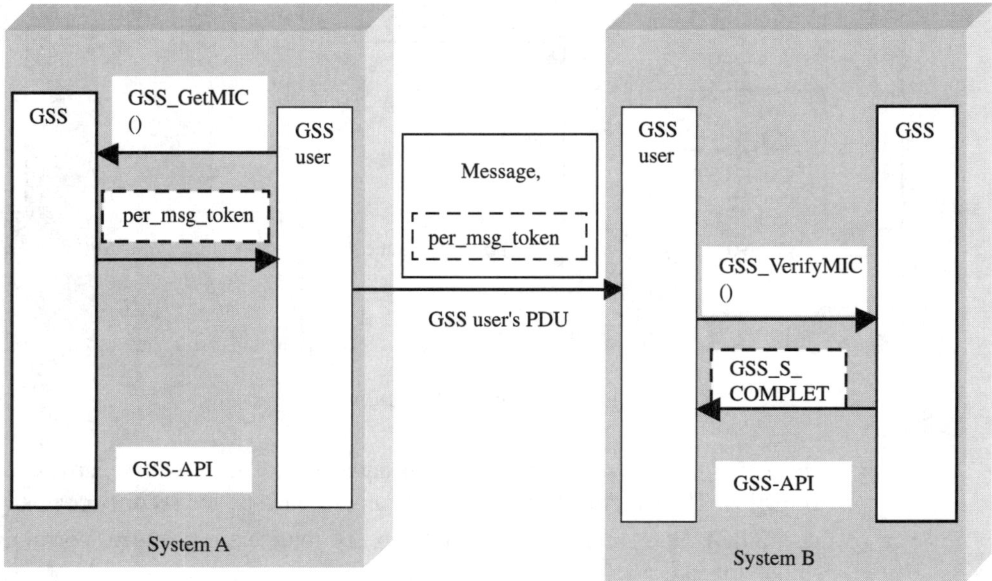

Figure 5.5 GSS-API secure transfer phase

1. A's GSS user issues a GSS_GetMIC() call. This call requests a Message Integrity Code (MIC) for a message that is included in the call as an octet string. The call may include additional parameters such as a context_handle and required Quality of Protection (QOP).
2. GSS returns a per_msg_token that includes the MIC for the message. In our simple example, it also returns a major_status code GSS_S_COMPLETE indicating that the MIC has been successfully computed and no further calls are expected regarding this message.
3. A's GSS user sends a PDU to its peer entity in target system B. The PDU includes a message along with the corresponding per_msg_token it just received.
4. B's GSS user issues a GSS_VerifyMIC() call to its GSS. The call parameters include the message it received along with the accompanying per_msg_token.
5. B's GSS verifies that the MIC in the per_msg_token correctly certifies the adjoining message and returns a major_status code GSS_S_COMPLETE, indicating that the MIC has been successfully verified and no further calls are expected regarding this message.

In addition to the two calls in the preceding example, GSS-API has two more per message calls: GSS_Wrap and GSS_Unwrap. They are quite similar to the two preceding calls, except that they can also support message encryption and decryption.

Table 5.2 provides the complete list of GSS-API secure transfer phase calls, grouped under the banner PER-MESSAGE CALLS

TABLE 5.2 Per-Message Calls

GSS_GetMIC	Apply integrity check, receive as token separate from message
GSS_Verify MIC	Validate integrity check token along with message
GSS_Wrap	Apply integrity check, optionally encrypt, encapsulate
GSS_Unwrap	Decapsulate, decrypt if needed, validate integrity check

5.6.3 GSS Administrative Interfaces

In addition to the calls for the handshaking and secure transfer phases, GSS-API has a rich set of administrative calls, grouped into two categories: credential management calls and support calls.

Credential management calls are typically invoked within a system (i.e., not between peer GSSs) before the handshaking phase. Their invocation provides GSS with necessary credentials (for example, passwords, encryption keys) for secure communications. After the secure transfer phase has been completed, credential management calls are invoked to remove any lingering credentials. The credential GSS-API management calls are listed in Table 5.3.

TABLE 5.3 Credential Management Calls

GSS_Acquire_cred	Acquire credentials for use
GSS_Release_cred	Release credentials after use
GSS_Inquire_cred	Display information about credentials
GSS_Add_cred	Construct credentials incrementally
GSS_Inquire_cred_by_mech	Display per-mechanism credential information

Support calls provide a broad range of management functions, they are listed in Table 5.4.

TABLE 5.4 Support Calls

GSS_Display_status	Translate status codes to printable form
GSS_Indicate_mechs	Indicate mech_types supported on local system
GSS_Compare_name	Compare two names for equality
GSS_Display_name	Translate name to printable form
GSS_Import_name	Convert printable name to normalized form
GSS_Release_name	Free storage previously allocated on behalf of a normalized-form name
GSS_Release_buffer	Free storage previously allocated on behalf of a printable name
GSS_Release_OID	Free storage previously allocated on behalf of an object identified by the OID
GSS_Release_OID_set	Free storage previously allocated on behalf of OID set object (i.e., an object corresponding to a set of OIDs)
GSS_Create_empty_OID_set	Create an empty OID set
GSS_Add_OID_set_member	Add a member to an OID set
GSS_Test_OID_set_member	Test if OID is member of OID set
GSS_OID_to_str	Display OID as string
GSS_Str_to_OID	Construct OID from string
GSS_Inquire_names_for_mech	Indicate name types supported by mechanism
GSS_Inquire_mechs_for_name	Indicate mechanisms supporting name type
GSS_Canonicalize_name	Translate name to per-mechanism form
GSS_Export_name	Externalize per-mechanism name
GSS_Duplicate_name	Duplicate name object

5.6.4 Status Reporting

Each GSS-API call provides two status return values. Major_status values provide a mechanism-independent indication of call status (for example, GSS_S_COMPLETE, GSS_S_FAILURE, GSS_S_CONTINUE_NEEDED), sufficient to drive normal control flow within the caller in a generic fashion. There are two types of GSS-API Major Status Codes: fatal error codes, listed in Table 5.5, and informatory status codes, listed in Table 5.6.

TABLE 5.5 Fatal Error Codes

GSS_S_BAD_BINDINGS	Channel binding mismatch
GSS_S_BAD_MECH	Unsupported mechanism requested
GSS_S_BAD_NAME	Invalid name provided
GSS_S_BAD_NAMETYPE	Name of unsupported type provided
GSS_S_BAD_STATUS	Invalid input status selector
GSS_S_BAD_SIG	Token had invalid integrity check
GSS_S_CONTEXT_EXPIRED	Specified security context expired
GSS_S_CREDENTIALS_EXPIRED	Expired credentials detected
GSS_S_DEFECTIVE_CREDENTIAL	Defective credential detected
GSS_S_DEFECTIVE_TOKEN	Defective token detected
GSS_S_FAILURE	Failure, unspecified at GSS-API level
GSS_S_NO_CONTEXT	No valid security context specified
GSS_S_NO_CRED	No valid credentials provided
GSS_S_BAD_QOP	Unsupported QOP value
GSS_S_UNAUTHORIZED	Operation unauthorized
GSS_S_UNAVAILABLE	Operation unavailable
GSS_S_DUPLICATE_ELEMENT	Duplicate credential element requested
GSS_S_NAME_NOT_MN	Name contains multi-mechanism elements

TABLE 5.6 Informatory Status Codes

GSS_S_COMPLETE	Normal completion
GSS_S_CONTINUE_NEEDED	Continuation call to routine required
GSS_S_DUPLICATE_TOKEN	Duplicate per-message token detected
GSS_S_OLD_TOKEN	Timed-out per-message token detected
GSS_S_UNSEQ_TOKEN	Reordered (early) per-message token detected
GSS_S_GAP_TOKEN	Skipped predecessor token(s) detected

Minor_status provides more detailed status information, which may include status codes specific to the underlying security mechanism. Minor_status values are not specified in RFC 2078.

5.7 GULS—SOARING SECURITY

Specifications for the seven layers of the OSI stack grew from the bottom up, starting with the physical layer and culminating with the application layer. Specification of security communications protocols for the various layers kept pace with this process. Publication of the Generic Upper Layers Security (GULS) standard marked the glorious achievement of OSI security standards. GULS is indeed the ultimate security communications protocol. It provides every function discussed in Section 3.3.7. In addition, it also supports selective field protection. Unique among security communications protocols, it even allows the explicit specification of encoding rules to be used on the data to be protected before any security transformations are applied, as well as specification of encoding rules for carrying over the wire the results of security transformations.

GULS consists of two parts: the handshaking function resides in the application layer, while the secure transfer function resides in the presentation layer. This configuration allows for optimum use of encoding/decoding in conjunction with security transformations.

GULS also broke new ground by being one of the first protocols to be expressed only in the 1994 ASN.1.

GULS quickly became the dream of security designers. Unfortunately, it remains a dream. Indeed, by the time GULS was published, other options, albeit with lesser functionality, were available. So GULS was evaluated based on its incremental benefits, not on its total capabilities. Use of the 1994 ASN.1 turned out to be a bit of a liability—few practitioners were as comfortable with the new version as with the older version of ASN.1. More to the point: compilers for the 1994 ASN.1 were harder to come by (though by now even compilers for the more advanced 1997 ASN.1 are commercially available). Finally, the considerable complexity needed to support the rich functionality scared away would-be developers.

Had GULS been commercially available, there would be little else in this book. Reality being what it is, this section is rather brief. A major reason for mentioning GULS is its use as a yardstick against which other security communications protocols can be measured.

5.8 SSL3—SAFE SURF

Secure Socket Layer version 3 (SSL3) was designed for securing Web surfing. It has been popularized by Netscape, and it is being standardized by the Internet Engineering task Force (IETF) as Transport Layer Security version 1 (TLS1). Its broad availability and ease of use make it an attractive candidate for securing any transactions over the Internet. In particular, it can be used for securing EDI- and CORBA-based network management transactions over TCP/IP.

It seems that SSL3 could just as well be used for protecting CMIP transactions over TCP. However, some current CMIP-based applications use X.25. Although TCP/IP is expected to become much more prevalent for the transport of CMIP transactions, it is not certain to become the only transport for CMIP. Since CMIP may be expected to operate over a variety of transport protocols, it is less appealing to provide its security in the transport layer. That is one reason for protecting CMIP in the application layer with STASE-ROSE rather than with SSL3 in the transport layer.

SSL3 consists of two parts. The first part is active during connection setup, it supports strong peer entity authentication and secure exchange of security information such as negotiation of which cipher suite shall be used and agreement on a secret key material. The second part secures the flow of bits between the communicating entities.

5.8.1 A Firm Handshake

Since SSL3 was designed for Web security, the initiator of a SSL3 session is called a client, while the responder is called a server. In the case of TMN applications, the initiator may well be a CMIP manager, and the responder will then be a CMIP agent. In talking about SSL3, we will stick to client and server.

As a paragon of civility, the SSL3 handshake protocol (see Figure 5.6) starts with a mandatory **ClientHello** to which the server responds with a mandatory **ServerHello.** There is a lot more than etiquette to the hello messages. They are the foundation of a secure session. Each contains the following information:

> **ProtocolVersion,** the version of SSL to be used. It consists of two integers (each represented by a single octet) denoting the major and minor version numbers. For SSL3 the value is {3,0}.

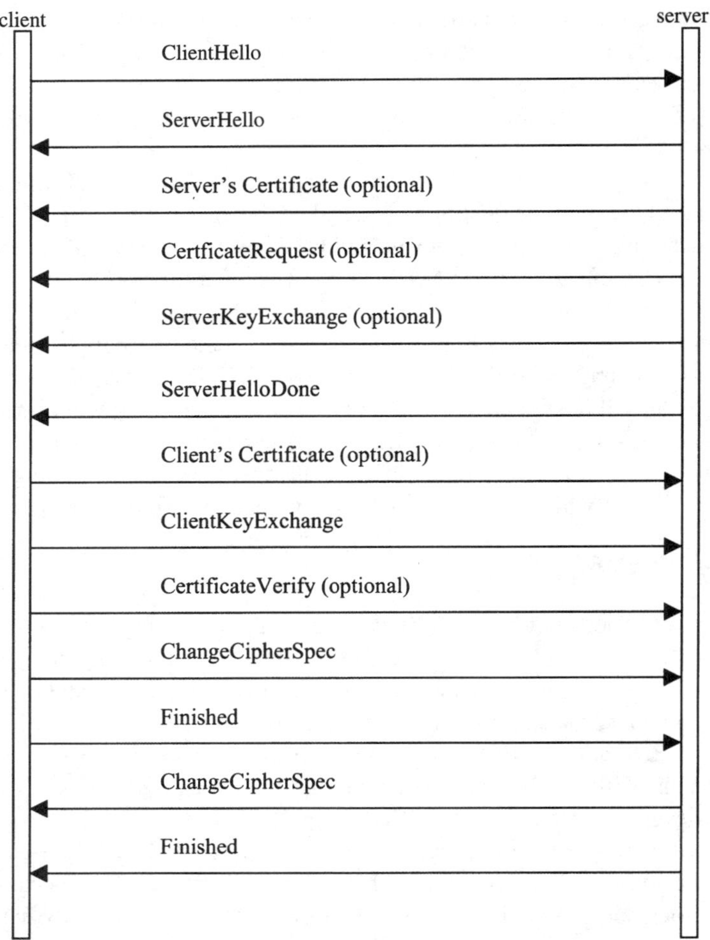

Figure 5.6 SSL3 handshake protocol

Random, which consists of the sender's internal clock value expressed in Unix 32-bit format followed by 28 octets generated by a secure random number generator.

SessionID consisting of 0 to 32 octets. The establishment of a SSL3 connection, with full negotiation of a security context, can be rather elaborate. Once two parties have agreed on a security context, they are quite likely to be happy with that context for other connections. The SessionID allows the two chummy parties to avoid renegotiating security contexts. Indeed, several connections between the same client-server pair may share the same SessionID. In that case, they all share the same security context without having to renegotiate it. Thus, if the SessionID corresponds to a previously established session and the server is willing to use the same security context, then the server can simply respond with the same SessionID and avoid the need for any further context negotiation. Otherwise, a new context must be negotiated. If the server responds with a different SessionID, this will be the identifier for the newly negotiated security context. If the server responds with a null SessionID, then the session being established cannot be resumed after it has been terminated (i.e., the security context negotiated for this session cannot be reused for a subsequent session without explicit negotiation).

CipherSuite is an integer that identifies the three encryption algorithms that will be used in the course of the session for public key encryption, symmetric key encryption, and hashing. At the beginning of the handshake protocol, the default CipherSuite denotes NULL_NULL_NULL, meaning that no encryption algorithms are used. The client can list several CipherSuites; the server must choose one of those listed by the client.

CompressionMethod is an integer that enumerates the compression method to be used in the course of the association. . At the beginning of the handshake protocol, the default is CompresionMethod null (0), meaning that no data compression is used. The client can list several CompresionMethods; the server must choose one of those listed by the client.

A server may despair of waiting for a client to call. Such an impatient server can prompt the client to send ClientHello by sending the client a **HelloRequest** message. This simple request has no parameters.

After a server has received a ClientHello and duly responded with the basic ServerHello, the server may elaborate with a few additional, optional messages. The server may find it helpful to send the client its own certificate, the server may ask the client to provide a certificate, or the server may find it appropriate to partake in key negotiation.

The **Certificate** consists of a sequence of X.509v3 certificates, starting with the sender's certificate and terminating with the root Certification Authority. The Certificate is needed for authentication of the sending party. For TMN applications each party must be authenticated, and therefore it must send its Certificate.

The **CertficateRequest** message allows the server to request the client's Certificate. It consists of two fields:

ClientCertficateType specifies the types of client certificate that the server would accept.

DistinguishedName is a list of the Certification Authorities that the server would accept for the client's certificate.

The server sends a **ServerKeyExchange** message only if it did not send a Certificate containing Diffie–Hellman parameters. If the RSA algorithm is used for public key encryption, which is presently the case for TMN applications, then this message contains the following two fields:

ServerRSAParams containing the RSA modulus and exponent of the server's temporary RSA public key, each represented as a string of octets. The server's temporary RSA public key would be used by the client for encryption; it is not used for signing.

Signature, when RSA is used, is the digital signature of (MD5(ClientHello.random + ServerHello.random + ServerRSAParams) + SHA(ClientHello.random + ServerHello.random + ServerRSAParams)). The signature is computed using PKCS#1 block type 1 and the server's private key corresponding to the public key in the server's Certificate.

When the server has finished sending its hello-related message, it sends a **ServerHelloDone** message; it has no parameters.

At this point, the client knows that it is exchanging messages with the desired server. It knows that the messages it has received from the server were indeed destined for it (and not messages destined for another client and replayed by an intruder), and it has a public key that it can use for sending secret information to the server.

The crucial message for subsequent security is the **ClientKeyExchange** message. It consists of a PreMasterSecret encrypted with PKCS#1 block type 2 and the server's public key from the server's Certificate or with the server's temporary public key from the ServerKeyExchange message. The **PreMasterSecret** consists of the ProtocolVersion followed by 46 random octets. Both the client and server compute the master_secret from the PreMasterSecret using the following formula:

master_secret =

MD5(PreMasterSecret + SHA('A' + PreMasterSecret + ClientHello.random + ServerHello.random)) +

MD5(PreMasterSecret + SHA('BB' + PreMasterSecret + ClientHello.random + ServerHello.random)) +

MD5(PreMasterSecret + SHA('CCC' + PreMasterSecret + ClientHello.random + ServerHello.random))

Both erase the PreMasterSecret after completing this computation. Next, each side derives shared key material from the master_secret. They start by computing

key_block =

MD5(master_secret + SHA('A' + master_secret + ServerHello.random + ClientHello.random)) +
MD5(master_secret + SHA('BB' + master_secret + ServerHello.random + ClientHello.random)) +

MD5(master_secret + SHA('CCC' + master_secret + ServerHello.random + ClientHello.random)) + [. . .]

This process is continued until enough key material has been obtained. The octets of the key_block are allocated as follows.

client_write_MAC_secret
server_write_MAC_secret
client_write_key
server_write_key
client_write_IV
server_write_IV

in that order. The number of octets for each of the above depends on the algorithm. For MD5 MAC (Message Authentication Code), each write MAC secret is 16 octets; for SHA it is 20 octets. For DES encryption, each write key is 7 octets and each IV is 8 octets. For exportable encryption algorithms, some additional processing is required.

At this point, the server still has no assurance that it is interacting with the purported client and not with an impostor. Indeed, although it has the client's Certificate, it does not have a single message signed by that client. To provide the necessary assurance that all the messages have been exchanged with the client identified in the client's Certificate, the client can send a **CertficateVerify** message. It consists of two digital signatures, computed with PKCS#1 and block type 1 and the client's private key. The first signature is of

MD5(master_secret pad2 + MD5(handshake_messages + master_secret + pad1))

And the second signature is of

SHA(master_secret + pad2 + SHA(handshake_messages + master_secret + pad1))

Where handshake_messages refers to all the messages exchanged so far, starting with the client's hello, up to but not including this message.

pad1 is the character 0x36 repeated 48 times for MD5 and 40 times for SHA.

pad2 is the character 0x5c repeated 48 times for MD5 and 40 times for SHA.

So far all the messages have been exchanged without any protection, except as provided through their own digital signatures and public key encryption. SSL3 was using the CipherSuite NULL_NULL_NULL as the default. Now, however, the two sides have agreed on a specific CipherSuite, and they share a common secret; they can rely on SSL3 to protect all subsequent messages. To start the process, the client sends a **ChangeCipherSpec** message. Although this message is not part of the handshake protocol (indeed, it can be used later on to change the ciphering strategy), it is conveniently used here. It consists of a single octet with the value 1. All messages following the ChangeCipherSpec message will be protected with the just established security context (i.e., the parameters exchanged in the initial hello messages and the master secret), until a new one is established. The ChangeCipherSpec message is immediately followed by a **Finished** message to formally announce the conclusion of the handshaking ceremony. The Finished message sent by the client consists of two message digests:

MD5(master_secret + pad2 + MD5(handshake_messages +
0x434C4E54 + master_secret + pad1))

SHA(master_secret + pad2 + SHA(handshake_messages +
0x434C4E54 + master_secret + pad1))

where handshake_messages refers to all the handshake protocol messages exchanged so far, starting with the client's hello, up to but not including this message. Note that the handshake protocol messages do not include the ChangeCipherSpec messages.

The server reciprocates by sending a ChangeCipherSpec message followed by the server's Finished message, consisting of two message digests:

MD5(master_secret + pad2 + MD5(handshake_messages +
0x53525652 + master_secret + pad1))

SHA(master_secret + pad2 + SHA(handshake_messages +
0x53525652 + master_secret + pad1))

Notice that both Finished messages are protected with the newly established cipher spec.

5.8.2 Secure Transfer

All SSL3 messages, including the handshake protocol messages, are carried over the SSL3 record layer. First, they are fragmented into records of 2^{14} octets or less. Short messages of the same type (for example, handshake protocol messages) may be grouped in a single record. Next, they are compressed. The default compression method, in force prior to the handshaking, is null compression (i.e., no compression). Finally, each record is protected for integrity and privacy. The default protection method, in force prior to the handshaking, is null protection (i.e., no protection).

As soon as security mechanisms (cipher suite, master secret, compression) for the session have been established, SSL3 starts doing its magic. Integrity is provided through a Message Authentication Code (MAC) computed as:

hash(MAC_write_secret + pad2 + hash(MAC_write_secret +
pad1 + seq_num + length + content))

where

> **hash** is the current hashing algorithm (MD5, SHA, or NULL).
>
> **MAC_write_secret** is derived from the shared master_secret.
>
> **seq_num** is the sequence number for this message; sequence numbers are generated independently by the client and the server with each keeping track of the most recent sequence number it has sent and the most recent sequence number it has received. The sequence number allows detection of deletion and/or reordering attacks.
>
> **length** is the length of the compressed record.
>
> **content** is the compressed record.

If an encryption algorithm is included in the current cipher spec, then it is applied to the record after it has been protected for integrity. For block ciphers (such as DES) in the CBC mode, the Initialization Vector (IV) for the first record is derived from the master_secret. The IV for each subsequent record is the last ciphertext block from the previous record.

5.8.3 Alerts

SSL3 defines a complete set of both **warning** and **fatal alerts.** When a user sends or receives a fatal alert, the user is required to terminate the session immediately and erase any information related to that session. That session is not resumable.

A **close_notify** message notifies the recipient that the sender will not send any more messages on that session. If a connection is terminated without a proper close_notify message with level equal to warning, then it is not resumable.

5.9 IPsec

IPsec provides strong one-way authentication or confidentiality for IPv4 and IPv6 packets [RFC2401]. As such it can protect all protocols riding on top of IP (for example, TCP, UDP). It is therefore a plausible candidate for protecting SNMP messages that use UDP.

IPsec consists of two distinct protocols: **Authentication Header** (AH) [RFC2402] and **Encapsulating Security Payload** (ESP) [RFC2406]. Using one neither requires nor prevents using the other. AH provides data integrity for the IP payload and the immutable fields of the IP header (i.e., IP header fields that cannot be changed in an unpredictable way by intermediate routers); ESP provides data integrity for the IP payload and/or encryption of the IP payload. Both are discussed in this section.

5.9.1 A Discrete Handshake

SSL3's firm handshake consumes about a dozen messages. Since IP is inherently connectionless, IPsec cannot use even a single message for handshaking. Its procedure for setting up the security context is stealthily done behind the scenes (i.e., it is outside the scope of IPsec) well ahead of any secure communications.

The two communicating parties have to agree in advance on the security services, corresponding algorithms, parameters, and shared secrets they will use. Different IP packets may receive different protection, though the difference, if any, may only depend on their IP header and some fields in the headers of higher protocols.

An IPsec security context consists of a single **Security Association** (SA) or a bundle of SAs. An SA defines a security context for IP packets in one direction only, from a fixed originator to a fixed receiver. Thus, at least two SAs are needed for secure two-way communications. An SA specifies which IPsec protocol is used: AH or ESP. If ESP is used, the SA specifies if it is used for integrity, confidentiality, or both (it cannot be neither). The SA further specifies which algorithm(s) are used by each of the selected protocols and with which parameters (for example, encryption keys). If an IP packet requires the services of both AH and ESP, it can be protected by two SAs (an SA bundle).

There are two types of SAs: **transport mode** and **tunnel mode.** Transport mode can be used only for host-to-host communications. It provides end-to-end security. Tunnel mode must be used if the IPsec processing at one or both ends of the communication is done by an intermediate security gateway (for example, a firewall). In this case, the IP packet will consist of an outer IP header addressing the security gateway (SG), an IPsec header to be used by the security gateway, and an inner IP header addressing the ultimate destination of the packet. Tunnel mode provides security between two security gateways or between a security gateway and an end-user. Figure 5.7 illustrates the use of both modes for a single packet.

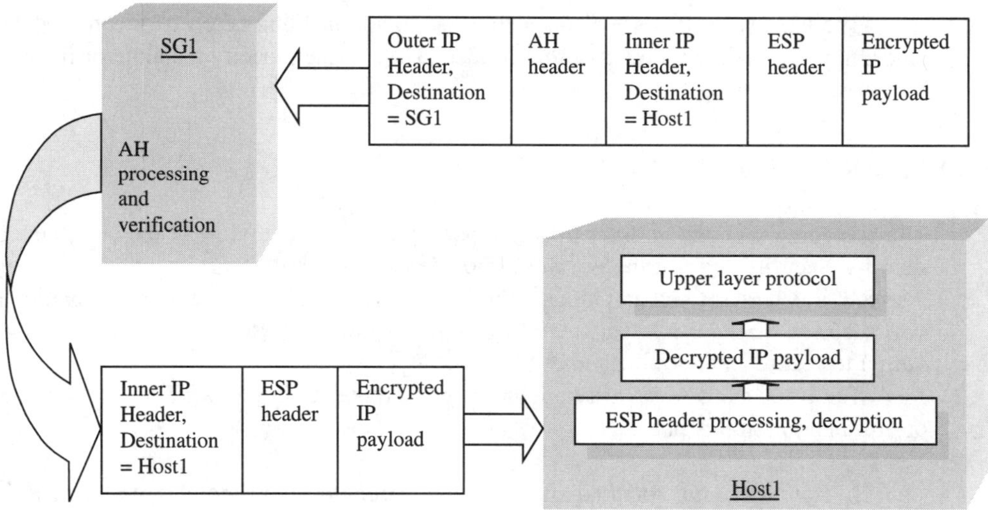

Figure 5.7 Example of tunnel mode authentication and transport mode encryption

If the final destination of an IP packet is a security gateway (for example, the packet carries an SNMP command to the security gateway), then the security gateway is considered a host and the SA type can be the transport mode.

An SA is identified by a remote IP address, an IPsec protocol (i.e., AH or ESP), and a locally unique (i.e., for the SA's remote address and protocol) **Security Parameter Index** (SPI).

An IPsec implementation maintains an **SA Database** (SAD) of all its SAs.

An IPsec implementation further maintains a **Security Policy Database** (SPD) for each interface. The SPD specifies the SA (bundle) corresponding to any incoming or outgoing IP packet based on the values of selected fields in the packet's IP header and upper layer protocols' headers. The SPD may specify that some packets (based on the values of fields in their headers) need no protection. It can also specify that some packets must be discarded. The entries in an SAD and an SPD represent the previously established security contexts with all potential interlocutors.

5.9.2 Secure Transfer

Both the AH and ESP headers contain an SPI and a sequence number. The SPI allows the receiver to associate the incoming packet with the correct SA. The sequence number can be used by the receiver to detect replay attacks. The sequence number is a monotonically increasing counter. The sender maintains one such counter for each SA. When the counter is about to reach its maximum value (for example, 2^{32}), it is reset to zero and a new key must be used (i.e., a new SA).

The real security protection of the AH header is a MAC (called Integrity Check Value, or ICV in IPsec). It is typically a keyed hash. It protects the IP packet payload (including the upper protocols' headers) as well as the fields of the IP header that may not be changed in an unpredictable way by intervening routers. Every IPsec implementation must support HMAC with MD5 [RFC2403] and HMAC with SHA1 [RFC2404]. An IPsec implementation may also support additional MAC algorithms published in RFCs.

The ESP header, or rather the ESP-encapsulated packet, has an SPI, a sequence number, payload data (encrypted if the SA specifies this service), and MAC (or ICV, if specified by the SA). The MAC protects only the IP packet payload. In addition to HMAC with MD5 and HMAC with SHA1, each ESP must support DES CBC encryption [RFC2405]; it can support additional cryptographic algorithms published in RFCs.

The AH and ESP RFCs also specify how to add padding octets to accommodate block algorithms and field alignment within headers They further specify a couple of fields needed for header processing.

5.10 SECURITY ENGINEERING

The preceding sections in this part of the book provide an overview of the basic concepts of security. Security engineering is a would-be science that is still an art, in putting it all together. As such, it is perhaps presumptuous to bundle security engineering with mathematically precise security mechanisms. However, a disciplined approach to security engineering is required to secure an enormous target like a TMN.

The basic steps of security engineering are quite straightforward and follow the order of presentation in this book:

1. Enumerate the potential attackers, their strengths, their motives, and their resources.
2. Identify vulnerabilities to the potential attackers.
3. Analyze the risks associated with the vulnerabilities to determine the security needs.
4. Select the security services, including security management, that provide the required protection.
5. Assemble an optimum set of security mechanisms to support the selected security services.
6. Deploy, activate, and monitor the security features.

Unfortunately, the preceding list is far from providing a simple recipe for security. It overlooks several challenges:

- Identifying most of the potential attackers is fairly easy; the list of the usual suspects in Section 3.1 provides a good starting point. The real challenge is to ferret out the more insidious intruders, especially crooks with advanced social engineering skills.
- The preceding sections provide several tables that help correlate security threats, risks, services, and mechanisms; such correlations assist in matching the security features to the perceived threats. However, those correlations are merely suggestive; different combinations of vulnerabilities may result in many different types of risks. Furthermore, different combinations of security services and mechanisms may protect against the same sets of threats.

- Selection of security mechanisms depends on several considerations: availability of commercial products, processing delay, transmission overhead, ease of deployment, ease of management (for example, of keys, certificates, audit trails), and transparency to users.
- Selection of security algorithms requires good luck as well good judgment: an algorithm that looks safe today may be broken tomorrow. It is therefore necessary to maintain the flexibility to quickly adapt to such developments.

Security engineering can be quite daunting indeed. However, within the context of a TMN it is easy to establish several simplifying guidelines:

- Unlike idle chat among friends, network management transactions are not used to exchange frivolous gossip. The corruption of any network management transaction has a potential for serious damage. Therefore, every TMN association setup requires **strong peer entity authentication,** and every TMN transaction requires cryptographic **integrity** protection. If you stumble upon a TMN transaction that does not merit such protection, then you can further streamline your TMN by eliminating this transaction (since its deletion by an intruder would be harmless anyway).
- All transactions with external entities (over the X interface) that impact substantial resources (for example, a service request for several broadband circuits) need **non-repudiation** protection. Non-repudiation may be valuable for other X-interface transactions as well, though perhaps with less urgency.
- In some cases, the law requires **encryption** of customer data. In more nebulous cases, the law specifies privacy of customer data, though whether this implies mandatory encryption is a matter of legal debate. In general, laws do not specify that any information specific to the telecom service provider should be protected. Therefore, each service provider needs to assess which of their own data, carried over its TMN, requires privacy protection.
- If a TMN involves more than a couple of hundred participants (network elements, operation systems, work stations, users), or if it is expected to grow to such size, then a **public key infrastructure** becomes a requirement for efficient, scalable key management.
- Nothing beats catching an intruder red-handed. Therefore, every TMN system should be able to issue **security alarms** when it suspects that it is under attack. Such alarms allow the security manager to isolate the TMN and the network from the attack, as well as to lure the intruder into a trap. It is still a matter of local security policy to determine the kinds of security alarms a system can issue and how to manage security alarms.
- Absolute security for a TMN, if it could be provided, would probably cost more than the TMN it protects. Recognizing that preventing all attacks is not practical, it is at least necessary to detect successful attacks. It is therefore necessary to maintain a log of all significant security events. It is a matter of local security policy to determine what kind of information should be contained in a security log and how to manage security logs.
- Employees must be educated about the need for security and how to use (or live with) the deployed security measures.

Within a structured framework, such as a TMN, it is also necessary to meld the security features with all the other components. The rest of the book addresses this issue.

PART III
SECURING THE TMN

This part addresses the security of the TMN. It shows how TMN messages, using any of the TMN protocols, can be protected. The coverage is not uniform. Long-standing, general-purpose security features (e.g., firewalls, SSL3) are not discussed in much detail since they are well covered by existing literature. Only their use within the TMN is discussed. Freshly minted TMN-specific features, on the other hand, are covered in glorious detail.

6

Security of OSI-Based TMN Protocols

If two entities are to communicate securely, they must be able to exchange security-related information, such as a digital signature, an authenticator, or a specification of which encryption algorithms are to be used. Several protocols used in the TMN provide the handles necessary for exchanging such information. This chapter describes the security features of relevant OSI-based TMN protocols. It focuses on CMISE and X.500, as well as protocols that support those Application Service Elements (ASEs), such as ACSE. This chapter does not address the security features of FTAM (which is included in the TMN communications protocol stack specified by the ITU-T) since it is not currently used in any TMN implementations.

6.1 IN THE BEGINNING—ACSE SECURITY

The Association Control Service Element (ACSE) is used to establish all OSI application associations, especially TMN associations. Therefore, it is crucial to the establishment of secure associations. Fortunately, ACSE provides several handles to support basic security features—identification, authentication, and ASE-specific security.

6.1.1 Identification

An OSI association is established between two application processes (APs). The APs perform network management functions that require interactions between systems. Two APs interact through a well-defined communication interface called an Application Entity (AE). Each AP can have several AEs, allowing it to have different interfaces with different remote APs. In most cases, however, an AP would have a single AE. An AP consists of one or more Application Service Elements (ASEs), a coordination function that manages interactions among the ASEs, and information processing functionality. Each ASE in one AP communicates only with a similar ASE in the remote AP through an ASE-specific protocol. For instance, two Common Management Information Service Element (CMISE) ASEs communicate through the Common Management Information Protocol (CMIP). An AE that includes CMISE must also include ACSE (for association setup) and the Remote Operations Service Element (ROSE) to sort out which reply corresponds to which request. In general, such an AE would also include one or more Functional Units (FUs) of some system management functions of the System Management Application Service Element (SMASE). (SMASE consists of several system management functions, such as security audit trail management and security alarm management; each system management function may contain several FUs, each of which can be implemented on its own merit. See Section 2.3.7.7.)

Two ACSE messages are exchanged to set up an association: the Application Association Request (AARQ) and the Application Association Response (AARE). The AARQ has fields for identifying both the calling AE and the called AE. The AARE has fields for identifying the responding entity (which may be different from the one specified in the AARQ). All of these fields are optional.

An AE is fully defined through the AE title, which consists of two fields in an ACSE PDU: the AP Title, which identifies a unique AP, and the AE Qualifier, which identifies a unique AE within the AP. To ensure the uniqueness of AE titles, two choices are possible:

- The syntax of AP Title can be chosen as OBJECT IDENTIFIER (see Section 2.3.7.8.1). In this case, the syntax of the AE Qualifier (if used) must be an INTEGER. This way their concatenation will result in an OBJECT IDENTIFIER. Or
- The syntax of AP Title can be chosen as directory DistinguishedName (see Section 2.3.7.8.2). In this case, the syntax of the AE Qualifier (if used) must be RelativeDistinguishedName, which uniquely identifies the AE under the AP.

DistinguishedName is an especially convenient choice because it can be used as a TMN directory entry, and it can be used in access control management to identify request initiators. Thus, it allows an AE to be identified by the same DistinguishedName in the TMN directory, in Access Control Lists (or other access control management mechanisms), and in the ACSE association setup PDUs.

For secure communications, each party must uniquely identify itself to the other party when an association is established. One way of doing so is to include the DistinguishedName of the calling AP Title and of the calling AE Qualifier in the AARQ, and the DistinguishedName of the responding AP Title and of the responding AE Qualifier in the AARE. Another way of providing unique identification is discussed under Authentication.

6.1.2 Authentication

Authentication uses secret information (e.g., a password, an encryption key) to corroborate the identity of the sender to the receiver. The format of the authentication information depends on the authentication mechanism that is used. For instance, if the authentication mechanism is a simple password exchange, the authentication value may be of type GraphicString. Other authentication mechanisms may include a time stamp, a digital signature, or other information. Such mechanisms will result in more complex structures for carrying authentication information.

If the two communicating entities have some agreement regarding the authentication mechanism, then this information need not be included in the ACSE PDUs. Otherwise, and for added flexibility, the authentication mechanism name can be included in the ACSE PDUs. Both scenarios are supported by ACSE, as described below.

ACSE contains an optional Authentication FU. The Authentication FU consists of three fields:

1. The ACSE Requirements field indicates which ACSE FUs are used. It must be present if the Authentication FU is used. In this case its value simply signals that the Authentication FU is used.

2. The Authentication-mechanism Name is an optional field, which can be used only if the ACSE Requirements field is present. This field specifies which mechanism is used for authentication. Although the use of this field is optional, it becomes mandatory if the authentication value (see next item) is of type ANY DEFINED BY (the Authentication-mechanism Name). The Authentication-mechanism Name is an OBJECT IDENTIFIER. It can be registered by anyone, though for the broadest interoperability it is ideally registered by ISO.
3. The Authentication-value, which can be used only if the ACSE Requirements field is present, provides the actual authentication information; its abstract syntax is a CHOICE among four possibilities:
   - GraphicString, which is suitable for exchanging passwords, for example.
   - BIT STRING, which is suitable for more general authenticators without explicit internal structure.
   - EXTERNAL, which allows any structure for the authentication value. It requires that the syntax for the authenticator be defined outside the ACSE standard and be properly registered (e.g., with ANSI, ITU-T). The abstract syntax must be part of the defined context set negotiated by the presentation layer during connection setup.
   - ANY DEFINED BY mechanism name provides the same flexibility as the EXTERNAL choice, though it requires that the Authentication-mechanism Name be present.

6.1.3 ASE-Specific Security

Some ASEs may require the exchange of some initialization information during association setup. To that end, the AARQ and AARE PDUs each contain a User Information field. This field is a SEQUENCE OF EXTERNAL, where typically one EXTERNAL may be used to carry initialization information specific to one ASE in the AE. In particular, the EXTERNAL may contain security-related information for the corresponding ASE.

For instance, the CMIP protocol which is used by the CMISE ASE defines the syntax for CMIPUserInfo, which is to be carried in the EXTERNAL allocated to CMIP in the ACSE User Information field. CMIPUserInfo includes an optional accessControl field. This field, which is of type EXTERNAL, can carry any security-related information that is specific to CMISE.

6.2 CMISE SECURITY

The CMIP protocol which is used by the CMISE ASE provides uneven security: the CMIP management operations PDUs (get, set, create, delete, and action) are each endowed with an (optional) access control field; CMIP responses, notifications, and confirmations, on the other hand, lack any security handles.

From a purist perspective, the access control field should carry only access control information. As such, it is meaningful only for CMIP operations and not for other CMIP messages. In practice, however, the CMIP access control field may be used to carry an authenticator that provides data origin authentication and that can allow detection of replay, deletion, delay, rerouting, and misrouting attacks. Such protection is often needed for CMIP responses and notifications. It can only be provided through other ASEs. As discussed in 6.1.3, CMIP access control information can also be carried at association setup time in the User Information field of the ACSE AARQ and AARE.

6.2.1 Electronic Bonding Authenticator—Homegrown Security

The first CMIP-based application over the X interface among Telecom companies in the United States was trouble administration (TA), the first in a series of TMN applications between Local Exchange Carriers (LECs) and Inter Exchange Carriers (ICs) collectively known as Electronic Bonding (EB). The security provided for TA [TR40] is not spectacular. Actually, it is rather timid and leaves a lot to be desired. Nonetheless, a whole section in this book is devoted to this topic because it provides a valuable insight into the process of security engineering: developing an acceptable security solution under draconian constraints, and squeezing the maximum amount of security out of the available resources. It is also far superior to security measures in earlier network management applications (basically, firewalls and passwords), and its use has spread to several other EB applications.

6.2.1.1 Historical Background. ICs lease access circuits from LECs in order to provide service to their customers. When the IC notices that one of these circuits breaks down (typically as a result of a customer complaint), a trouble report is entered into the IC's Operation System (OS) for TA. The OS helps track the trouble report and its resolution. To resolve the problem it must be reported to the LEC that provides the circuit. The traditional way of exchanging the corresponding information between different companies is rather laborious, even if each of the two companies is highly automated: an IC technician calls a LEC technician to report the problem. Further telephone conversations are required for information to be requested by the IC and provided by the LEC about the progress of the trouble resolution. With EB TA the trouble reports are conveyed electronically from the IC OS to the LEC OS. The potential for introducing human errors is much reduced, the productivity of maintenance technicians is streamlined, and the quality of service perceived by the customers is improved since troubles get fixed faster.

When EB was first proposed, it was clear to all parties that a standard interface would be desirable. It would allow each participant to develop a single interface for interacting with all its EB partners. The Electronic Communications Implementation Committee was thus formed to develop uniform implementation guidelines.

When TA EB was first proposed, it faced a mixed reception: it was much welcomed thanks to the benefits described above, yet some managers responsible for TA OSs, especially on the LECs' side, were horrified at letting strangers into their OSs. Some security measures were obviously needed to assuage such concerns. Unfortunately, there were many different views as to the appropriate level of security. Some participants felt that passwords at association setup time were quite adequate, since this was the traditional approach. At the other extreme, some participants worried about opening their OSs and their internal operations to outside parties and insisted on comprehensive protection, including non-repudiation, for every EB message.

The result was, as usual, a compromise. Early on, all parties agreed to the need for Closed User Groups (CUGs) and strong, cryptographic peer entity authentication. The CUGs were easy since the underlying X.25 transport networks supported them. Strong authentication provided an opportunity for lively discussions. The first mechanism proposed for the authenticator was based on the Enigma algorithm, simply because it was available on several work stations. During World War II Alan Turing, armed with little more then paper and pencil, succeeded in breaking this algorithm, heavily used by German U-boats. RSA public key encryption was proposed as a much superior alternative. However, several participants had no prior experience with this technology; they were concerned about its performance and were reluctant to pay royalties for use of a patented technology. Both Enigma and RSA were therefore off-limits for EB security (for excellent and poor reasons, respectively).

Some participants proposed to use hardware encryptors at each EB termination. Certainly, use of such encryptors would not put any burden on the processors doing TA and would not require any programming effort from the developers. However, hardware encryptors from different suppliers do not interoperate. Therefore, hardware encryption could not offer a standard interface, unless everyone agreed to use products from the same supplier. While this remains an option, there was still a need for a truly interoperable, supplier-independent solution.

The next proposal was to base EB security on the recently developed Generic Upper Layers Security (GULS) [X.830, X.831, X.832, X,833] standard. Support for GULS must be provided from within the OSI communications protocol stack. All the suppliers of such protocol stacks to EB participants made it clear that GULS, though very powerful, is much too complicated to be seriously considered. No other standard solutions were available at the time for security services within the protocol stack. It therefore became clear that EB application developers would be saddled with security. Project managers, facing tight deadlines and budgetary constraints, were understandably reluctant to commit scarce resources to introduce security features into their applications.

The absence of security support within the communication stack was a double whammy: in addition to imposing new tasks on application developers, it effectively precluded any hope of providing security for whole messages. Indeed, any security transformations such as providing a Message Integrity Code (MIC) to ensure message integrity, must be computed over the message exchanged between the two systems. The application does not see that message. It is generated (on the sending side) and decoded (on the receiving side) by the presentation layer.

Meanwhile, the deadline for deployment of TA EB became ever more urgent. The prospects for securing EB became gloomy. Only the constraints imposed on it emerged with blinding clarity:

- No public key encryption
- No support from stack suppliers
- No chance for whole message protection
- Solution needed yesterday
- Must be quick and easy to develop and use
- Must be frugal with processing and transmission overhead

Altogether, these constraints are familiar to most system engineers and developers.

6.2.1.2 The Authenticator. Authentication [TR40] is accomplished through the exchange of a time stamp encrypted with DES in the CBC mode and use of a key known only to the two communicating systems. It protects against replay (each time stamp is unique), reordering (the time stamp is monotonically increasing), rerouting or misrouting (only the intended recipient can correctly decrypt the authenticator), and it detects unexpected long delays. DES is unencumbered by patents, it is widely available, and it is pretty fast (compared to public key encryption). That's about as much security as possible within the design constraints; it's a good start.

An encrypted time stamp is meritoriously brief—unfortunately too brief to be useful. A few more tidbits are needed for proper security and ease of processing: Proper DES encryption requires an 8-byte initialization vector (IV) that must accompany the encrypted string. Proper security requires occasional changes of the encryption key, so a key identifier must be added to specify which key, from a previously exchanged list, is now being used. A popular system is likely to have dealings with several other systems. It shares lists of secret keys with each of its partners. Therefore, including the identity of the sender will help the receiver retrieve the key from the correct list. By now the authenticator has four components:

- Sender identifier
- Initialization vector
- Key identifier
- Encrypted time stamp

The four fields are captured in the following ASN.1 expression:

```
AccessControl ::= SEQUENCE {

            entityIdentifier        [0] IMPLICIT GraphicString (SIZE (1..64)),

            initializationVector    [1] IMPLICIT OCTET STRING (SIZE (8)),

            keyIdentifier           [2] IMPLICIT INTEGER,

            encryptedString         [3] IMPLICIT OCTET STRING (SIZE(8..64))

            }
```

The reason the sequence is called AccessControl will become clear shortly.

To ensure interoperability of the format of the time stamp, the GMT system time must be specified. The GeneralizedTime (version b, the Coordinated Universal Time [UTC] option, defined in ASN.1) format is Y2K proof and is therefore an obvious choice. Since an encoder of ASN.1 expressions is not usually available to application programmers, it is further necessary to specify how the GeneralizedTime is to be represented. Since the EB participants were using ASCII-based computers, the choice was easy: represent GeneralizedTime as a string of ASCII characters.

6.2.1.2.1 Vulnerabilities. Although this authenticator seems pretty solid, it is not. Its vulnerabilities, discussed below, are as serious as they are subtle.

6.2.1.2.1.1 Immediate Replay. DES is used in the CBC mode with a new Initialization Vector (IV) for every PDU that contains the access control field. The IV is an arbitrary string of 64 bits chosen by the sender. The IV is transmitted, in clear text, in the access control field as shown in the ASN.1 syntax above.

Using an arbitrary IV seems to enhance security (that's the intent), but actually it introduces a vulnerability: An intruder that captures the AARQ PDU can use the encrypted time stamp and IV in the captured message to form a valid-looking authenticator in a spoofed AARE. The receiver of the AARE would not necessarily reject a time stamp identical to the one it just sent, since the clocks of both systems are allowed some drift (two minutes) from the true GMT. This attack is thwarted through a simple requirement on the IV: Each IV must be different from the IV in the previous PDU sent (which contains an access control field) and different from the IV in the most recent PDU received (which contains an access control field). The GeneralizedTime itself (in clear text) is not transmitted.

6.2.1.2.1.2 Delayed Replay. If the same encryption key is used for several days, then IV introduces yet another vulnerabilty: An intruder can make copies of EB authenticators transmitted in the course of one or more days on a given association. For each copy the intruder notes the time when the authenticator has been transmitted and the ID of the encryption key. On any subsequent day, while the same encryption key is used, the intruder can replay a captured authenticator at the exact time of day when it was initially transmitted, with an appropriately designed IV as follows: For the initial authenticator, the intruder knows the IV and the first 8 bytes of GeneralizedTime: YYYYMMDD (i.e., 4 bytes for the year, 2 for the month, and 2 for the day). Therefore, the intruder also knows IV XOR YYYYMMDD (the exclusive OR of the IV and the first 8 bytes of the GeneralizedTime that went into the DES encryption in the first round of the CBC mode). When the intruder replays the authenticator on date YYYYMMDD′, the IV will be changed to

$$\text{IV}' = (\text{IV XOR YYYYMMDD}) \text{ XOR YYYYMMDD}'$$

This way

$$\text{IV}' \text{ XOR YYYYMMDD}' = \text{IV XOR YYYYMMDD}$$

Thus, the input into the first round of DES encryption is the same in both cases. All communications after the first day the association is established, and for the duration of the usage of the key, are vulnerable to this attack. The defense against this attack is simple: change the key at least once every 24 hours. Since an EB association can go on for several months, it is necessary to change the key without interrupting the association. That's why a key identifier is included in the authenticator. The authenticator is thus defined. Now it must be woven into the protocol data units (PDUs).

6.2.1.2.2 Protocol Implications. Peer entity authentication in the OSI environment is accomplished by including authentication information in the two ACSE (Association Control Service Element) messages used to set up the association. The obvious way of exchanging the authenticator during association setup would be through the Authentication Functional Unit (FU) of the ACSE. This FU, however, is optional. Most OSI stack providers did not support it. The solution to this new challenge may look contrived, but it is actually rather efficient: Since the association is being set up for a CMISE-based application, the user information field in the ACSE association setup messages can carry CMIP-specific data. CMIP operation messages include an access control field. This field got commandeered to support authentication (which is needed for access control anyway). So the authenticator is assigned to the CMIP access control field, and at association setup (before any CMIP messages are being exchanged) the ACSE user information field transports the payload of the CMIP access control field. Placing the authenticator in the CMIP access control field is not an undue burden. It provides data origin authentication (though without integrity protection) for the CMIP operations PDUs that are exchanged in the course of the association. It turns out that the absence of support for the ACSE Authentication FU did not matter after all.

Within the ACSE user information field, the CMISE-related data is of type EXTERNAL, which means that it must be defined somewhere outside of ACSE. The ASN.1 expression for AccessControl above is that definition.

6.2.1.2.3 Operations Implications. The manager initiates the association request by creating and sending to the agent an AARQ (application association request) PDU message. Within this message the manager includes the authenticator in the portion of the ACSE user information field allocated to CMISE. The manager starts a (tunable) timer (with a default expiration time of two minutes). If this time expires before the AARE (application association response) PDU is received, the manager will terminate the association attempt.

The agent validates the manager's request. If the request is valid, the agent responds with an AARE PDU. This PDU will contain the DES-encrypted GeneralizedTime in the AccessControl field as discussed above.

If the agent is not able to validate the manager's association request, it issues an

A-ABORT without any additional information. The A-ABORT (ABRT) prevents intruders from tying up resources like the underlying X.25 connections.

Authentication information is crucial in implementing access control—that is, checking if the requester has the right to establish an association, and if the requester has the right to request the various operations, with their associated parameters, in the course of the association. Access control is of the utmost importance to EB. It is needed to ensure that one EB entity does not obtain or modify information that it is not entitled to (thereby compromising the privacy or integrity of someone else's service).

6.2.1.2.4 DES Padding. The DES algorithm takes as input clear text whose length is an integer multiple of 8 bytes (and an encryption key). If the length of the clear text is not an integer multiple of 8 bytes, additional padding bytes must be added to comply with the DES specifications. The convention adopted for TA EB is that the last byte of text denotes how many of the last 8 bytes are relevant data bytes; the remaining bytes; if any (i.e., after the last relevant data byte and before the last byte of the text), are arbitrary. The length of GeneralizedTime is 17 bytes. For its length to be a multiple of 8 bytes (i.e., 24 bytes), 6 arbitrary padding bytes are added, followed by the value 1, indicating that only the first byte of the last 8 is relevant.

If the original text were an exact multiple of 8 bytes (which is not the case for GeneralizedTime but may be the case if additional information is added) the padding would consist of 7 arbitrary bytes followed by a 0, indicating that none of the last 8 bytes is relevant.

6.2.1.2.5 Key Management. Before the EB cutover date, the Agent provides the Manager with a list of DES keys. The default size of the list is 1,000 keys. However, the Manager and the agent may agree on a different size list. The keys are numbered, and thus identified, from 1 to 1,000. The list is in mutually agreed, electronically readable form. It is exchanged manually.

The Manager changes keys at random time intervals that are not to exceed one day. The change of keys is accomplished by a change in the keyIdentifier field of the AccessControl structure. The keyIdentifier is used as an index into the list of keys. The indexed key is used to decrypt the encryptedString field of the AccessControl structure. The AccessControl structure is sent not only with the AARQ PDU and RESPONSE, but also with m-Get, m-Set, m-Create, m-Delete, and m-Action. This allows keys to be changed in mid-association. The Manager normally makes Key changes. Should the Agent desire a key change, it will issue an A-ABORT; then in the next association establishment, it will send a new key in the AccessControl structure of the AARE PDU.

Keys are not reusable. If a key identifier is reused or invalid, the behavior is at the discretion of the recipient.

In the event that an Agent becomes uneasy about the goodness of a particular key, the Agent may telephone the Manager and request a key change, or issue an A-ABORT and specify a new key as above.

In the event that a Manager becomes uneasy about the goodness of a particular key, it will disable that key so that it will never enter that key's index in the keyIdentifier field. A backup list (of the same number of keys as in the initial list) of keys will be kept on hand by both the Manager and the Agent for use in case it is suspected that the entire key list has been compromised.

After a major portion of the keys (e.g., 700 of the 1,000) keys have been used, or after a year has passed, whichever comes first, the Agent will provide the Manager with a new list of keys. The numbering of the keys in each subsequent list will be a continuation of the numbering of the previous list. Every key will have a unique keyIdentifier (e.g., 1–1,000, 1,001–2,000, 2,001–3,000, etc.).

The first list of keys must be exchanged with utmost precaution—manually between two persons who know each other, for example. Each subsequent list is triple DES encrypted with three keys from the current list that have not been used before. The three keys are then destroyed. The triple DES encrypted list can be transmitted electronically or by any other convenient means.

6.2.1.2.6 Future Proofing. It is usually foolhardy to claim that a security measure is immune to any future developments. It would be downright preposterous to make such claims about the simple EB authenticator. Surprisingly, however, it has fared pretty well in the face of recent turmoil involving DES.

The Technology Frontier Foundation (TFF) has constructed a DES cracker for under $500,000. It can crack a DES key in a couple of days. This development casts fresh doubt on the usefulness of DES for protecting highly confidential information. Indeed, an intruder may copy encrypted messages as they zoom over the network and then decrypt them in a couple of days. In EB, however, DES is used for authentication, not for confidentiality protection. The DES key used for encrypting the time stamp is changed at least once a day. So by the time this new DES cracker has cracked a key, it is probably no longer in use. Changing the key more often would further strengthen the security of EB. This can be done easily without any changes to the EB interface. The most the intruder can get out of this exercise is the value of the time stamp—very slim pickings.

EB uses triple DES encryption to protect new lists of DES keys. Even with the TFF DES cracker, it would take billions of years to break triple DES. However, Eli Biham has recently proposed a fiendishly clever attack on triple DES that is no more demanding than a brute force attack on DES. Combining Biham's cryptanalysis with TFF's cracker seems to threaten EB's key exchange procedure. However, Biham's attack requires a vast amount (billions of bytes) of encrypted text. EB key exchange uses three DES keys to encrypt only a few thousand fairly random bytes. Since these keys are never reused, the procedure is immune to Biham's attack.

It is most comforting to notice that EB security can withstand attacks that were not anticipated when EB started. Well, it better be pretty strong. Indeed, although this scheme was designed for TA, it is now used for all other CMIP-based EB applications. Following the law of unintended consequences, EB implementers are now rather lax about the need for more comprehensive security, such as integrity protection.

6.2.2 Selective Field Protection—If ABSolutely Necessary

The second CMIP-based EB application was PIC (Primary Inter Exchange Carrier). As end-users change their PIC, they notify either their LEC or their new IC of the change. The party that receives this information from the end-user must notify the other, as well as that customer's previous IC. As churn in PIC selection increased, this process became an obvious candidate for EB automation.

PIC exchanges may include the customer's name, address, phone number, and social security number. Several federal and state laws proclaim that this information is private. Some attorneys interpret those laws narrowly as forbidding public dissemination of the protected information. Since the information is only exchanged between two parties that are entitled to it, no further protection is needed. Another school of thought interprets the laws more broadly as requiring encryption of customer-specific data as it traverses the network. EB participants, armed with their respective legal counsels' opinions, could not settle the issue, especially since different states have different privacy laws. The only agreement was that the PIC application should be able to support privacy protection of customer data for those who want it.

The PIC application faced the same constraints as the TA application—most glaringly the lack of security functionality within the communication protocol stacks. Therefore, to the great consternation of application developers, the application had to provide privacy protection.

The solution that was developed [T1.254] provides Application-Based Security (ABS) for selective field protection. At the time it was the only option for providing any sort of privacy protection. As security features (in particular, STATE-ROSE) are being built into OSI protocol stacks, there is little need for selective field protection within applications. The only case where selective field protection might, theoretically, be of any value is for very strong privacy protection (e.g., using triple DES encryption) for just a few fields within very large PDUs.

Ideally, privacy is provided by encrypting messages after they have been encoded at the presentation layer into common transfer syntax. Unfortunately, most protocol stacks do not offer application programmers any access to the output of the encoder. The only solution then is to agree on a common representation of individual fields and then encrypt each field separately; for example, all character strings would be encoded as ASCII strings before encryption.

The approach chosen for application-based selective field protection is illustrated with the simple case of encryption of the ASCII representation of a PrintableString, as depicted in Figure 6.1.

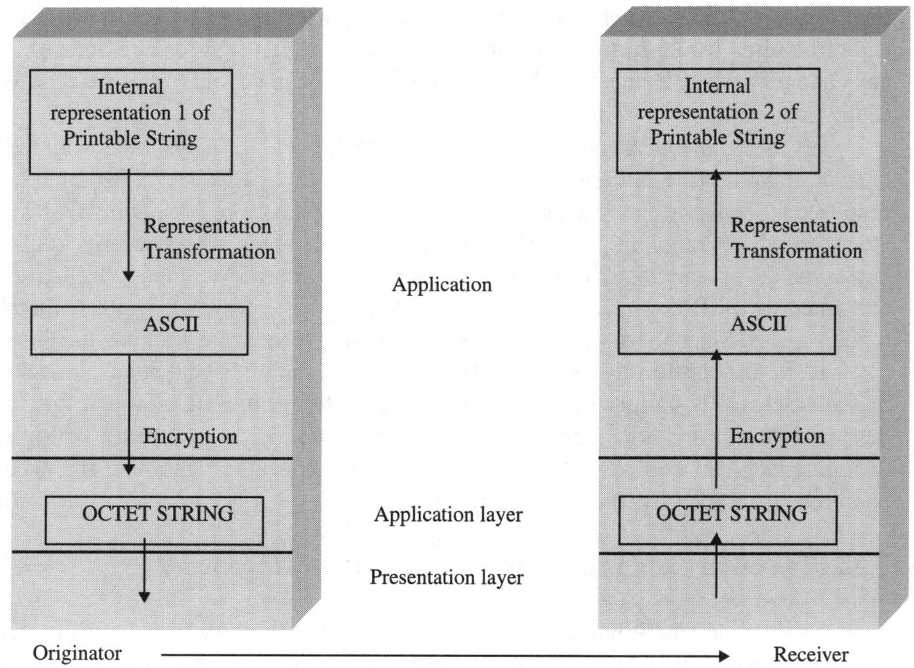

Figure 6.1 Encryption in the application

As shown in Figure 6.1, a PrintableString, in the originator's internal representation, is transformed into an ASCII string, which is then encrypted and passed to the presentation layer as an OCTET STRING. This is done by the application that requires security protection. The OCTET STRING is then encoded as such in the presentation layer, using, for example., the Basic Encoding Rules (BER). On the receiving side, the decoded OCTET STRING is passed from the presentation layer to the application layer where it is decrypted to recover the ASCII string. The syntax for this field specifies that it is a PrintableString. Therefore, the receiving application "knows" to decode the ASCII string into a PrintableString in its own internal representation. A slight generalization of this concept has resulted in the Application-Based Security (ABS) standard. ABS rests on some basic ideas:

- Every simple (as opposed to constructed) field can be represented as a string of bits, octets, or ASCII characters.

- Once a field is represented in the canonical form, it is fair game for any security transformations (e.g., hashing, signing) in addition to encryption.
- If a field is the subject of security transformations (ST), then there must be a way of communicating to the receiver the nature of those transformations and which security parameters were used (e.g., which algorithm, which key).

6.2.2.1 Data Representation. This standard specifies that security transformations (STs) shall be performed in the application on specific representations of the items to be protected. This standard is limited to the simple types listed in this section. This limitation is intended to simplify the implementation of the standard. (Generalizing the standard to include all ASN.1 types would require capabilities that are similar to a full ASN.1 Distinguished Encoding Rules (DER) encoder. See Section 2.3.6.1.) This section specifies representations used for various ASN.1 types for the purpose of STs.

6.2.2.1.1 CHARACTER STRINGS. ABS addresses security transformations for strings of characters that are members of the ASCII character set. Other character strings are not covered by ABS. Security transformations on such character strings (e.g., printable strings, visible strings, IA5 strings, numeric strings) are done on ASCII representations of these strings.

6.2.2.1.2 TIME. GeneralizedTime and UniversalTime are represented as character strings that are encoded as ASCII characters.

6.2.2.1.3 OCTET STRINGS. Security transformations on an OCTET STRING are done on its representation, as specified below, by the application. The representation of the (clear text) OCTET STRING for the purpose of ST is defined by the following conventions (based on Basic Encoding Rules (BER) conventions:

- The OCTET STRING shall not be split into smaller strings.
- The leftmost octet (the first octet that would be transmitted if clear text were chosen) shall be the high-order octet.
- Within each octet the leftmost bit (the first bit that would be transmitted if clear text were chosen) shall the high-order bit.

6.2.2.1.4 BIT STRINGS. Security transformations on a BIT STRING are done on its representation, as specified below, by the application. The representation of the (clear text) BIT STRING for the purpose of ST is defined by the following conventions (based on BER conventions):

- The BIT STRING shall not be split into smaller strings.
- The bits are ordered such that the leftmost bit (the first bit that would be transmitted if clear text were chosen) shall be the high order-bit.
- The BIT STRING shall be represented as a series of octets.
- A header octet shall be inserted in front of the resulting octet string and shall indicate how many bits in the last octet are unused.

6.2.2.1.5 BOOLEAN. Security transformations on a BOOLEAN are done on its representation, as specified below, in the application layer. The representation of the (clear text) BOOLEAN for the purpose of ST is defined by the following conventions (based in part on BER conventions):

- The value FALSE is represented as an octet with all bits set to zero.
- The value TRUE is represented as an octet with all bits set to one.

6.2.2.1.6 Integers. For the purpose of ST, a (signed) integer is represented as an optional sign followed by decimal digits, treated as a CharacterString and encoded using ASCII representation.

6.2.2.1.7 Real Numbers. For the purpose of ST, a (signed) real number is represented as an optional sign followed by a string of decimal digits, including (optionally) one decimal point, and treated as a CharacterString that is encoded using ASCII representation.

6.2.2.1.8 Object Identifier. An OBJECT IDENTIFIER is a sequence of integers. For the purpose of ST it shall be represented as a sequence of integers, separated by blanks, where each integer is represented as a string of decimal digits as in Section 6.2.2.1.6.

6.2.2.1.9 Object Descriptor. For the purpose of ST, an OBJECT DESCRIPTOR is represented as a GraphicString and treated as a CharacterString in Section 6.2.2.1.1.

6.2.2.2 Syntax for Security Transformations. In a CMIP PDU, fields that need protection are typically values of attributes of managed object (MO) instances. The syntax of such attributes is defined with the help of Abstract Syntax Notation 1 (ASN.1). Thus, the syntax for customer name may be defined as being of type PrintableString, while the syntax for the number of phone lines may be defined as being of type INTEGER. If either field is encrypted, the result will be of type OCTET STRING. If an OCTET STRING value is transmitted for the customer name instead of PrintableString, the recipient will interpret this field as being an error. In order to allow a field like a customer name to be transmitted either as PrintableString or OCTET STRING, the ASN.1 module of the corresponding information model must be revised. The required change is introduced with a simple example.

For instance, if CustomerName is of type PrintableString and may need privacy protection, its syntax shall be changed from

```
CustomerName ::= PrintableString

to:

CustomerName ::= CHOICE

         {clear                 PrintableString,

          simpleConfidential    OCTET STRING

         }
```

This allows the customer name to be transmitted in the clear as PrintableString or to be transmitted encrypted as OCTET STRING.

If some encryption parameters (a key identifier and/or an IV) need to be conveyed along with the ciphertext, a better rewrite of the customer name syntax will be:

```
CustomerName::= CHOICE

         {clear          PrintableString,

          confidential   Enciphered

         }
```

Where Enciphered is defined as:

```
Enciphered ::= SEQUENCE
```

```
{
encrypted OCTET STRING,
encryptionParameters EncryptionParameters OPTIONAL
}
```

And EncryptionParameters in turn is defined as:

```
EncryptionParameters ::= SET
    }
    keyId                  KeyId                 OPTIONAL,
    initializationVector   OCTET STRING (SIZE(8)) OPTIONAL
    }
KeyId ::= INTEGER
```

Although the new syntax allows the exchange of some encryption parameters, it comes at a price: If there is no need to transmit any encryption parameters, they can be skipped since the EncryptionParameters field in Enciphered is OPTIONAL. Nevertheless, since Enciphered is of type SEQUENCE, when it gets encoded into the transfer syntax a few extra bytes will still be required. Fortunately, in this case we can have both flexibility and transmission efficiency. We can define the syntax for CustomerName as:

```
CustomerName::= CHOICE
    {clear                 PrintableString,
    simpleConfidential     OCTET STRING,
    confidential           Enciphered
    }
```

In this case, both PrintableString and OCTET STRING would be transmitted with no undue overhead, while Enciphered would be used only when there is a need to convey some encryption parameters.

The preceding example illustrates the main concepts of ABS. A more general case, addressing all possible STs for simple fields, is presented next.

For simplicity of presentation, and without loss of generality, the rest of this section focuses on printable strings. The generalization to other types is fairly straightforward. Mostly it involves replacing occurrences of PrintableString with the name of the appropriate type. To make it easier to spot these occurrences, **PrintableString** is shown in bold below.

The ASN.1 changes presented in this section support any ST. In some applications, it is sufficient to support only a few of the several possible STs. In such cases, the syntax can be substantially simplified, as illustrated by the example at the beginning of this section. Let Item1 be of type PrintableString; that is, it is defined in some ASN.1 module as:

```
Item1 ::= PrintableString
```

In order to allow any STs of Item1 values, the definition of Item1 in the ASN.1 module shall be changed to:

```
Item1 ::= CHOICE

        { clear                   PrintableString,

          simpleConfidential      [1] OCTET STRING,

          confidential            [2] Enciphered ,

          simplePublicEnciphered [3] INTEGER,

          publicEnciphered        [4] PublicEnciphered ,—encrypted using
                                      PKCS#1

          hashed                  [5] HashedPrintableString ,

          simpleHash              [6] OCTET STRING,

          hash                    [7] Hash ,

          sealed                  [8] SealedPrintableString ,

          simpleSeal              [9] OCTET STRING,

          seal                    [10] Seal ,

          signed                  [11] SignedPrintableString ,

          simpleSignature         [12] SEQUENCE (SIZE(1..4)) OF INTEGER,

          signature               [13] Signature,

          confidentialSigned      [14] ConfidentialSigned,

          ...

        }
```

And the following type definitions shall be either included in the revised ASN.1 module or imported from a module that contains these definitions:

```
Enciphered ::= SEQUENCE

               {encrypted                OCTET STRING,

                encryptionParameters    EncryptionParameters OPTIONAL

               }
```

Both Enciphered (with the "confidential" label) and simpleConfidential are included because the latter requires zero transmission overhead if no EncryptionParameters need to be exchanged. Indeed, the transfer syntax for Enciphered may require a few octets to denote a type SEQUENCE and its length even if the OPTIONAL EncryptionParameters is not present. Notice also that if the option labeled "clear" is chosen, the **PrintableString** value of Item1 is transmitted in the exact same transfer syntax as if the original ASN.1 were used, with no additional transmission overhead.

```
PublicEnciphered ::= SEQUENCE

                 {publicEncrypted         INTEGER,

                  encryptionParameters   EncryptionParameters OPTIONAL

                 }
```

This option is used when the encryption is done with a public key encryption algorithm. Again, simplePublicEnciphered is present in the CHOICE in order to minimize transmission overhead when the OPTIONAL EncryptionParameters is not present.

```
Hash ::= SEQUENCE
        {hashValue                OCTET STRING (SIZE(8..64)),
         encryptionParameters     EncryptionParameters OPTIONAL
        }
```

Hash represents the message digest resulting from hashing the clear text. simpleHash is present in the CHOICE in order to minimize transmission overhead when the OPTIONAL EncryptionParameters is not present.

```
HashedPrintableString ::= SEQUENCE
        {clear       PrintableString,
        hash         CHOICE {simpleHash OCTET STRING (SIZE (8..64)),
                             hash       Hash
                            }
        }
```

Hashed**PrintableString** represents both the clear text and its message digest. The message digest can be represented by the "hash" choice or by the simpleHash choice to minimize transmission overhead when the OPTIONAL EncryptionParameters is not present.

```
Seal ::= SEQUENCE
        {sealValue                OCTET STRING (SIZE(8..128)),
         encryptionParameters     EncryptionParameters OPTIONAL
        }
```

Seal represents a digital seal of the clear text. simpleSeal is present in the CHOICE in order to minimize transmission overhead when the OPTIONAL EncryptionParameters is not present.

```
SealedPrintableString ::= SEQUENCE
        {clear       PrintableString,
         seal        CHOICE {seal       Seal,
                            simpleSeal OCTET STRING (SIZE(8..128))
                            }
        }
```

Sealed**PrintableString** represents both the clear text and its seal. The seal can be represented by the "seal" choice or by the simpleSeal choice to minimize transmission overhead when the OPTIONAL EncryptionParameters is not present.

```
Signature ::= SEQUENCE
        {signatureValue          SEQUENCE (SIZE(1..4)) OF INTEGER,
         encryptionParameters    EncryptionParameters OPTIONAL
        }
```

Signature represents a digital signature of the clear text. simpleSignature is present in the CHOICE in order to minimize transmission overhead when the OPTIONAL EncryptionParameters is not present.

```
SignedPrintableString ::= SEQUENCE
    {clear       PrintableString,
     signature  CHOICE {signature     [1] Signature,
                     simpleSignature [2] SEQUENCE (SIZE(1..4)) OF INTEGER
                       }
    }
```

Signed**PrintableString** represents both the clear text and its signature. The signature can be represented by the "signature" choice or by the simpleSignature choice to minimize transmission overhead when the OPTIONAL EncryptionParameters is not present.

```
ConfidentialSigned ::= SEQUENCE
    { encrypted     OCTET STRING,
      signature     CHOICE {signature   [1] Signature,
                        simpleSignature [2] SEQUENCE (SIZE(1..4)) OF INTEGER
                          }
    }
```

ConfidentialSigned represents both the encrypted text and its signature. The signature can be represented by the "signature" choice or by the simpleSignature choice to minimize transmission overhead when the OPTIONAL EncryptionParameters is not present.

EncryptionParameters is an extensible type that is used as a catch-all for any parameters that may be used by any of the STs. In most applications only a small number, if any, of the components of EncryptionParameters will be used.

```
EncryptionParameters ::= SET
        {symmetricKeyId       [0] KeyID                          OPTIONAL,
        publicKeyId           [1] KeyID                          OPTIONAL,
        sealKeyId             [2] KeyID                          OPTIONAL,
        signatureKeyId        [3] KeyID                          OPTIONAL,
        passwordId            [4] KeyID                          OPTIONAL,
        initializationVector  [5] OCTET STRING (SIZE(8))         OPTIONAL,
        feedBackBits          [6] INTEGER (SIZE (1..63))         OPTIONAL,
```

```
      —for k-bit output feedback mode or k-bit cipher feedback mode of DES
        symmetricAlgorithm     [7] OBJECT IDENTIFIER         OPTIONAL,
        publicKeyAlgorithm     [8] OBJECT IDENTIFIER         OPTIONAL,
        signatureAlgorithm     [9] OBJECT IDENTIFIER         OPTIONAL,
        sealAlgorithm          [10] OBJECT IDENTIFIER        OPTIONAL,
        hashAlgorithm          [11] OBJECT IDENTIFIER        OPTIONAL,
        keyDigest              [12] OCTET STRING (SIZE(8..64)) OPTIONAL,
                                     —for verification of public keys
        blockSize              [13] INTEGER                  OPTIONAL,
                                  —for square mod-n hashing
        keySize                [14] INTEGER                  OPTIONAL,
                                  —for RSA
        publicKey              [15] SEQUENCE
                                   {modulus       INTEGER,
                                   exponent       INTEGER
                                   }                         OPTIONAL,
        sequenceNumber         [16] INTEGER                  OPTIONAL,
        timeStamp              [17] GeneralizedTime          OPTIONAL,
        ...
        }
KeyId ::= CHOICE  {identifier  OBJECT IDENTIFIER,
                   name        GraphicString,
                   number      INTEGER
                  }
```

The revised syntax allows the value of Item1 to be sent in clear text as a **printable string,** or in transformed form. In the latter case, it further allows the communication of any of several parameters that the receiver may need to decrypt the transformed Item1 value. If no such parameters need to be transmitted, this syntax provides for the efficient transmission of just the transformed value, without the additional transfer syntax overhead for SEQUENCE type.

This rather elaborate discussion addresses only the easiest case of simple fields. Nevertheless, it demonstrates the main features of ABS:

- It allows transmission of protectable fields in the clear without additional overhead.
- It allows different STs for different fields in the same PDU.

- It allows different STs for a given field in different PDUs.
- It supports the exchange of field-specific security parameters.

The price to pay for such flexibility is quite heavy: all the security work must be done by the application, and the ASN.1 syntax in the underlying information model must be redrawn.

ABS also addresses more complex issues, though the underlying principles are the same. The reader is referred to T1.254 for the complete definition of ABS. In addition to simple fields, T1.254 also covers more esoteric topics:

- Size restrictions on values of attributes
- ASN.1 constructor types
- STs including sequence numbers and time stamps
- Sharing of encryption parameters among fields in a PDU and across PDUs
- Secure binding of independently encrypted fields (to ensure they all belong to the same PDU)

6.2.2.3 DES Padding for ABS. The basic DES algorithm encrypts a block of 8 octets of clear text into ciphertext of 8 octets. Typically, when DES is used in the CBC mode and the clear text is not a multiple of 8 octets, additional octets are added to have a multiple of 8. The last byte indicates how many bytes of the last 8 are meaningful (the remaining are padding bytes). If the clear text is a multiple of 8 octets, then 8 padding octets are added. This approach adds, on the average, 4.5 bytes of overhead per encrypted field. That's the padding specified for the EB authenticator where only a single time stamp is encrypted. In the case of ABS, the padding penalty may be much heavier: if several fields are to be encrypted, each one will exact an extra penalty of 4.5 bytes. Therefore, ABS adopts an alternative, zero padding overhead approach, as the default.

If the clear text is a multiple of 8 octets, it shall be encrypted in CBC mode without any padding. Otherwise, if the length of the clear text is $n*8 + k$ octets ($n > 0$, $0 < k < 8$), then the first $n.8$ octets are encrypted in DES CBC, the last 8 octets of ciphertext are transmitted, as usual, then they are encrypted again, the first k octets of this last encryption operation are XORed with the remaining k octets of the original clear text, and are transmitted. If the clear text is less than 8 octets (i.e., $n = 0$), then a confounder may be used (8 random octets in front of the clear text). If a confounder is not used but an initialization vector (IV) is used (the two provide the same function), then the IV (which must be known or transmitted in clear text) is encrypted, and the first k octets of the resulting ciphertext is XORed with the k octets of clear text. If neither an IV nor a confounder is used, one should encrypt the key itself and then XOR the first k octets of the encrypted key with the clear text. (This solution may be more vulnerable to replay, known clear text attack, and electronic codebook attack.)

If the default procedure for zero overhead padding is not used, the communicating parties must agree on a different approach by means outside the scope of ABS. This could be done, for instance, by using the EncryptionParameters or by entering a joint implementation agreement.

6.3 STASE-ROSE

Neither ABS nor EB authentication can provide two essential security services: whole PDU integrity protection and non-repudiation. Both impose an undue burden on application programmers. Both were intended to be stopgap measures, while waiting for GULS. GULS never came, and, based on comments from OSI stack suppliers, it never will. STASE-ROSE was specifically designed to remedy this sad situation. In particular, it provides a full range of whole PDU security services from within the communications protocol stack. Furthermore, in view of the humbling GULS nonevent, it was clear that in order to have the flimsiest chance, any alternative must be very easy to implement. This goal resulted in four design considerations:

- Several TMN platform suppliers buy their underlying OSI protocol stack, up to and including the presentation layer from other parties. Such TMN platform suppliers have no access to the presentation layer; they can implement any security features only in the application layer.
- Several TMN platform suppliers integrate various ASEs in the application layer within a single module. Therefore, any security functionality could not be implemented between any two ASEs (e.g., between CMISE and ROSE). The only practical solution was to provide security for ROSE PDUs, just before they were being transferred to the underlying presentation layer.
- Several security features are desirable in addition to whole PDU protection: peer entity authentication, negotiation of security parameters, and dynamic update of security parameters. All such embellishments are explicitly made optional, so it is possible to provide whole PDU protection without having to carry extra luggage.
- The 1988 version of ASN.1 includes constructs, such as MACRO and ANY, that cause much grief to builders of ASN.1 compilers. To remedy this defect, a more sophisticated version of ASN.1 was defined in 1994. The new version was promptly used in GULS. Some insiders claim that this was one reason for the neglect of GULS: the new ASN.1 is less transparent to novices, and at the time there were few, if any, compilers for the new ASN.1. To avoid such pitfalls STASE-ROSE uses only the portion of ASN.1 that is common to both the old and new versions. Thus, it sacrifices elegant sophistication for down-to-earth simplicity.

This section starts, innocuously enough, in plain English for all to see and understand—at least in presenting what STASE-ROSE does. Then, in talking about how it does it, the language gradually shifts into the more esoteric formulation of abstract service definition and ASN.1.

STASE-ROSE is an ASE that exists solely for the purpose of providing security. It is designed to protect whole ROSE PDUs and thereby any ASEs that use ROSE, such as CMISE and X.500. The functionality of STASE-ROSE can be divided into four distinct areas:

- Security transformations on whole ROSE PDUs
- Peer entity authentication during association setup
- Negotiation of security algorithms and parameters during association setup
- Dynamic update of security parameters during the association

6.3.1 Security Transformations on ROSE PDUs

The core functionality of STASE-ROSE resides between the presentation layer and ROSE. It provides for security transformations on whole ROSE PDUs. The security transformations can be any of the following:

- No security transformation applied, this is a pass-through mode
- Encryption with a symmetric algorithm for confidentiality
- Encryption with an asymmetric algorithm for confidentiality and non-repudiation
- Hashed or double hashed for integrity protection
- Sealed for integrity protection
- Signed for integrity and non-repudiation protection

- Encrypted with a symmetric encryption algorithm for confidentiality and signed for integrity and non-repudiation protection
- Encrypted with a symmetric encryption algorithm for confidentiality and hashed for integrity protection
- Encrypted with a symmetric encryption for confidentiality algorithm and sealed for protection

Only some of these options are needed to protect network management transactions. Selecting which transformations shall be supported depends on the security needs of the application to be protected. All network management applications need integrity protection and should be able to exchange unprotected PDUs (with obvious risks) if security mechanisms become temporarily disabled. Therefore, the following two options need to be supported for all network management applications:

- No security transformation applied (this is a pass-through mode)
- Hashed or double hashed for integrity protection

In addition, if confidentiality is required, then the following need to be supported:

- Encrypted with a symmetric encryption algorithm for confidentiality
- Encrypted with a symmetric encryption algorithm for confidentiality and hashed for integrity protection

If non-repudiation is required, most likely over an X interface to an external entity, then the following transformations need to be supported:

- Signed for integrity and non-repudiation protection
- Encrypted with a symmetric encryption algorithm for confidentiality and signed for integrity and non-repudiation protection

6.3.2 Peer Entity Authentication

The STASE-ROSE standard specifies abstract syntax to be used in the Authentication-value field of the ACSE Authentication FU. Use of the authenticator, as specified in STASE-ROSE, is optional. Actually, STASE-ROSE offers two abstract syntax possibilities: one for symmetric encryption and one for asymmetric encryption. Only the authenticator based on asymmetric encryption needs to be supported for most TMN applications. It consists of:

- The sender's unique identifier
- The receiver's unique identifier
- A time stamp, consisting of GeneralizedTime
- Optionally, a symmetric encryption key that the sender will use during the association, encrypted with the receiver's public key
- A digital signature of the preceding fields, signed with the sender's private key
- Optionally, the sender's public key certificate

If this authenticator is used, then it is not necessary, for security purposes, to use the Calling AP Title and Calling AE Qualifier in the ACSE AARQ or the Responding AP Title and Responding AE Qualifier in the ACSE AARE. Indeed, the sender's ID and the receiver's ID can each be the corresponding AE-title.

6.3.3 Negotiation of Security Algorithms

In order for two parties to communicate securely, both must use the same set of security algorithms. In general, such agreement is negotiated at association setup. However, STASE-ROSE allows two communicating entities to agree on a set of default algorithms before any communications are initiated. For instance, the two parties may agree that RC5 will be used for symmetric encryption. Such an agreement can simplify the setup procedure by eliminating the need for negotiating security algorithms. In order to prime the pump, STASE-ROSE proposes a default set of algorithms and conventions: unless the communicating parties decide otherwise, their defaults are the ones specified by STASE-ROSE.

6.3.3.1 Defaults. The defaults specified by STASE-ROSE are:

- The default encryption algorithm for symmetric key encryption is the Digital Encryption Standard (DES) in the Cipher Block Chaining (CBC) mode.
- If triple DES is needed, the default is Encryption Decryption Encryption (EDE) in the CBC outer feedback mode with three different DES keys.
- If no Initialization Vector (IV) is specified, then the first IV of the association shall consist of 64 zero-valued bits; every subsequent IV shall consist of the last 8 bytes of the previous encrypted ROSE PDU.
- The default public key encryption algorithm is RSA.
- The default hashing algorithm is MD5.
- The default MAC (for keyed hashing) is HMAC.
- The default seal is the MD5 hash of the DER encoded ROSE PDU encrypted with DES.
- The default digital signature is the MD5 hash of the DER encoded ROSE PDU encrypted with RSA and the user's private key.
- Peer entity authentication shall occur at association setup time. The optional authentication Functional Unit (FU) of the ACSE is used. Authentication information is carried in the calling-authentication-value and responding-authentication-value fields of the authentication FU of the ACSE AARQ and AARE PDUs, respectively. The bit strings for the sender-acse-requirements and responder-acse-requirements fields of the authentication FU are set to include the authentication FU. The calling-authentication-value and responding-authentication-value fields are of type Authentication-value, which is further defined in ISO 8650 as a CHOICE. The CHOICE for the Authentication-value is EXTERNAL. The presentation context shall include a reference to the abstract syntax that is used for the EXTERNAL. When the default values and conventions are used, it is not necessary to use the (optional) mechanism-name field of the authentication FU of the ACSE PDUs.
- If peer entity authentication with public key encryption is required, it shall consist of:
  — The sender's unique identifier
  — The receiver's unique identifier
  — A time stamp, consisting of GeneralizedTime
  — Optionally, a symmetric encryption key that the sender will use during the association, encrypted with the receiver's public key
  — A digital signature of the preceding fields, signed with the sender's private key
  — Optionally, the sender's public key certificate
- Unless otherwise agreed upon between the communicating entities, the digital signature will be computed using MD5 for hashing and RSA for public key encryption. The (optional) publicly encrypted symmetric encryption keys in the AARQ and

TABLE 6.1 Negotiated Algorithms

Set of Acceptable Algorithms in AARE				Set of acceptable algorithms in AARQ			
				Present			Absent
				Nonempty		Empty	
				2 or more Elements	1 Element	(NULL)	
Present	Nonempty (subset of AARQ set)	2 or more elements	Does not include predefined default	The AARE algorithms, no default	Error	Error	Error
			Includes predefined default	The AARE algorithms, with predefined default	Error	Error	Error
		1 element		The AARE algorithm is the default	The selected algorithm, is the default	Error	Error
	Empty (Null)			None	None	None	None
Absent				None	None	None	Only the predefined default, if there is no predefined default then none

AARE messages can be different, allowing the initiator and responder of the association to use different keys in the course of the association.

- Systems clocks may be set back if they move too fast, or they may be set back due to some malfunction. Regardless of such events, consecutive time stamps produced by a system must be monotonically increasing. (See Section 4.7.2 for construction of monotonically increasing time.)

6.3.3.2 Negotiation. For most TMN applications, the defaults specified by STASE-ROSE can be used, and there is no need to use the (optional) capability to negotiate any such algorithms. Some of the defaults specified in STASE-ROSE, however, may not always be well suited for some applications. In particular, any of these algorithms may become vulnerable to as yet unknown attacks. Therefore, STASE-ROSE allows any two communicating entities, or any community of users, to agree on a different set of default algorithms. As long as such an agreement exists, there is no need to use the STASE-ROSE capability for negotiating security algorithms. If the communicating parties have not agreed up front on default algorithms and parameters, they can do so at association setup time. Furthermore, they can negotiate an agreement that several algorithms of the same type (e.g., RC4 [RC4], RC5 [RC5], and DES for symmetric encryption) will be supported in the course of the association.

The negotiation procedure is straightforward: the association initiator can submit a list of proposed security algorithms in the AARQ. The association responder can respond with a similar list in the AARE. The algorithms listed in the AARE, if any, must be a subset of those listed in the AARQ. Table 6.1 shows how to interpret the results of the negotiation under all possible conditions. In particular, it shows how default algorithms for the current association become established.

The negotiation of algorithms is illustrated with a simple example. Figure 6.2 illustrates a possible scenario for the negotiation of security algorithms.

At the end of the exchange illustrated in Figure 6.2, the initiator may decide that the sets of algorithms that the responder is ready to support for the proposed association are unac-

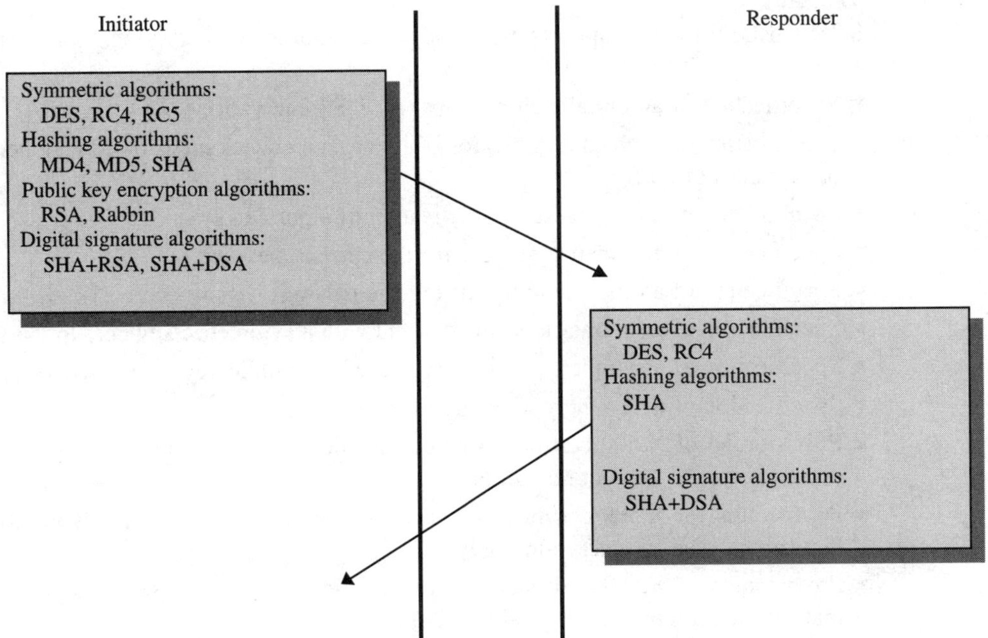

Figure 6.2 Negotiation of security algorithms

ceptable. In that case the initiator will reject the association. If the initiator is willing to accept the sets of algorithms proposed by the responder, then the association will proceed with the following defaults:

- Symmetric algorithm: DES, since it is the default specified in this standard and it is among the negotiated options for symmetric algorithms
- Hashing algorithm: SHA, since it is the only mutually acceptable hashing algorithm
- Public key encryption algorithm: no public key encryption algorithm can be used during this association since none has been agreed upon
- Digital signature algorithm: SHA + DSA, since it is the only mutually acceptable digital signature algorithm
- Digital seal algorithm: MD5 + DES, since this is the default specified in this standard and the negotiation did not involve digital seal algorithms

In addition to negotiation of security algorithms, STASE ROSE also supports the negotiation of various encryption parameters:

- Identification of symmetric encryption keys that can be used
- Identification of public keys that can be used
- Identification of seal keys that can be used
- Identification of password IDs that can be used
- Specification of sizes of public keys that can be used
- Specification of public keys that can be used
- Sender's secret key

Unlike encryption algorithms, STASE-ROSE does not provide any default values for any of these parameters.

STASE-ROSE further supports the conveying of additional encryption parameters:

- Specification of an initialization vector (for DES encryption)
- Specification of the feedback bits for *k*-bit output feedback mode or *k*-bit cipher feedback modes of DES
- Specification of a key digest for verification of a public key
- Specification of a sequence number for the current message
- Specification of a time stamp for the current message
- Specification of a symmetric key encrypted with a symmetric key encryption key
- Specification of a symmetric key encrypted with a public key (of the receiver)
- Specification of a key encryption key identifier
- Provisioning of X.509 certificates or certification paths of the sender's public keys that can be used without restrictions
- Provisioning of X.509 certificates or certification paths of the sender's public keys that can be used for encryption only
- Provisioning of X.509 certificates or certification paths of the sender's public keys that can be used only for digital signatures
- Specification of a symmetric session key encrypted with the receiver's public key and signed with the sender's secret key

Either party can provide the values of these parameters during association setup, but they are not subject to negotiation. For instance, each party can provide its own public key certificate(s) to the other party.

By the end of the negotiation phase (i.e., association setup), the two entities have an agreement as to which STs and which algorithms for those STs they will support. In some cases, but not necessarily all, they also have an agreement on the default algorithms for some or all of the STs they have agreed to support. In some cases they may also have an agreement on the values of some or all of the security parameters.

More formally, the negotiation information is carried in the user information field of the AARQ and AARE using EncryptionParametersSelection defined with the following syntax:

```
DEFINITIONS IMPLICIT TAGS ::= BEGIN
–EXPORTS everything
IMPORTS
SenderId, ReceiverId, Signature, KeyID, PublicKeyCertificate,
EncryptionCertificate,
SignatureCertificate, EncryptedAuthenticatedSymmetricKey
FROM ST-CMIP-PCI {itu recommendation q(17) 813(813) stase(1) stase-pci(1)
asn1module(4) version1(1)}

EncryptionParametersSelection ::= SET
       {symmetricKeyIds          [0] SET OF KeyID                OPTIONAL,
        publicKeyIds             [1] SET OF KeyID                OPTIONAL,
        sealKeyIds               [2] SET OF KeyID                OPTIONAL,
       signatureKeyIds           [3] SET OF KeyID                OPTIONAL,
       passwordIds               [4] SET OF KeyID                OPTIONAL,
       initializationVector      [5] OCTET STRING (SIZE(8))      OPTIONAL,
       feedBackBits              [6] INTEGER (SIZE (1..63))      OPTIONAL,
       –for k-bit output feedback mode or k-bit cipher feedback mode of DES
       symmetricAlgorithms       [7] SET OF OBJECT IDENTIFIER    OPTIONAL,
       publicKeyAlgorithms       [8] SET OF OBJECT IDENTIFIER    OPTIONAL,
       signatureAlgorithms       [9] SET OF OBJECT IDENTIFIER    OPTIONAL,
       sealAlgorithms            [10] SET OF OBJECT IDENTIFIER   OPTIONAL,
       hashAlgorithms            [11] SET OF OBJECT IDENTIFIER   OPTIONAL,
       keyDigest                 [12] OCTET STRING (SIZE(8..64)) OPTIONAL,
       –for verification of public keys
       blockSize                 [13] INTEGER                    OPTIONAL,
       –for square mod-n hashing
       keySizes                  [14] SET OF INTEGER             OPTIONAL,
       –for RSA
       publicKeys                [15] SET OF SEQUENCE
                                          {modulus     INTEGER,
                                          exponent     INTEGER
                                          }                      OPTIONAL,
       sequenceNumber            [16] INTEGER                    OPTIONAL,
       timeStamp                 [17] GeneralizedTime            OPTIONAL,
       encryptedKey              [18] OCTET STRING (SIZE(64..128)) OPTIONAL,
       – symmetric session key, encrypted with Key-Encryption-Key
       encryptedSymmetricKey     [19] INTEGER                    OPTIONAL,
       – symmetric session key, encrypted with the receiver's public key
       keyEncryptionKey          [20] SEQUENCE (SIZE (1..3)) OF KeyID OPTIONAL,
       – one to three symmetric keys used for encrypting a session key
       keyListIds                [21] SET OF KeyListId           OPTIONAL,
       – list of encryption keys that can be used during the association
       publicKeyCertificate      [21] SET OF PublicKeyCertificate  OPTIONAL,
–X.509 certificates or certification paths of the sender's public keys with no
usage restrictions–
       encryptionCertificate     [22] SET OF EncryptionCertificate OPTIONAL,
```

```
—X.509 certificates or certification paths of the sender's public keys used for
encryption only
signatureCertificate                [23] SET OF SignatureCertificate   OPTIONAL,
—X.509 certificates or certification paths of the sender's public keys used for
digital signatures only —
encryptedAuthenticatedSymmetricKeys [24] SET OF EncryptedAuthenticatedSymmetricKey
OPTIONAL,
—symmetric session key, encrypted with the receiver's public key and signed with
sender's key—
macAlgorithms                       [25] SET OF OBJECT IDENTIFIER      OPTIONAL,
        ...
        }
— EncryptionParametersSelection is optionally used during association setup to
negotiate which algorithms and other encryption parameters will be supported dur-
ing the association. It is not used in STASE-ROSE PDUs. —
KeyListId ::= CHOICE {identifier OBJECT IDENTIFIER,
                      name       GraphicString,
                      number     INTEGER
                      }

END
```

6.3.4 Dynamic Update of Security Parameters

If several algorithms for the same security transformation have been negotiated during association setup time, it may be necessary to specify which of these algorithms is used for any particular PDU, especially if the selected algorithm for the PDU is not the default algorithm. This optional capability is not needed if only the defaults are used (which is expected to be the most common case) since the defaults specify only one algorithm per security transformation. However, this capability is needed to specify other security parameters that vary among PDUs. Such security parameters may include a time stamp, a sequence number, and an Initialization Vector. For the purpose of TMN applications, however, only one security parameter might need dynamic updating: the symmetric encryption key (if any) used for confidentiality protection. It may therefore be necessary to support STASE-ROSE's capability to exchange a symmetric encryption key, encrypted with the receiver's public key and signed with the sender's private key.

6.3.5 STASE-ROSE Services

An ASE's definition includes the specification of the services it provides to other ASEs. For STASE-ROSE this is simple: it provides only one service: SR-TRANSFER, to one user: ROSE.

The related service structure consists of two service primitives, as illustrated in Figure 6.3. On the sender's side, ROSE uses the SR-TRANSFER request to signal STASE-ROSE that it has a ROSE PDU that needs protection. On the receiving side, STASE-ROSE uses the SR-TRANSFER indication to signal to ROSE that it has just recived a PDU from a remote system's ROSE.

At first, it may seem a feat of magic that the rich functionality of STASE-ROSE, outlined over the preceding few pages, can be provided through a single, fairly simple service. The trick is not one of magic but rather of brazen leveraging of ACSE services.

6.3.5.1 SR-TRANSFER Parameters. The SR-TRANSFER service has three parameters: ROSE-PDU, Encryption-Type, and Encryption-Parameters.

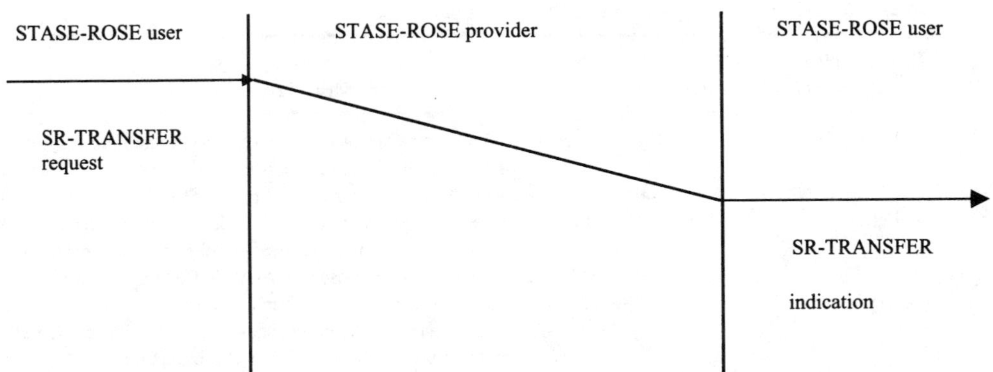

Figure 6.3 SR-TRANSFER Service Primitives

6.3.5.1.1 ROSE-PDU. This parameter identifies the ROSE PDU to be transferred. It has to be supplied by the requester of the service, and the data values must conform with the ROSEapdus definition in X.229.

6.3.5.1.2 ENCRYPTION-TYPE. This parameter identifies the type of STs desired by the service user for the current ROSE PDU. The valid values for the type are listed in Table 6.2.

6.3.5.1.3 ENCRYPTION-PARAMETERS. This service parameter identifies the parameters to be used for the security transformations. The presence of this parameter is dependent on the Encryption-Type selected by the user (as described in the previous section). It is seman-

TABLE 6.2 Encryption-Type Values

Encryption-Type Value	Meaning
clear	No ST is desired
simpleConfidential	Whole PDU privacy protection using the defaults supported by the service provider
confidential	Whole PDU privacy protection using the parameters provided in the Encryption-Parameters
simplePublicEnciphered	Whole PDU privacy protection using default public key supported by the service provider
publicEnciphered	Whole PDU privacy protection using the public key provided in the Encryption-Parameters
simpleHashed	Hashing-based MAC of the PDU using the default values
hashed	Hashing-based MAC of PDU using the parameters provided in the Encryption-Parameters
simpleSealed	Sealing of the PDU using the default values
sealed	Sealing of the PDU using the parameters provided in the Encryption-Parameters
simpleSigned	Digital signature of the PDU using the default values
signed	Digital signature of the PDU using the parameters provided in the Encryption-Parameters
simpleConfidentialSigned	Whole PDU privacy protection and digital signature of the PDU using the default values
confidentialSigned	Whole PDU privacy protection and digital signature of the PDU using the parameters provided in the Encryption-Parameters
simpleConfidentialMAC	Whole PDU privacy protection and MAC of the PDU using the default values
confidentialMAC	Whole PDU privacy protection and MAC of the PDU using the parameters provided in the Encryption-Parameters
simpleConfidentialSealed	Whole PDU privacy protection and seal of the PDU using the default values
confidentialSealed	Whole PDU privacy protection and seal of the PDU using the parameters provided in the Encryption-Parameters

TABLE 6.3 Components of EncryptionParameters

EncryptionParameters Components	Meaning
symmetricKeyIds	The set of key identifiers for the keys to be used on this association for symmetric encryption. The application-responder responds with the same set or a subset thereof if the symmetricKeyIds parameter is present in A-ASSOCIATE indication.
publicKeyIds	The set of key identifiers for the keys to be used on this association for public key encryption. The application-responder responds with the same set or a subset thereof if the publicKeyIds parameter is present in A-ASSOCIATE indication.
sealKeyIds	The set of key identifiers for the keys to be used on this association for sealing. The application-responder responds with the same set or a subset thereof if the sealKeyIds parameter is present in A-ASSOCIATE indication.
signatureKeyIds	The set of key identifiers for the keys to be used on this association for digital signature. The application-responder responds with the same set or a subset thereof if the signatureKeyIds parameter is present in A-ASSOCIATE indication.
passwordIds	The set of password identifiers for the passwords to be used on this association. The application-responder responds with the same set or a subset thereof if the parameter is present in A-ASSOCIATE indication.
initializationVector	The initialization vector (IV) to be used for DES encryption in Ciphered Block Chaining (CBC) mode. Each party can use a different IV for the messages it sends.
feedBackBits	The feedback bits to be used for DES encryption in *k*-bit cipher feedback or *b*-bit output feedback modes. Each party can use a different value for the feedback bits for the messages it sends.
symmetricAlgorithms	The set of symmetric algorithms the association-initiator is capable of supporting. The association-responder responds with same set or a subset thereof if the symmetricAlgorithms parameter is present in the A-ASSOCIATE indication.
publicKeyAlgorithms	The set of public key algorithms the association-initiator is capable of supporting. The association-responder responds with the same set or a subset thereof if the publicKeyAlgorithms parameter is present in the A-ASSOCIATE indication.
signatureAlgorithms	The set of digital signature algorithms that the association-initiator is capable of supporting. The association-responder responds with the same set or a subset thereof if the signature Algorithms parameter is present in the A-ASSOCIATE indication.
sealAlgorithms	The set of sealing algorithms that the association-initiator is capable of supporting. The association-responder responds with the same set or a subset thereof if the sealAlgorithms parameter is present in the A-ASSOCIATE indication.
hashAlgorithms	The set of hashing algorithms that the association-initiator is capable of supporting. The association-responder responds with the same set or a subset thereof if the hashAlgorithms parameter is present in the A-ASSOCIATE indication.
keyDigest	The message digest (fingerprint) of a public key, used to verify the validity of a public key.
blockSize	The block size to be used for square mod-*n* hashing. Each party can choose a different block size for the messages it sends.
keySize	The key size for the RSA encryption algorithm. The association-responder responds with the same or a different key size if the keySize parameter is present in the A-ASSOCIATE indication.
publicKeys	The set of publicKey to be used by the sender on this association. The association-responder may respond with the set of the responder's public keys.
sequenceNumber	The starting sequence number for ROSE PDUs, if the association should be protected against replay and deletion attacks. Each party can choose to start with a different sequence number. If this parameter is present, the STASE-ROSE-provider of any party that supplies it must assign a sequence number for each STASE-ROSE APDU sent on the application association.
timeStamp	The time at which the A-ASSOCIATE request was initiated by the association-initiator. The interpretation of this parameter is implementation-dependent and is outside the scope of this standard. If the association-responder sends the time stamp, it's value must correspond to the time at which the A-ASSOCIATE response primitive was issued.

continued on next page

TABLE 6.3 Continued

EncryptionParameters Components	Meaning
encryptedKey	A symmetric key used for (part of) the association and encrypted with a symmetric Key Encryption Key (KEK).
encryptedSymmetricKey	A symmetric key used for (part of) the association and encrypted with the receiver's public key.
keyEncryptionKey	Identifies one to three symmetric keys to be used as the symmetric KEK.
keyListIds	The set of identifiers of lists of symmetric encryption keys that the association-initiator proposes to use. The association-responder responds with the same set or a subset thereof if the keyListIds parameter is present in the A-ASSOCIATE indication.
publicKeyCertificates	X.509 Certification Path certifying the sender's public key.
encryptionCertificates	X.509 Certification Path certifying the sender's public key which can be used only for encryption.
signatureCertificates	X.509 Certification Path certifying the sender's public key which can be used only for digital signatures.
encryptedAuthenticated SymmetricKeys	A symmetric key used for (part of) the association and encrypted with the receiver's public key, followed by a time stamp (GeneralizedTime), the sender's ID, receiver's ID, and a signature computed over the ASCII representation of these four fields using the sender's private key.
macAlgorithms	The set of MAC algorithms that the association-initiator is capable of supporting. The association-responder responds with the same set or a subset thereof if the macAlgorithms parameter is present in the A-ASSOCIATE indication.

tically equivalent to the EncryptionParameters type specified earlier as part of the STASE-ROSE protocol. The components of EncryptionParameters are listed in Table 6.3.

6.3.6 Interaction between Application Service Elements

The previous section mentioned that STASE-ROSE provides its services to ROSE by leveraging on ACSE services. This section unveils the wheeling and dealing among the ASEs. In particular, it describes the interactions between the application, ACSE, ROSE, STASE-ROSE, the ROSE user (e.g., CMISE) and presentation layer services. It covers the various stages of communication between two application entities. Other interactions, which result in the same set of messages exchanged between the communicating systems and the same functionality, are possible. Selection of such interactions is a local matter.

6.3.6.1 Association Establishment. Figure 6.4 illustrates the sequence of interactions between the application, various ASEs and the presentation provider during association establishment.

6.3.6.1.1 Association Initiator. The interactions on the association initiating side in Figure 6.4 are:

1. The application of the application entity involving STASE-ROSE issues an A-ASSOCIATE request to ACSE to establish an application association. If peer entity authentication is desired, the application process provides ACSE with the value of the authenticator to be carried in the Authentication-value field of the AARQ PDU (using the ACSE Authentication FU). During the same phase, the application may also inform STASE-ROSE and the ROSE user ASE(s) (e.g., CMISE) of the requested association and provide STASE-ROSE with any proposed values for encryption parameters.

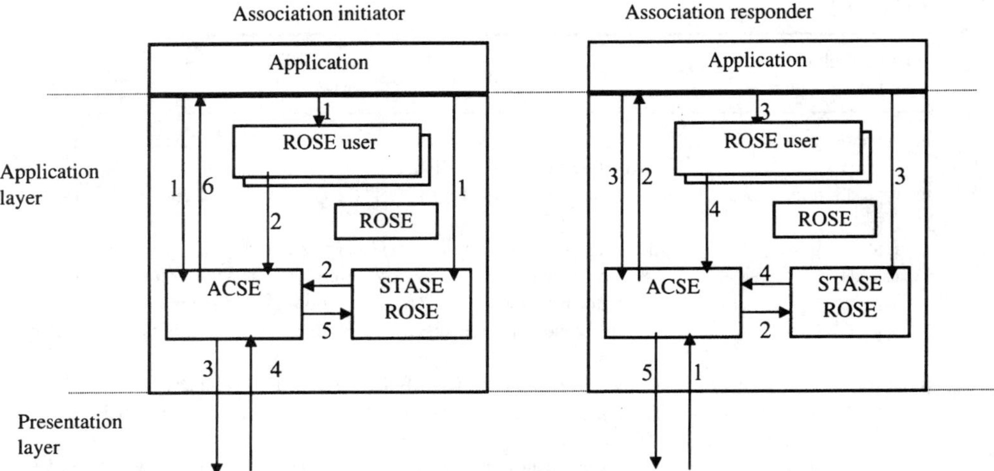

Figure 6.4 Interaction during association establishment

2. STASE-ROSE provides ACSE with any proposed values for encryption parameters. The mechanism by which STASE-ROSE informs ACSE is an implementation matter. This information is carried in the ACSE user information field using the EncryptionParametersSelection parameter. During the same phase, the ROSE user ASE(s) may also provide ACSE with information relevant to that ASE(s). All such information is carried in the ACSE user information field. The ACSE user information field defined in ITU-T Rec. X.227 | ISO 8650 consists of a SEQUENCE OF EXTERNAL. The information for STASE-ROSE, if any, is carried in the first EXTERNAL. The application context must specify which EXTERNAL(s) carry information for each of its other ASE(s). If the application context includes CMISE, its initialization information is carried in the second EXTERNAL.
3. ACSE issues a P-CONNECT request to the presentation provider to establish an application association. The presentation service provider then transfers the P-CONNECT request and receives a response (not shown).
4. The presentation provider issues a P-CONNECT confirmation primitive to ACSE, confirming the establishment of a presentation connection.
5. ACSE informs STASE-ROSE that a new application association has been established and provides STASE-ROSE with the values (if any) of the encryption parameters. The mechanism by which ACSE informs STASE-ROSE is a local implementation matter.
6. ACSE issues an A-ASSOCIATE confirm primitive to the application, confirming establishment of the association. ACSE provides the application with the values (if any) of the encryption parameters. The mechanism by which ACSE addresses the application is a local implementation matter.

6.3.6.1.2 Association Responder. The following describes the interactions on the association responding side in Figure 6.4. The presentation service provider receives a connect request from the remote presentation provider.

1. The presentation provider issues a P-CONNECT indication primitive to ACSE, announcing a remote service-user's interest in establishing an association.
2. ACSE issues an A-ASSOCIATE indication primitive to the application. During the same phase, ACSE informs STASE-ROSE that establishment of a new application as-

sociation has been requested and provides STASE-ROSE with the proposed values (if any) of the encryption parameters. ACSE further provides ASE-specific information (carried in the user information field), if any, to the ROSE users. The mechanism by which ACSE informs STASE-ROSE is a local implementation matter.

3. The application issues an A-ASSOCIATE response primitive to ACSE accepting or rejecting the application association. If the A-ASSOCIATE indication contains proposed values for (any of) the encryption parameters, the application process may indicate to STASE-ROSE which values of the encryption parameters should be accepted. The mechanism by which the application process informs STASE-ROSE is a local implementation matter. During this phase, the application may also inform the ROSE user ASE(s) of the association.
4. STASE-ROSE provides ACSE with the accepted values, if any, for the encryption parameters in the AARQ. The mechanism by which STASE-ROSE informs ACSE is an implementation matter and is a local implementation matter. This information will be carried in the ACSE user information field using the EncryptionParametersSelection parameter. During the same phase, the ROSE user ASE(s) may also provide ACSE with information relevant to the corresponding remote ASE(s). The ACSE user information field defined in ITU-T Rec. X.227 | ISO 8650 consists of a SEQUENCE OF EXTERNAL. The information for STASE-ROSE, if any, is carried in the first EXTERNAL. If CMISE is present in the application context, then its initialization information is carried in the second EXTERNAL. The application context specifies which EXTERNAL(s) shall carry information for each of its other ASE(s).
5. ACSE issues a P-CONNECT response primitive to the presentation provider accepting or rejecting establishment of association.

6.3.6.2 Association Release. Figure 6.5 illustrates the sequence of interactions between the application process, various ASEs, and the presentation provider during association release.

6.3.6.2.1 SENDER. The interactions on the sending side in Figure 6.5 are:

1. The application of the application context involving STASE-ROSE issues an A-RELEASE request to ACSE to release an application association.

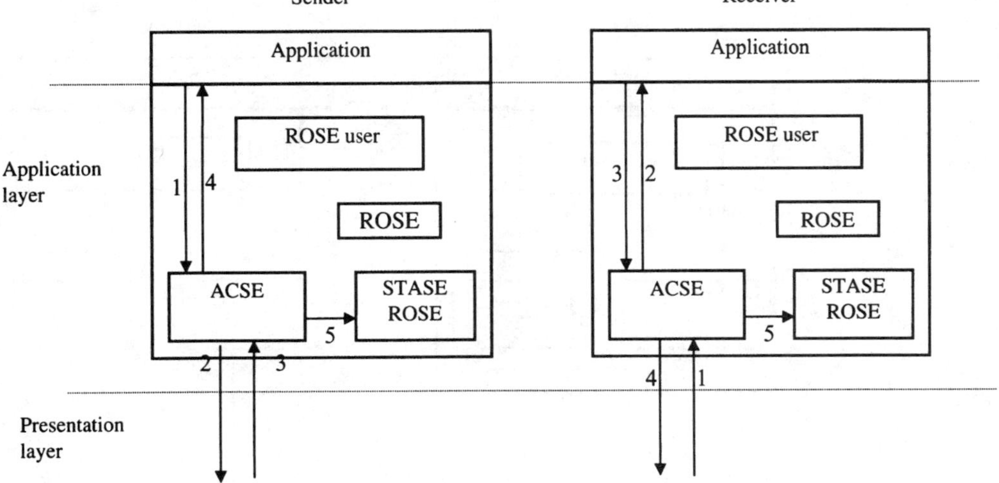

Figure 6.5 Interaction during association release

2. ACSE issues a P-RELEASE request to the presentation provider to release an application association. The presentation service provider then transfers the P-RELEASE request to the peer application-entity and receives a response (not shown).
3. The presentation provider issues a P-RELEASE confirmation primitive to ACSE, confirming the release of a presentation connection.
4. ACSE issues an A-RELEASE confirmation primitive to the application process confirming release of the application association.
5. ACSE informs STASE-ROSE and other ASEs (not shown in the figure) that the application association has been released. The mechanism by which ACSE informs STASE-ROSE and other ASEs is a local implementation matter.

6.3.6.2.2 RECEIVER. The interactions on the receiving side in Figure 6.5 are as follows. The presentation service provider receives a release request from the remote presentation provider.

1. The presentation provider issues a P-RELEASE indication primitive to ACSE, reporting that a remote service-user is interested in releasing an association.
2. ACSE issues an A-RELEASE indication primitive to the application process.
3. The application process issues an A-RELEASE response primitive to ACSE accepting release of the application association.
4. ACSE issues a P-RELEASE response primitive to the presentation provider accepting release of association.
5. ACSE informs STASE-ROSE and other ASEs about the release of the application association. The mechanism by which ACSE informs STASE-ROSE and other ASEs is a local implementation matter.

6.3.6.3 Association Abort. Figure 6.6 illustrates the sequence of interactions between the application process, various ASEs, and the presentation provider during association aborting.

6.3.6.3.1 SENDER. The interactions on the sending side in Figure 6.6 are:

1. The application using STASE-ROSE issues an A-ABORT request to ACSE to abort an application association.

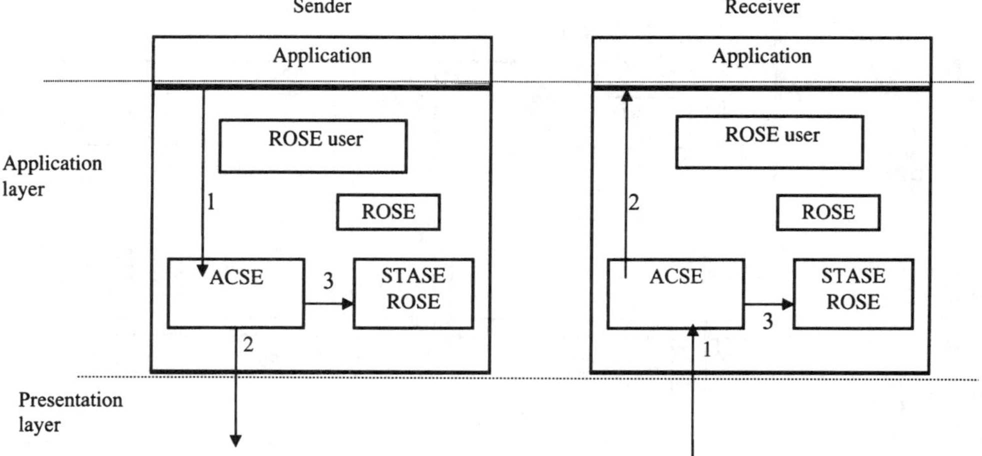

Figure 6.6 Interaction during association abort

2. ACSE issues a P-ABORT request to the presentation provider to abort the presentation connection.
3. ACSE informs STASE-ROSE and other ASEs about the aborting of the application association. The mechanism by which ACSE informs STASE-ROSE and other ASEs is a local implementation matter.

6.3.6.3.2 RECEIVER. The interactions on the receiving side in Figure 6.6 are:

1. The presentation service provider detects the aborting of a presentation connection.
2. The presentation provider issues a P-ABORT indication primitive to ACSE, announcing that an application connection has been aborted.
3. ACSE issues an A-ABORT indication primitive to the application process.
4. ACSE informs STASE-ROSE and other ASEs that the application association has been aborted. The mechanism by which ACSE informs STASE-ROSE and other ASEs is a local implementation matter.

6.3.6.4 Data Transfer. Figure 6.7 shows the interaction between the application, various ASEs, and the presentation provider during the data transfer phase.

6.3.6.4.1 SENDER. The interactions on the sending side in Figure 6.7 are:

1. The ROSE user, prompted (not shown) by the application, issues a ROSE request or response primitive to ROSE, requesting transfer of data.
2. ROSE issues the SR-TRANSFER request primitive to STASE-ROSE, requesting secure transfer of data.
3. STASE-ROSE performs the required encoding and security transformation on the ROSE PDU (present in the request parameters) and issues a P-DATA request primitive to the presentation provider.

6.3.6.4.2 RECEIVER. The interactions on the receiving side in Figure 6.7 are:

1. The presentation provider issues a P-DATA indication primitive to STASE-ROSE, announcing the arrival of data from the peer application entity on an application connection.

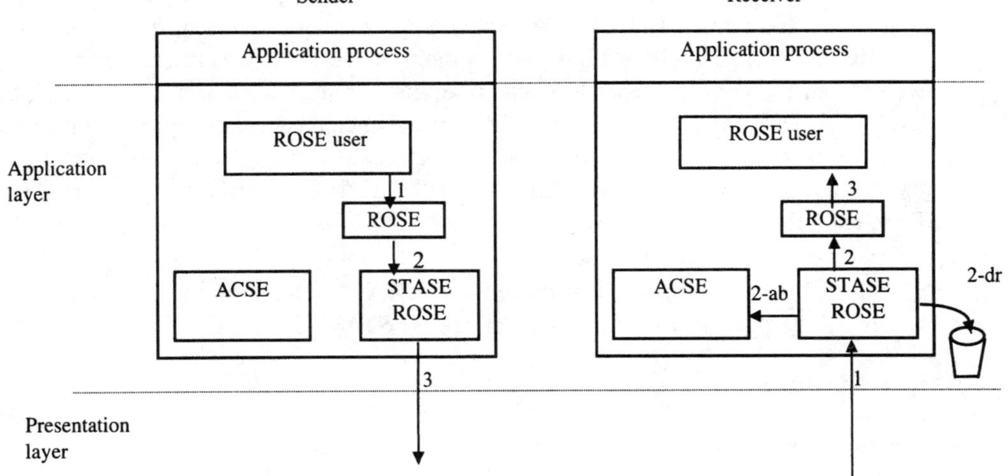

Figure 6.7 Interaction during data transfer

2. STASE-ROSE performs the reverse security transformations on the incoming data, checks for the validity of the PDU (e.g., validity of the seal or signature, currency of the time stamp, value of the sequence number), and if valid issues a SR-TRANSFER indication primitive to ROSE.

   ROSE issues a ROSE indication or confirmation primitive to the ROSE user. The ROSE user then informs (not shown) the application of the arrival of data from a peer application entity.

The receiving STASE-ROSE may find the incoming APDU unacceptable (e.g., decryption fails). In such a case the action to be taken by STASE-ROSE is a local matter. However, this standard recommends the following two alternatives.

- The receiving STASE-ROSE implementation may drop the incoming APDU as shown by **2-dr** in Figure 6.7.
- The receiving STASE-ROSE may issue an A-ABORT primitive to ACSE as shown in **2-ab** in Figure 6.7.

In either case, it is recommended (not shown) that the event be reported to the user element, that the event be logged in a security audit trail, and that a security alarm be issued to the local security administrator.

6.3.7 STASE-ROSE Protocol

The previous two sections described the services provided by STASE-ROSE and how the ASEs residing within a system cooperate to establish secure communications. In order to deliver the services, as advertised, STASE-ROSE must communicate with its remote counterpart using a well-defined protocol. This protocol, specified in this section, supports the STASE-ROSE services described earlier. It consists of defining the syntax of STASE-ROSE PDUs using ASN.1. As mentioned previously, STASE-ROSE uses the presentation layer P-DATA service for the transfer of secure ROSE PDUs.

The STASE-ROSE-protocol-machine (SRPM) communicates with ROSE by means of the SR-TRANSFER service primitives described previously. The SRPM is driven by service requests from ROSE and by indication primitives from the presentation-service. The SRPM in turn issues indication primitives to ROSE and request primitives to the presentation-service. P-DATA request and P-DATA indication presentation service primitives are used.

Reception of a STASE-ROSE service primitive or reception of a presentation-service primitive and generation of the dependent actions are local implementation matters. During the exchange of APDUs, the existence of an application association between the peer AEs is presumed.

Note: Each application association may be identified in an end system by an internal, implementation-dependent mechanism so that the STASE-ROSE service-user (ROSE) and SRPM can refer to it.

6.3.7.1 Abstract Syntax Definition of APDUs. In addition to the ASN.1 types defined in X.229, the following types are defined for STASE-ROSE.

```
DEFINITIONS IMPLICIT TAGS ::=  BEGIN

— EXPORTS everything

IMPORTS
```

```
ROSEapdus
FROM Remote-Operations-ADPUs {joint-iso-ccitt remote-operations(4) apdus(1)}

AE-title
FROM ACSE-1 {joint-iso-ccitt association-control (2) abstract-syntax(1) apdus(0)
version(1)}

DistinguishedName
FROM InformationFramework {joint-iso-ccitt ds(5) modules(1) informationFramework
(1)}

Certificate, CertificationPath
FROM AuthenticationFramework {joint-iso-ccitt ds(5) modules(1)
authenticationFramework(7)};

SR-APDU ::= CHOICE
      { clear                      [0] ROSEapdus,
        simpleConfidential         [1] OCTET STRING
        confidential               [2] Enciphered ,
        simplePublicEnciphered     [3] SimplePublicEnciphered,
        publicEnciphered           [4] PublicEnciphered ,
        hashed                     [5] HashedROSEpdu ,
        sealed                     [6] SealedROSEpdu ,
        signed                     [7] SignedROSEpdu ,
        confidentialSigned         [8] ConfidentialSigned,
        confidentialMAC            [9] ConfidentialMAC,
        confidentialSealed         [10] ConfidentialSealed,
        gssToken                   [11] GssToken,

        ...
        }
Enciphered ::= SEQUENCE
               {encrypted                  OCTET STRING,
                encryptionParameters       EncryptionParameters OPTIONAL
               }

- encrypted represents the DER encoded and encrypted ROSE PDU.
- encryptionParameters represents the parameters used for encryption.

SimplePublicEnciphered ::= CHOICE
                           { integers    SEQUENCE OF INTEGER,
                             string      OCTET STRING
                           }
-SimplePublicEnciphered represents the DER encoded and public key encrypted ROSE
PDU.
- A large PDU may be broken into smaller blocks, each of which may be encrypted
- as an INTEGER. The size of such blocks depends on the public key encryption al-
gorithm
- used and on the size of the public key; specification of such block sizes is
outside the
- scope of this standard.
-In some cases the result of public key encryption may be represented as an OCTET
STRING.

PublicEnciphered ::= SEQUENCE
                     {publicEncrypted        SimplePublicEnciphered,
                      encryptionParameters          EncryptionParameters OPTIONAL
                     }

-publicEncrypted represents the DER encoded and public key encrypted ROSE PDU.
```

–encryptionParameters represents the parameters used for encryption.

```
Hash ::= SEQUENCE{
            hashValue                  OCTET STRING (SIZE(8..64)),
            encryptionParameters       EncryptionParameters OPTIONAL
           }
```

–hashValue represents the message digest resulting from hashing the DER encoded
–ROSE PDU.
–encryptionParameters represents the parameters used for the hashing algorithm.

```
HashedROSEpdu ::= SEQUENCE
            {data     OCTET STRING,
             hash     CHOICE { hashHash,
                               simpleHash  OCTET STRING (SIZE (8..64))
                             }
            }
```

–data represents the DER encoded ROSE PDU
–hash represents the hash value either as a simple OCTET STRING or as a Hash
–structure defined above.

```
Seal ::= SEQUENCE
           {sealValue                OCTET STRING (SIZE(8..128)),
            encryptionParameters     EncryptionParameters OPTIONAL
           }
```

–sealValue represents the seal value for the DER encoded ROSE PDU
–encryptionParameters represents the parameters used by the seal generation algorithm

```
SealedROSEpdu ::= SEQUENCE
           {data       OCTET STRING,
            seal       CHOICE {sealStructure Seal,
                               simpleSeal    OCTET STRING (SIZE(8..64))
                               }
           }
```

–data represents the DER encoded ROSE PDU
– seal represents the seal value either as a simple OCTET STRING or as a Seal structure
– defined above.

```
Signature ::= SEQUENCE
            {signatureValue          SEQUENCE (SIZE(1..4)) OF INTEGER,
             encryptionParameters    EncryptionParameters  OPTIONAL
            }
```

–signatureValue represents the signature for the DER encoded ROSE PDU
–encryptionParameters represents the parameters for the signature algorithm.

```
SignedROSEpdu ::= SEQUENCE
      {data       OCTET STRING,
       signature CHOICE {signatureStructure [1] Signature,
                         simpleSignature    [2] SEQUENCE (SIZE(1..4)) OF INTEGER
                         }
      }
```

–data contains the DER encoding of the ROSE PDU
–signature represents the signature of the DER encoded ROSE PDU, either as a simple sequence

```
—of INTEGER or the Signature structure defined above.

ConfidentialSigned ::= SEQUENCE
       { encrypted OCTET STRING,
         signature CHOICE {signature          [1] Signature,
                          simpleSignature     [2] SEQUENCE (SIZE(1..4)) OF INTEGER
                          }

—encrypted represents the encryption of the DER encoded ROSE PDU.
—signature represents the signature of the DER encoded ROSE PDU in either a sim-
ple form
— or as Signature type defined above.

ConfidentialMAC ::= SEQUENCE
       { encrypted     OCTET STRING,
         mac           CHOICE {mac             [1] Hash,
                              simpleMAC        [2] OCTET STRING (SIZE (8..64))
                              }
       }

—encrypted represents the encryption of the DER encoded ROSE PDU.
— mac represents the MAC of the DER encoded ROSE PDU in either a simple form
— or as Hash type defined above.

ConfidentialSealed ::= SEQUENCE
       { encrypted     OCTET STRING,
         seal          CHOICE {sealed          [1] Seal,
                              simpleSealed     [2] OCTET STRING (SIZE (8..64))
                             }
       }

—encrypted represents the encryption of the DER encoded ROSE PDU.
— seal represents the seal of the DER encoded ROSE PDU in either a simple form
— or as Seal type defined above.

EncryptionParameters ::= SET
       {symmetricKeyId          [0] KeyID                          OPTIONAL,
        publicKeyId             [1] KeyID                          OPTIONAL,
        sealKeyId               [2] KeyID                          OPTIONAL,
        signatureKeyId          [3] KeyID                          OPTIONAL,
        passwordId              [4] KeyID                          OPTIONAL,
        initializationVector    [5] OCTET STRING (SIZE(8))         OPTIONAL,
        feedBackBits            [6] INTEGER (SIZE (1..63))         OPTIONAL,
        —for k-bit output feedback mode or k-bit cipher feedback mode of DES
        symmetricAlgorithm      [7] OBJECT IDENTIFIER              OPTIONAL,
        publicKeyAlgorithm      [8] OBJECT IDENTIFIER              OPTIONAL,
        signatureAlgorithm      [9] OBJECT IDENTIFIER              OPTIONAL,
        sealAlgorithm           [10] OBJECT IDENTIFIER             OPTIONAL,
        hashAlgorithm           [11] OBJECT IDENTIFIER             OPTIONAL,
        keyDigest               [12] OCTET STRING (SIZE(8..64)) OPTIONAL,
        —for verification of public keys
        blockSize               [13] INTEGER                       OPTIONAL,
        —for square mod-n hashing
        keySize                 [14] INTEGER                       OPTIONAL,
        —for RSA
        publicKey               [15] SEQUENCE
                                        {modulus     INTEGER,
                                         exponent    INTEGER
                                        }                          OPTIONAL,
        sequenceNumber          [16] INTEGER                       OPTIONAL,
```

```
        timeStamp               [17] GeneralizedTime             OPTIONAL,
        encryptedKey            [18] OCTET STRING (SIZE(64..128)) OPTIONAL,
        —symmetric session key, encrypted with Key-Encryption-Key
        encryptedSymmetricKey   [19] INTEGER                     OPTIONAL,
        —symmetric session key, encrypted with the receiver's public key
        keyEncryptionKey        [20] SEQUENCE (SIZE (1..3)) OF KeyID
                                                                 OPTIONAL,
        —one to three symmetric keys used for encrypting a session key
        publicKeyCertificate    [21] PublicKeyCertificate        OPTIONAL,
—X.509 certificate or certification path of the sender's public key with no usage
restrictions
        encryptionCertificate   [22] EncryptionCertificate       OPTIONAL,
—X.509 certificate or certification path of the sender's public key used for en-
cryption only
        signatureCertificate    [23] SignatureCertificate        OPTIONAL,
—X.509 certificate or certification path of the sender's public key used for dig-
ital signatures only
—
encryptedAuthenticatedSymmetricKey [24] EncryptedAuthenticatedSymmetricKey
                                                                 OPTIONAL,
—symmetric session key, encrypted with the receiver's public key and signed with
sender's key—
        macAlgorithm            [25] OBJECT IDENTIFIER           OPTIONAL,

                ...
                }

—EncryptionParameters is an extensible type that is used as a catch-all for any
—parameters that may be used by any of the STs. In most applications only a small
—number, if any, of the components of EncryptionParameters will be used.

KeyId ::= CHOICE        {
                         name        GraphicString,
                         number      INTEGER
                        }

PublicKeyCertificate ::= CHOICE {certificate           [0] Certificate,
                                 certificationPath     [1] CertificationPath
                                 }

EncryptionCertificate ::= CHOICE {certificate          [0] Certificate,
                                  certificationPath    [1] CertificationPath
                                  }

SignatureCertificate ::= CHOICE {certificate           [0] Certificate,
                                  certificationPath    [1] CertificationPath
                                  }

EncryptedAuthenticatedSymmetricKey ::= SEQUENCE {
                                   encryptedSymmetricKey          INTEGER,
               — symmetric session key, encrypted with the receiver's public key
                                   time                GeneralizedTime,
                                   sender              SenderId,
                                   receiver            ReceiverId,
                                   signature           Signature
—the signature is computed over ASCII representation of the preceding four fields
with the sender's private key
                                       }

SenderId ::= CHOICE {
```

```
                    identifier      [1] DistinguishedName,
                    name            [2] GraphicString,
                    application     [3] AE-title
                    }

ReceiverId ::= CHOICE {
                    identifier      [1] DistinguishedName,
                    name            [2] GraphicString,
                    application     [3] AE-title
                    }

GssToken::= CHOICE {
                  micToken      [1] MicToken ,
                  wrapToken     [2] OCTET STRING ,
                  }
MicToken::= SEQUENCE {
                rosePDU         [1] OCTET STRING ,
                token           [2] OCTET STRING ,
                }
END
```

6.3.8 Use of GSS API with STASE-ROSE

GSS-API [RFC2078] (Generic Security Services API) is a high-level API for integration of communication security services.

Use of GSS-API can give several benefits to OSI-stack suppliers that want to implement STASE-ROSE. GSS-API is a high-level API; it provides a very simple way of integrating security services for application implementers. Use of GSS-API for STASE-ROSE can ensure that security algorithms or even entire security mechanisms can be changed without having to modify STASE-ROSE.

This section describes how STASE-ROSE can realize its cryptographic functionality (security transformations upon ROSE PDUs) through the use of GSS-API. This section describes the use of GSS-API during the different phases of communication. The figures used in this section are based on figures used in Section 6.3.6 that describe interactions between application service elements independent of GSS-API. The numbers along the arrows in these figures correspond to interactions discussed in Section 6.3.6 and not to GSS-API interactions.

Figure 6.8 indicates a possible way of using GSS-API during the association establishment phase. It shows that both the application process and STASE-ROSE will need to have access to the same supporting GSS-API cryptographic module. The application process will use the GSS-API module for authentication purposes during association establishment, whereas STASE-ROSE will use the GSS-API module during the data transfer phase. The interaction between the different components in the figure at the initiator side (left) and the responder side (right) is described next.

GSS-Specific Interaction of the Association Initiator

a. The application performs GSS_aquire_cred() to acquire its credentials from the GSS-API module.

b. The application calls GSS_init_sec_context() to initiate a security context with a specified association responder. The application must decide on a set of security parameters for the association (e.g., whether to apply unilateral or mutual authentication, whether message sequence protection and replay protection is required). The GSS-API module returns an initial_context token to the application.

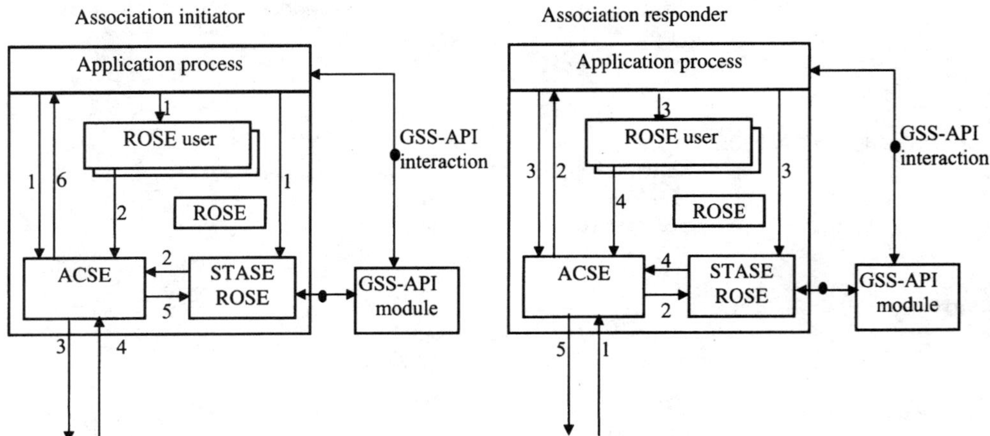

Figure 6.8 Use of GSS-API with STASE-ROSE at association establishment time

c. The application invokes an ACSE A-ASSOCIATE request to the association responder, providing ACSE with the initial_context token to be carried in the Authentication-value field. The token structure syntax should in this case be supported by ACSE. At the same time, the application may provide STASE-ROSE with information relevant to data protection during the data transfer phase. (As a minimum, the GSS-API context_handle parameter should be provided as a security context reference.)
d. The remaining steps are the same as without GSS-API.

GSS-Specific Interaction of the Association Responder

a. Steps 1 and 2 are the same as without GSS-API.
b. When the application receives an A-ASSOCIATE indication, the application calls GSS_aquire_cred() to acquire its credentials from the GSS-API module (if the application has not already acquired its credentials at this point in time). The application thereafter calls GSS_accept_sec_context() with the token received from the initiator (within the ACSE Authentication-value field) as one of the input parameters. The GSS-API module authenticates the initiator by verifying that the token is valid. If the initiator requests mutual authentication, the application receives from the GSS-API module a second token that must be transmitted back to the initiator.

Steps 3, 4, and 5 are the same as without GSS-API. As part of the A-ASSOCIATE response provided in step 3, the application will provide the second token (in the case of mutual authentication) to be carried in the Authentication-value field. Again the token structure syntax should be supported by the ACSE A-ASSOCIATE response Authentication-value parameter.

6.3.8.1 Security Context Negotiation. As part of the initial-context token exchange at association establishment time, the initiator and target GSS-API modules negotiate (transparent to STASE-ROSE) a common valid set of integrity and confidentiality algorithms for the established security association. By default, the negotiated valid set of algorithms will always be the largest common set of algorithms that is supported by both parties. This algorithm negotiation is done automatically by the communicating GSS-API modules without being controlled by the GSS-API user (application). This level of negotiation is sufficient for interoperability reasons. However, it does not give the communicating applications any flexibility to restrict, for example, the number of valid algorithms any further.

To allow for a more flexible security policy, an option would therefore be to add a second negotiation mechanism at the STASE-ROSE level (external to the GSS-API modules), using the negotiation parameters defined by STASE-ROSE. In particular, the following negotiation parameters would be relevant:

- signatureAlgorithms (negotiating non-repudiation algorithms for this association)
- sealAlgorithms (negotiating integrity algorithms for this association)
- hashAlgorithms (negotiating hash algorithms for this association)

This negotiation facility would mean that two STASE-ROSE entities can agree upon use of a set of algorithms that is a subset of the set of algorithms that is being negotiated by the GSS-API modules. A prerequisite in this case is that the STASE-ROSE entities know what algorithms their local GSS-API modules support. The principles for negotiation are the same as those for STASE-ROSE in general.

6.3.8.2 Data Transfer Phase. Figure 6.9 illustrates the use of GSS-API during the data transfer phase. The interactions between the different components at the initiator side (left) and the responder side (right) will be the same as for STASE-ROSE without GSS-API.

During this phase, each STASE-ROSE ASE interfaces with its local GSS-API module by using the GSS-API primitives GSS_wrap() and GSS_get_mic() to provide security transformations. At this point in time, each application must have provided to their local STASE-ROSE ASE the GSS-API context handle that is needed as an input parameter to all GSS_wrap()/GSS_get_mic() function calls. In addition, the communicating GSS-API modules and the STASE-ROSE ASEs (if additional context negotiation was used during the asso-

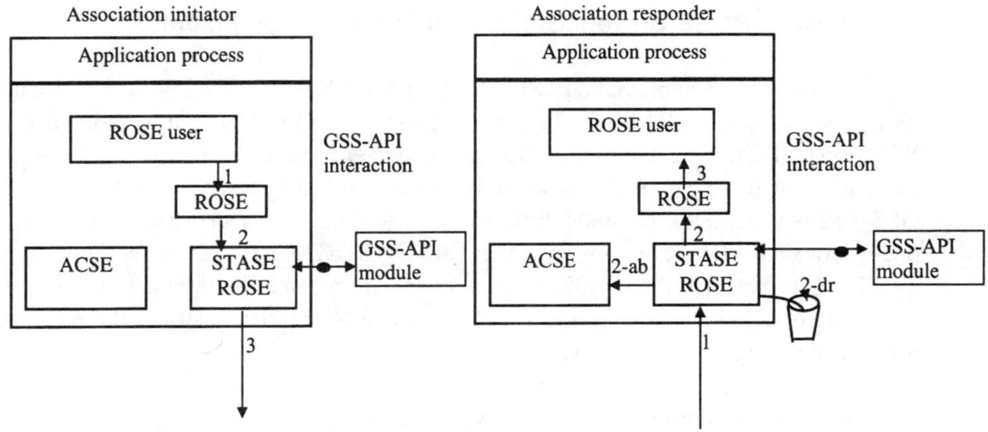

Figure 6.9 Use of GSS-API with STASE-ROSE during data transfer

ciation establishment phase) have already negotiated the set of alternative algorithms for the association and thereby the possibilities for Quality of Protection (QoP).

For each message that is sent, ROSE, or some other ASE within the application process, requests a certain level of protection from STASE-ROSE. How an ASE decides the necessary level of protection is not discussed here. (It should preferably get this information from the local application.)

At a minimum, the ASE must tell STASE-ROSE what type of encryption to use, if any. STASE-ROSE receives this information through the SR-TRANSFER parameter "Encryption-Type." These requests must be mapped onto a gss_wrap()/gss_get_mic() function call. Each time some kind of confidentiality protection (with or without integrity protection) is requested, STASE-ROSE must use the gss_wrap() function. Each time integrity or non-repudiation protection is requested without any confidentiality protection, STASE-ROSE must use the gss_get_mic() function.

It is not sufficient for the entity using STASE-ROSE to know whether to use gss_wrap() or gss_get_mic(). The initiating STASE-ROSE entity must also know what quality of protection level (QoP-level) to ask for when calling these functions (qop_req input parameter of gss_wrap() and gss_get_mic()). In principle, STASE-ROSE can decide the quality of protection level in two ways:

1. ROSE (or some other ASE) tells STASE-ROSE what quality of protection is required through the SR-TRANSFER parameter "Encryption-Parameters."
2. If the SR-TRANSFER "Encryption-Parameters" parameter is not used, the default level of protection for "Encryption-Type" indicated by SR-TRANSFER is used. In this case, it is assumed that the STASE-ROSE entities have already agreed upon a default level of protection at context establishment time.

Whatever solution is used, it is important to ensure that STASE-ROSE selects a QoP level that stays within the limits of protection that was negotiated at association establishment time. The implementers of STASE-ROSE must therefore decide what should happen if, for example, the SR-TRANSFER primitive from ROSE requests a higher Quality of Protection level than the established context is able or allowed to provide. After a gss_wrap() token or a gss_get_mic() token has been created, it is inserted into the STASE-ROSE protocol using the gssToken protocol field. The receiving entity verifies the tokens through calls to gss_unwrap() or gss_verify_mic().

6.3.9 STASE-ROSE Current Status and Future Developments

Committee T1 approved STASE-ROSE as ANSI Standard T1.259. It was also approved by Study Group 4 (SG4) of the ITU-T as ITU-T Rec. Q.813. The international version of STASE-ROSE provides the option of using the Generic Security Service (GSS) Application Programming Interface (API) for actually carrying out the security transformations. When GSS-API is used, both the authenticator exchanged during association setup and the subsequent protected PDUs are represented as OCTET STRINGs that are only "understandable" by GSS. It is expected that T1.259 will be amended to match the international standard.

The optional inclusion of GSS-API in STASE-ROSE demonstrates three valuable features of the design of STASE-ROSE:

- The incorporation of GSS-API was fairly simple.
- Any implementation of the current T1.259 would be compliant with Q.813.
- When the GSS-API option is used, STASE-ROSE does not impose any additional overhead.

It is expected that at least one of the following steps will be taken:

- The definition of STASE-ACSE which protects ACSE PDUs in the same way that STASE-ROSE protects ROSE PDUs. Additional variations may be developed for other ASEs.
- The modification of STASE-ROSE into STASE which applies to any ASE; this would require more clever tweaking of the standard.

It is expected that either course could be achieved in a manner that would be backwards compatible with the current T1.259. That is, implementations of the current T1.259 would be compliant with the newer versions when only ROSE PDUs are being protected.

6.4 Q3 SECURITY

T1.261 specifies how various security services for the Q3 interface are to be provided. It is based largely on existing security standards. T1.261 also specifies that peer entity authentication at association setup time is to be accomplished as described in T1.259. That is, the default authenticator specified in STASE-ROSE is exchanged in the authentication FU of ACSE.

Data origin authentication without integrity protection follows the procedure currently used in Electronic Bonding: the authentication information is carried in the CMIP access control field. It provides for a simple authenticator consisting of the sender's ID, the receiver's ID, and an encrypted time stamp, as well as a more general authenticator that can be used to exchange a symmetric key encrypted with the receiver's public key.

Whole PDU protection is provided by using STASE-ROSE with the default algorithms. The following security transformations are included:

- Clear (i.e., no security transformations)
- Simple confidential (i.e., encrypted; not including update of the symmetric encryption key)
- Confidential (including update of the symmetric encryption key)
- Hashed
- Signed
- Confidential and signed
- Confidential and hashed

Access control management is based on ITU-T Rec. X.741 (Objects and attributes for access control). It supports both the Access Control List (ACL) and the Access Control Certificate (ACC) approaches.

T1.261 supports effective and efficient management of access control information. It allows the grouping of several initiators into classes with specific access privileges. Each member of a class has all the access privileges associated with that class. It also allows the grouping of target systems into categories that offer similar access privileges. Thus, T1.261 supports the definition, for example, of a class of "central district switch testers"; each member of this class has a specific set of privileges for accessing any switch in the central district. Management of access control then consists of adding/removing members of the class of central district switch testers, adding/removing members of the category of central district switches, and/or changing the privileges that central district switch testers have in accessing central district switches. T1.261 allows one initiator to be a member of any number of access classes, and it allows each target system to be a member of any number of access categories.

6.5 X.500

The X.500 directory standard provides ample provisions for security. Many of those security provisions have been (adapted and) reused by other standards, such as STASE-ROSE. The level of security for any instance of the directory depends on the sensitivity of the information it contains. If the information is not confidential, then it need not be encrypted as it is exchanged over an interface, and it may not be necessary even to control who can request this information (unless there is a risk that the directory will be overwhelmed by requests from undesirable sources). In all cases, however, it is necessary to ensure the integrity of the messages exchanged with the directory. This section provides an overview of the X.500 security features that have been specified in the ANSI T1.252-1996 standard for secure TMN directory.

An association with an X.500 directory is established through the directoryBind operation, as defined in X.511. It includes an OPTIONAL Credentials field. This field is mandatory for the secure TMN directory.

The Credentials field is a CHOICE among SimpleCredentials, StrongCredentials, or an EXTERNAL procedure. For the secure TMN directory, the choice is StrongCredentials. StrongCredentials contains a CertificationPath field. For interactions over Q3 interfaces, this field can be a single certificate.

StrongCredentials also contains a Token, which is a signed sequence consisting of an AlgorithmIdentifier, a DistinguishedName, a UTCTime (a time stamp using the ASN.1 format for coordinated universal time), and a random BIT STRING. For the secure TMN directory, the UTCTime is a time stamp showing the time when the message was constructed. Since this time stamp is unique, there is no need for the random BIT STRING, which therefore has zero bits.

X.511 defines the Registration request protocol (RRP). However, the RegistrationBindArgument of the RRP does not provide any security features. T1.252 therefore defines the SecureRegistrationBindArgument, which includes strong credentials with the certification path.

Directory operations PDUs include the CommonArguments, which in turn include OPTIONAL SecurityParameters. The SecurityParameters contain five OPTIONAL elements: CertificationPath, DistinguishedName, UTCTime, a random BIT STRING, and ProtectionRequest, which specifies whether the response from the directory should be signed by the directory. For the secure TMN directory, the CertificationPath and UTCTime must be present. The UTCTime is a time stamp showing the time when the message was constructed. Since this time stamp is unique, there is no need for the random BIT STRING, which is therefore absent.

6.6 X.25

Most TMN security is provided in the TMN applications and in the Application Layer where detailed fine-tuning of the security measures can customize the services provided to the needs of each application. Such security measures, however, cannot prevent an intruder from using up network resources while attempting to establish an unauthorized connection. Even if the connection request is rejected, some network resources would have been wasted. Further resources would have been wasted by the targeted system to examine the request and reject it. An intruder with no hope of successfully establishing a connection with the targeted system may still cause disruption simply by overloading that system with connection requests. The best place to provide protection against such attacks is in the network layer. Ideally, the intruder would be stopped at the first network node. Such security can be provided for X.25 networks through use of a Closed User Group (CUG).

The purpose of a CUG is to hinder access to TMN entities by unauthorized entities. Any set of systems, within and outside a TMN, can form a CUG. A system can be part of several CUGs. Typically, if a system is part of one or more CUGs, it can be accessed only by systems that are members of at least one of these CUGs.

If a TMN entity is a member of two or more CUGs, the X.25 network must support CUG with outgoing access selection, allowing the call originator to specify the desired CUG.

Although access to a CUG member is tightly controlled, access by a CUG member can be more relaxed. Some CUGs allow any member of the CUG to establish connections with any other system, regardless of CUG membership. Others limit CUG members to establish connections only within a CUG. A CUG that allows any of its members gateways to send messages to any Data Terminal Equipment within or outside the CUG is a CUG with outgoing access.

If several X.25 networks are used instead of a single X.25 network, then the X.75 interfaces among those X.25 networks must support CUGs.

7

EDI-Based TMN Security

EDI messages are usually stuffed in electronic mail envelopes. This store-and-forward transfer is not fast enough for TMN real-time interactions; TMN needs speedier delivery of EDI messages. Thus, a special-purpose Interactive Agent (IA) has been crafted to map EDI messages (the output of an EDI translator) into the underlying transport facility for fast delivery. (For TMN applications, only TCP/IP is used for the transport of EDI messages.) The IA also provides flow control.

Since TMN EDI messages are to be carried over TCP/IP and since SSL3 is broadly available to secure TCP data streams, SSL3 was selected to secure EDI transactions. SSL3 provides strong peer entity authentication, confidentiality, and integrity services. However, SSL3 does not, and cannot, provide non-repudiation. This service has to be provided prior to any fragmentation that is done by SSL3. The IA, which is invoked prior to any such fragmentation, has been drafted to provide non-repudiation. When TLS1, the standardized version of SSL3 becomes commercially available, it is expected to supersede SSL3 in TMN applications.

7.1 TLS1 FOR EDI

The initial usage of TLS1 to secure EDI TMN messages is based on the following guidelines:

- Strong peer entity authentication, based on public key encryption, shall be provided for all associations. (This precludes interoperability with SSL2.)
- Session secrets shall be encrypted with the receiver's public key.
- Message encryption is optional.
- SHA1 shall be used for integrity by TLS1.
- If privacy protection by TLS1 is chosen, DES (Data Encryption Standard) in the CBC (Cipher Block Chaining) mode shall be used for symmetric key encryption.
- Every participant is required to obtain a public key certificate from a CA acceptable to the communicating parties.
- Message integrity shall be computed on clear text (unencrypted) messages.
- Entity public key size shall be at least 768 bits.

- CA's public key size shall be at least 1,024 bits.
- Certificates shall be X.509 version 3.
- The following cipher suites will be supported (a SSL3 cipher suite specifies a public key algorithm, a symmetric encryption algorithm, and a hashing algorithm in this order):
  - RSA, NULL (meaning no algorithm), SHA1 (if no privacy protection is desired)
  - RSA, DES–CBC, SHA1 (if privacy protection is desired)

SSL3 allows sessions to be terminated and then resumed with the previously established security parameters. For EDI-based applications over the TMN X interface, resumable sessions do not present any additional threat.

SSL3 does not offer integrity based on double hashing, such as HMAC. Integrity protection based on a single hash (using SHA1 in our case) has some known vulnerabilities. Such vulnerabilities are eliminated if both encryption and integrity are used.

7.2 INTERACTIVE AGENT

We now examine the IA in some detail. Figure 7.1 illustrates the flow of IA messages.

7.2.1 Message Formatting

In order to meet both the business and security needs of companies, five message types have been defined:

- Basic Security (plain text or message privacy encryption only)
- Enhanced Security with message integrity support
- Enhanced Security with non-repudiation (includes message privacy and message integrity)
- Interactive Agent status
- Message receipt (optional)

Each of these is defined using ASN.1 and encoded using the definite length method of ASN.1's Distinguished Encoding Rules (DER).

Although these five formats are treated distinctly here, all are variants based on secure messaging services and protocols expressed using ASN.1 notation. The basic structure of the four mandatory messages (i.e., excluding the optional receipt message format) is illustrated in Figure 7.2.

The message formats are presented twice. The first presentation provides a conceptual overview. It is intended mostly for EDI users who do not care much about ASN.1. The second presentation, in Section 7.2.2, provides a formal ASN.1-based description of the messages, and it is intended mostly for implementers who care a lot about such technicalities.

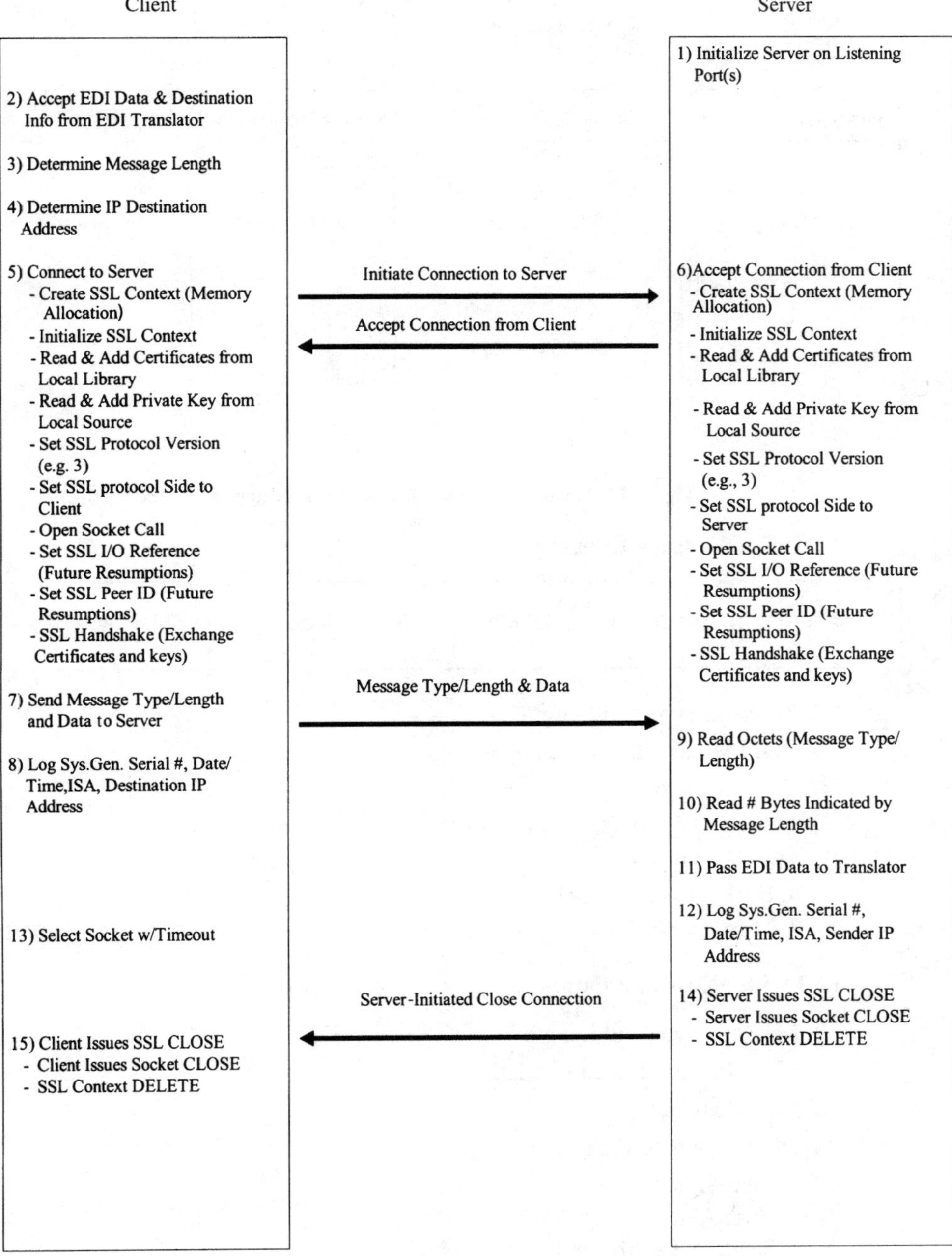

Figure 7.1 IA Client-server interaction

Basic EDI

Header (1)	EDI Message (2)

(1) Message Type and Length
(2) EDI Message

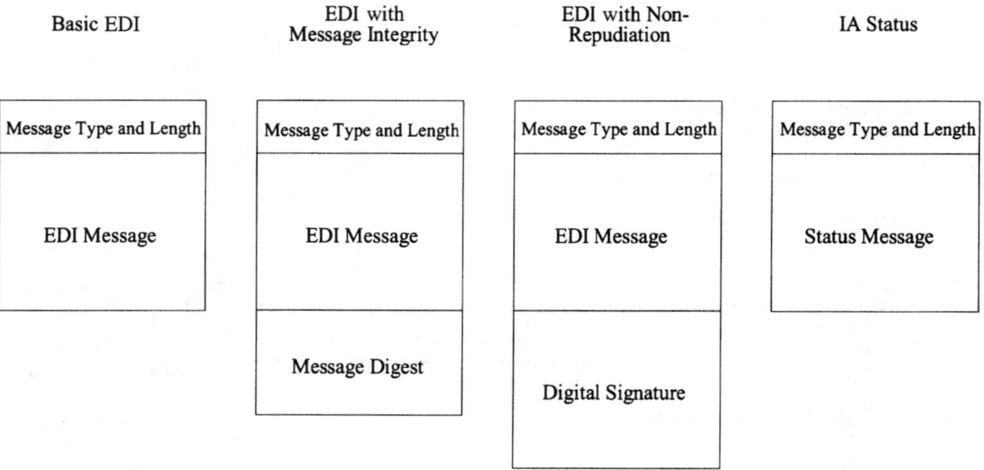

Figure 7.2 Message format architecture for mandatory messages

EDI with Message Integrity

Header (1)	Header(2)	EDI Message (3)	Header (4)	Message Digest (5)

(1) Message Type and Length
(2) EDI Message Tag and Length
(3) EDI Message
(4) Message Digest Tag and Length
(5) Message Digest
- Hash Algorithm (SHA1)
- Message Digest

EDI with Non-Repudiation

(1) Message Type and Length
(2) EDI Message Tag and Length
(3) EDI Message
(4) Digital Signature Tag and Length
(5) Digital Signature
- Hash Algorithm (SHA1)
- Sender's Certificate Info
- Hash Encryption Algorithm
- Message Digest encrypted using sender's private key

IA Status

Header (1)	IA Status (2)

(1) Message Type and Length
(2) Encoded IA Status

7.2.1.1 IA Status Message Detail Format. The IA Status message consists of an identifying header and four octets containing status information, structured as follows:

The **first octet** is utilized for status conditions pertaining to the wide area interface of the IA (i.e., TCP/IP and SSL3). This octet is to be interpreted as 2 hexadecimal values in the range 00 to FF. The currently defined values include:

00 = NULL (no error or action)
08 = Request for a peer to cease transmissions on inbound data stream due to problems with WAN interface

All other values are currently undefined.

The **second octet** is utilized for status conditions pertaining to the downstream systems on the local interface (i.e., the EDI translator or other related, downstream system). This octet is to be interpreted as 2 hexadecimal values in the range 00 to FF. The currently defined values include:

00 = NULL (no error or action)
08 = Request for peer to cease transmissions on inbound data stream due to problems with IA communicating with EDI translator or other downstream processing systems.

All other values are currently undefined.

The **third and fourth octets** are available for pairwise definitions. Parties to a pairwise agreement may define the values in these octets in any manner they choose. It is recommended that the values be encoded as 2 hexadecimal values per octet. These octets may be utilized to convey their own independent meanings in those cases where the first two octets contain 00 00, or they may be used as amplifying identifiers in those instances where either of the first two octets is nonzero.

A message containing four octets of hexadecimal 00 00 00 00 is defined as a NULL message that may be utilized as a **test message.** No actions are taken in response to receipt of such a message.

Only **one action or status command may be encoded per message.** It is not valid to combine operations within a single message. It is therefore a requirement that either octet one or octet two (or both) contain 00.

Status Message Examples

00 08 00 00—Message sent to peer requesting no further transmissions due to downstream problem

00 00 00 00—Test message, no action required

08 00 00 00—Problems with inbound WAN; peer should cease transmissions

00 00 06 00—Message defined in pairwise agreement, interpretation at discretion of the parties

7.2.1.2 Optional Message Receipts. As part of a pairwise agreement, companies may choose to enable Interactive Agent "positive receipts." Three different Receipt types are defined. This allows Receipt types to be matched with the three existing IA message formats. Although the receipt type selected need not match the format of the original data message, it is recommended that the two formats match.

Basic Receipt

Header (1)	IA Receipt (2)

(1) Message Type and Length
(2) Receipt

Receipt with Message Integrity

(1) Message Type and Length
(2) Receipt
(3) Message Digest Tag and Length
(4) Message Digest
- SHA-1 Message Digest Algorithm

Receipt with Digital Signature (Non-Repudiation)

(1) Message Type and Length
(2) Receipt
(3) Digital Signature Tag and Length
(4) Digital Signature
- SHA-1 Message Digest Algorithm
- RSA Digital Signature Algorithm

When Basic Receipts are utilized, the EDI message is parsed to extract the ISA segment. A date/time stamp is appended to the ISA forming the Receipt. This information is formatted in accordance with Section 7.2.2.5 and returned to the originator. Receipts with Message Integrity extend the Basic Receipt by the addition of a Message Digest of the original EDI data. Receipts with Digital Signatures similarly extend the Basic Receipt by adding a Digitally Signed Acknowledgment of the original EDI message.

7.2.2 Message Syntax Definitions

This section provides a more formal description of the IA messages using ASN.1. The following OBJECT IDENTIFIER has been registered for the transport of ASC X12 EDI data via Interactive Agent Transfer Protocol (IATP):

```
eciaAscX12Edi OBJECT IDENTIFIER ::= {

          iso(1) org(3) dod(6) internet(1) private(4) enterprises(1)

          mciwcom(3576) 7 }
```

The general syntax used for the IA is based on the structure described by RSA Data Security, Inc. in PKCS #7 (IETF RFC 2315). It associates a content type with content. The general syntax has ASN.1 type ContentInfo:

```
ContentInfo ::= SEQUENCE {

 contentType ContentType,

 content [0] EXPLICIT ANY DEFINED BY contentType OPTIONAL }

ContentType ::= OBJECT IDENTIFIER

Useful types include:
```

```
AlgorithmIdentifier ::{= SEQUENCE {

      algorithm OBJECT IDENTIFIER,

      Parameters NULL

}
```

Note: The following two OBJECT IDENTIFIERs have been registered for the transport of EDIFACT and non-EDI data, respectively:

```
eciaEdifact OBJECT IDENTIFIER ::= {

         iso(1) org(3) dod(6) internet(1) private(4) enterprises(1)

         mciwcom(3576) 8 }

eciaNonEdi OBJECT IDENTIFIER ::= {

         iso(1) org(3) dod(6) internet(1) private(4) enterprises(1)

         mciwcom(3576) 9 }
```

7.2.2.1 ASN.1 Syntax for Basic EDI Messages. The plain EDI message content has ASN.1 type PlainEDImessage:

```
PlainEDImessage ::= SEQUENCE {

 plainEDImessage OBJECT IDENTIFIER,

 ediContent [0] EXPLICIT OCTET STRING

 }

PlainEDImessage references the following object identifier:

plainEDImessage OBJECT IDENTIFIER ::= { eciaAscX12Edi 1 }

ediContent consists of an ASC X12 formatted EDI message.
```

7.2.2.2 ASN.1 Syntax for EDI with Message Integrity. The integrity EDI message content has ASN.1 type IntegrityEDImessage:

```
IntegrityEDImessage ::= SEQUENCE {

integrityEDImessage OBJECT IDENTIFIER,

 integrityContent [0] EXPLICIT SEQUENCE {

  integrityVersion INTEGER,

  integrityAlgorithmIdentifier AlgorithmIdentifier,

  innerContent SEQUENCE {

     plainEDImessage OBJECT IDENTIFIER,

     ediContent [0] EXPLICIT OCTET STRING
```

```
      },
   integrityDigest OCTET STRING
   }
}
```

EDI with Message Integrity references the following object identifiers:

```
integrityEDImessage OBJECT IDENTIFIER ::= { eciaAscX12Edi 5 }
integrityAlgorithmIdentifier OBJECT IDENTIFIER ::={
     iso(1) org(3) oiw(14) secsig(3) algorithm(2) 26 }       —SHA-1
```

integrityVersion is { 0 } for this revision of the syntax.
ediContent consists of an ASC X12 formatted EDI message.
integrityDigest consists of the SHA-1 digest of the ediContent.

7.2.2.3 ASN.1 Syntax for EDI with Non-Repudiation. The signed EDI message content has ASN.1 type SignedEDImessage:

```
SignedEDImessage ::= SEQUENCE {
signedEDImessage OBJECT IDENTIFIER,
 signedContent [0] EXPLICIT SEQUENCE {
   signedVersion INTEGER,
   signedDigestAlgorithms SET OF AlgorithmIdentifier,
   innerContent SEQUENCE {
      plainEDImessage OBJECT IDENTIFIER,
      ediContent [0] EXPLICIT OCTET STRING
      }.
   signerInfos SET OF SEQUENCE {
     signerVersion INTEGER,
     issuerAndSerialNumber SEQUENCE {
       issuerCountry SEQUENCE OF SET OF SEQUENCE {
         country OBJECT IDENTIFIER,
         countryValue PrintableString
         },
       issuerOrg SEQUENCE OF SET OF SEQUENCE {
         organizationName OBJECT IDENTIFIER,
```

```
          organizationValue PrintableString

          },

        serialNumber INTEGER

        },

      signedDigestAlgorithm AlgorithmIdentifier,

      digestEncryptionAlgorithm AlgorithmIdentifier,

      encryptedDigest OCTET STRING

      }

    }

  }
```

A note to the (rightly) puzzled reader: the construct "SEQUENCE OF SET OF SEQUENCE" in the preceding ASN.1 is the result of some simplification of the original syntax in PKCS #7 where it is useful for a more general case.

EDI with Non-repudiation references the following object identifiers:

```
signedEDImessage OBJECT IDENTIFIER ::= { eciaAscX12Edi 2 }

signedDigestAlgorithms OBJECT IDENTIFIER ::=  {

        iso(1) org(3) oiw(14) secsig(3) algorithm(2) 26 }     —SHA-1

country OBJECT IDENTIFIER ::= { 2 5 4 6 }

organizationName OBJECT IDENTIFIER ::= { 2 5 4 10 }

signedDigestAlgorithm OBJECT IDENTIFIER ::= {

        iso(1) org(3) oiw(14) secsig(3) algorithm(2) 26 }      —SHA-1

digestEncryptionAlgorithmIdentifier OBJECT IDENTIFIER ::= {

  iso(1) member-body(2) us(840) rsadsi(113549) pkcs(1) 1 1 }    —RSA
```

EDI with Non-repudiation Value Information:

signedVersion is { 0 } for this version of the syntax.
ediContent consists of an ASC X12 formatted EDI message.
signerVersion is { 0 } for this version of the syntax.
countryValue is the two-character representation of the issuer's country name.
organizationValue is the certificate issuer's organization name.
serialNumber is a unique number assigned to the certificate by the issuer.
encryptedDigest consists of a DER encoding of the SHA-1 digest of the ediContent, RSA encrypted with the sending party's private key. The RSA Digital Signature Algorithm converts the INTEGER data type to an OCTET STRING.

Notes pertaining to EDI with Non-Repudiation:

signedDigestAlgorithms—single member only in the set (AlgorithmIdentifier for the message digest algorithm). Must be the same as signedDigestAlgorithm within signerInfos.

signerInfos—single member only in the set. The private/public key pair associated with the digital certificate used for authentication is to be used for signing/verifying.

issuerCountry—single member only in the SEQUENCE OF and SET OF.

issuerOrg—single member only in the SEQUENCE OF and SET OF.

7.2.2.4 ASN.1 Syntax for IA Status. The IA Status Message content has ASN.1 type IaStatusMessage:

```
IaStatusMessage ::= SEQUENCE {
  iaStatusMessage OBJECT IDENTIFIER,
  iaStatus [0] EXPLICIT BIT STRING ( SIZE (32) )
  }
```

IA Status messages reference the following object identifier:

iaStatusMessage OBJECT IDENTIFIER ::= { eciaAscX12Edi 97 }

iaStatus consists of an IA formatted status message.

7.2.2.5 ASN.1 Syntax for Optional IA Receipts. The IA Receipt Message content has ASN.1 type IaReceiptMessage:

```
IaReceiptMessage ::= SEQUENCE {

 iaReceiptMessage OBJECT IDENTIFIER,

  receiptContent [0] EXPLICIT SEQUENCE {

    isaSegment OCTET STRING,

    dateTimeStamp PrintableString ( SIZE(15) ),

    enhancements Enhancements OPTIONAL

    }

  }

Enhancements :: = CHOICE {

 withDigest [1] WithDigest,

 withDigSig [2] WithDigSig

 }

WithDigest ::= [1] EXPLICIT SEQUENCE {

 receiptDigestAlgorithm OBJECT IDENTIFIER,

 receiptMessageDigest OCTET STRING

 }
```

```
WithDigSig ::= [2] EXPLICIT SEQUENCE {
 receiptSignatureAlgorithm OBJECT IDENTIFIER,
 receiptDigitalSignature OCTET STRING
 }
```

IA Receipt messages reference the following object identifiers:

```
iaReceiptMessage OBJECT IDENTIFIER ::= { eciaAscX12Edi 65 }
receiptDigestAlgorithm OBJECT IDENTIFIER ::= {
      iso(1) org(3) oiw(14) secsig(3) algorithm(2) 26 }          —SHA-1
receiptSignatureAlgorithm OBJECT IDENTIFIER ::= {
         iso(1) member-body(2) us(840) rsadsi(113549) pkcs(1) 1 5 }
```

isaSegment consists of the first 105 octets of an EDI data message.

dateTimeStamp are formatted as follows: *CCYYMMDDhhmmssz*
where: *CC* = century, *YY* = year, *MM* = month, *DD* = day, *hh* = hour, *mm* = minute, *ss* = second, *z* = alpha time zone indicator. A blank character indicates *local observed time.* The date and time in the time stamp should be the time that the receiving party received the complete message being receipted. The use of *coordinated universal time* (UTC) in the receipt is recommended to avoid any time zone ambiguities.

receiptMessageDigest is the digest of the message being receipted. If the message is of the EDI with Message Integrity format, this digest is the one received with the message and verified by the recipient. If the original message is of any other data type, this digest is calculated by the recipient using the EDI message data as input to the specified message digest algorithm.

receiptDigitalSignature is computed by formatting a concatenation of the ISA segment of the EDI message (105 octets), the dateTimeStamp of the receipt (15 octets), and the Digital Signature (96 octets, assuming a 768-bit key) received with the original message. This 216-octet structure is then *signed* by encrypting it in accordance with the RSA digital signature algorithm using SHA-1 as the digest algorithm and the receipt generating party's private key. The party receiving the receipt should construct a concatenation of the ISA segment of 105 octets (from either the transmitted message or the body of the receipt), the dateTimeStamp of 15 octets (from the body of the receipt), and the digital signature from the **original message** (96 octets if a 768-bit key was used). This concatenated structure, together with the digital signature received in the receipt, should be passed to the digital signature verification algorithm. The receipting party's public key is used to verify the signature. This receipt type requires that the original message being receipted for must have been of the type EDI with Non-Repudiation.

7.2.3 Client Specifications

Upon receipt of data from the EDI translator, the Interactive Agent client performs the following functions:

1. **Determine IP Destination Address.**

Based on the trading partner identity and any other relevant information, the corresponding IP address and port number is determined for routing the data to the appropriate server.

2. **Connect to a Server.**
   a. **Create SSL Context (Memory Allocation).**
   An opaque data structure known as a *Context* is used to store the cryptographic data associated with the individual connection, certificates, references to callbacks, and various other data. An independent context must be maintained for each connection.
   b. **Initialize SSL Context.**
   This action initializes the internal values of the SSL Context structure and must be called before any of the other *SSL Set* functions.
   c. **Read and Add Certificates from Local Library.**
   This function adds to the chain of certificates used when authenticating an SSL peer. A variation of this function adds *Trusted Certificates* to complete certificate chains.
   d. **Read and Add Private Key from Local Source.**
   Specify the private key used for signing and nonexport key exchange. This key must match the installed public key.
   e. **Set SSL Protocol Version.**
   This parameter specifies what versions of the SSL protocol can be used for the connection. This will normally be set to SSL Version 3 Only.
   f. **Set SSL Protocol Side.**
   This SSL instance is set to be the client side of the connection. Clients may only connect to servers.
   g. **Open Socket.**
   A client application creates a socket and then issues a connect to a service specified in a sockaddr_in structure.
   The result returned is a file descriptor that is connected to a server process, a communications channel on which one can conduct an application specific protocol.
   h. **Set SSL I/O Reference.**
   This function specifies the reference parameter passed by the library to the I/O call back functions. It configures the SSL context structure for later use.
   i. **Set SSL Peer ID.**
   This function uniquely identifies the SSL peer associated with this SSL connection. It is required in those instances where session resumption capabilities are specified.
   j. **SSL Handshake.**
      - The client performs the *Client Hello* operation, which consists of sending the *challenge,* session ID (if any), and its cipher specifications.
      - After receiving the *Server Hello,* the client performs the *Send Client Master Key* operation. (This step consists of sending the client's master key encrypted by the server's public key.) Note that this step is performed only if this is an initial session and is not done for a *resumed session.*
      - The client then performs the *Client Finish* operation. (This consists of transmitting the connection ID encrypted by the client write key); or
      - The server responds with the *Server Verify* function followed by a request for the client's certificate.
      - The client responds with its certificate data (encrypted by the client write key). The server then executes the *Server Finish* function, and the connection is ready to pass data.

3. **Send Data to Server.**
   After the connection to the server is established, the client DER encodes the message and transmits it to the server as a data stream. This process is accomplished using the *SSLWrite ()* function.
4. **Log Transmissions.**
   Log entries are created in response to transmission of data to the server. The minimal set of data to be logged consists of the following:
   - Date/time
   - ISA segment
   - Remote IP address and port number
   - Success/fail indicator with interfaces
5. **Wait for Server Disconnect.**
   This process provides for the determination of a successful receipt of the data stream by the destination server. The client performs the *Select Socket w/Timeout,* which allows the server to gracefully initiate the disconnect. The *Timeout* ensures complete message transfer to the server—followed by a proper closing sequence. Current vendor and toolkit implementations recommend a default value for the timeout of two (2) minutes.
6. **Client Disconnect.**
   If the server has initiated its disconnect sequence before the end of the timeout, the client performs the following steps:
   - Execute SSL CLOSE.
     This function closes the SSL session with the peer. No further communications are expected. If the server has initiated its disconnect sequence before the end of the timeout, the client will perform the following steps:
   - Execute Socket CLOSE.
     The client closes the socket connection with the *close()* function, passing the socket descriptor as a parameter.
   - Execute SSL Context DELETE.
     This function typically releases the resources utilized by the SSL connection. Note that this is performed only if this session is to be terminated and is not done for sessions that may later be resumed. Depending on implementation, the memory used by the context structure may have to be deallocated after the Context Delete.

If the server has not initiated its disconnect sequence and the end of the timeout has occurred, the client will log the error condition and perform the following steps:

- Execute SSL CLOSE.
- Execute Socket CLOSE.
- Execute SSL Context DELETE.

7.2.4 Server Specifications

This section defines the processing associated with the Interactive Agent server process. Upon startup of the server process, the server performs the following:

Initialize server.
To accept connections, a socket is created and bound to a service port; a queue for incoming connections is specified, and then the connections are accepted. When the

process and the port are bound, the server listens to the port for connection requests. After a connection request is received, the socket is created. When using a multiprocessing scheme, connections are made and the process forks a child process to handle that service request. The parent process continues to listen for and accept further service requests.

Accept connection from client.
Based on the identity of the trading partner and other relevant information, which could include the remote IP address and local port number, routing is determined for the data to the appropriate EDI translator.

Message read setup.
The following steps, summarized below, comprise this operation.

- Create SSL Context (Memory Allocation).
- Initialize SSL Context.
- Bind SSL Context to the socket.
- Read and Add Certificates from Local Library.
- Read and Add Private Key from Local Source.
- Set SSL Protocol Version (e.g., 3).
- Set SSL protocol side to = SERVER.
- Set SSL I/O Reference (required if sessions are to be resumed).
- Set SSL Peer ID (required if sessions are to be resumed).
- SSL Handshake (Exchange certificates and keys).

When an inbound connection request is received, the server listening process passes the connection request to the connection handling software. This software performs the following:

Create SSL context.
An opaque data structure known as a **Context** is used to store the cryptographic data associated with the individual connection, certificates, references to callbacks, and various other data. An independent context must be maintained for each connection.

Initialize SSL context.
This action initializes the internal values of the SSL context structure and must be called before any of the other *SSL Set functions.*

Read and add certificates from local library.
This function adds to the chain of certificates used when authenticating an SSL peer. A variation of this function adds *Trusted Certificates* to complete certificate chains.

Read and Add Private Key from Local Source.
This function specifies the private key used for signing and nonexport key exchange. This key must match the installed public key.

Set SSL Protocol Version.
This parameter specifies what versions of the SSL protocol can be used for the connection. This will normally be set to SSL Version 3 Only.

Set SSL Protocol Side.
This SSL instance is set to be the server side of the connection. Servers may only accept connections from clients.

Set SSL I/O Reference.
The function specifies the reference parameter passed by the library to the I/O call back functions. This function configures the SSL context structure for later use.

Set SSL Peer ID.
This function uniquely identifies the SSL peer associated with this SSL connection.

This function is required in those instances where session resumption capabilities are specified.

SSL Handshake.

- After receiving the *Client Hello,* the server sends the *Server Hello* consisting of the *Connection ID* and either *Session ID Hit* or the server certificate and cipher specs (depending on whether this is a resumed session).
- The server then waits for the *Client Finish* message to be received. The server responds with the *Server Verify* consisting of a challenge encrypted with the server write key.
- The server then *Requests Certificate* from the client by sending an *Auth Type Challenge* encrypted by the server write key.
- The client responds with its client certificate.
- The server sends the *Server Finish* message consisting of the *Session ID* encrypted by the server write key. The connection is now ready to pass data.

7.2.4.1 SSL Read Processing. This process requires that an initial SSL3 *read* of two octets be performed. The first octet contains the DER representation of an ASN.1 tag identifying the *type* of this IA message. The second octet will either be a DER short-form definite length or the first octet of a DER long-form definite length, enumerating the number of octets contained in the remainder of the length field.

If the second octet is a short-form DER length, it represents the length of the remainder of the current IA message. If the second octet is a long-form DER length, a second *read* of the indicated number of octets is required to determine the remaining number of octets comprising the long-form definite length field. These remaining octets enumerate the total length of the remainder of the IA message.

A final *read* should then be initiated to *read* all of the octets comprising the remainder of the IA message. This third read may be done either as a single *read* operation or as a series of partial *reads* whose sum equals the total value. All *reads* in this section are accomplished using the SSLRead () function.

7.2.4.2 Route Data to Translator. After the read is completed, the server will decode the message and transfer the EDI content to the translator.

7.2.4.3 Receipt Logging. Log entries are created in response to receipt of data by the server. The minimal set of data to be logged consists of the following:

- Date/time
- ISA segment
- Remote IP address and local port number
- Success/fail indicator with interface

7.2.4.4 Message Validation. In the event of a validation failure, internal security policies are to be followed as this specification does not provide a mechanism for negative acknowledgments.

7.2.4.4.1 MESSAGE INTEGRITY. After receiving a message of the type EDI with Message Integrity, the recipient should calculate a message digest of the EDI data portion of the message, using the algorithm specified in the message. The calculated message digest should then be compared with the message digest received with the message and an exact match verified. If they do match, the EDI data should be forwarded to the EDI translator, and the fact that a successful reception has occurred should be written to the log. If the digests do not match, some type of abnormal situation exists. The EDI data should be stored in a suspense file, and both digests should be saved for analysis. An error condition should be noted in the log.

7.2.4.4.2 NON-REPUDIATION. After receiving a message of the type EDI with Non-Repudiation, the recipient should calculate a message digest of the EDI data portion of the message, using the algorithm specified in the message.

Next, the recipient should decrypt the message digest received from the sender, utilizing the decryption algorithm and the key referenced in the message. The calculated message digest should then be compared with the message digest received with the message and an exact match verified. If they do match, the EDI data should be forwarded to the EDI translator and the fact that a successful reception has occurred written to the log. If the digests do not match, some type of abnormal situation exists. The EDI data should be stored in a suspense file, and both digests should be saved for analysis. An error condition should be noted in the log.

7.2.4.5 Server Disconnect. The server executes the following procedure after completing the SSL read processing:

- SSL3 CLOSE
  This function closes the SSL3 session with the peer. No further communications are expected.
- Socket CLOSE
  The server closes the socket connection with the *close()* function, passing the socket descriptor as a parameter.
- SSL Context DELETE
  This function typically releases the resources utilized by the SSL3 connection. This is performed only if this session is to be terminated, and it is not done for sessions that may later be resumed. Depending on implementation, the memory used by the context structure may have to be deallocated after the *context delete.*

7.2.4.6 Example of Parsing the Received Message. This section presents an illustrative example of how incoming messages might be parsed by an Interactive Agent. The English text describing a field with known value in this example is followed by the hexadecimal representation of that value.

The initial section of all IA messages identifies the message type via a unique Object Identifier.

```
read first byte (OID tag)          : 06

read second byte (OID length) (L)  : 09

read (L) bytes (OID VALUE):         .. .. .. .. .. .. .. .. ..

look up OID to get message type

 if Basic go to step 10.

 if MessageIntegrity go to step 20.

 if NonRepudiation go to step 30.

 if IAstatus go to step 40.
```

```
if IAreceipt go to step 50.

if none of the above, error = "Unknown message type"

Log event, save message, exit
```

7.2.4.6.1 Basic EDI Message

```
10. read next tag (EXPLICIT):      a0

    read next length:              82 .. ..

    read next tag (OCTETSTRING):   04

    read next length (LL):         82 .. ..

    read next (LL) bytes:          .. .. .. .. .. .. .. .. .. .. ..
    (EDI Message)

11. Log message received

    check if receipting enabled,

       if enabled, format receipt and pass to output routine

    pass EDI data to translator

    Log if successful/unsuccessful

    exit
```

7.2.4.6.2 Message Integrity

```
20. read next tag (EXPLICIT):      a0

    read next length:              82 .. ..

    read next tag (SEQUENCE):      30

    read next length:              82 .. ..

    read next tag (VERSION):       02

    read next length: (LL)         01

    read (LL) bytes VERSION #:     00

21. read next tag (SEQUENCE):      30

    read next length:              09

    read next tag (OID):           06

    read next length (LL):         05

    read (LL) bytes OID:           2a 0e 03 02 1a   (Digest Algorithm = SHA1)

    read next tag (NULL):          05

    read next length:              00
```

```
22. read next tag (SEQUENCE):     30
    read next length:             82 .. ..
    read next tag (OID):          06
    read next length (LL):        09
    read (LL) bytes OID:          2a 86 48 86 f7 0d 01 07 01  (Plain Data)
    read next tag (EXPLICIT):     a0
    read next length (LL):        82 .. ..
    read next tag (OCTETSTRING):  04
    read next length (LL):        82 .. ..
    read next (LL) bytes:         .. .. .. .. .. ..      (EDI Message Content)
23. read next tag(OCTETSTRING):   04
    read next length (LL):        14
    read (LL) bytes:              .. .. .. .. ..         (SHA1 Message Digest)
24. Calculate Local Message Digest
    using EDI message received in step 22 and algorithm
    specified in step 21
25. Compare Remote (Step 23) and Local (Step 24) Message Digests
    if digests match:
      LOG message acceptance
      save received message digest
      check if receipting enabled,
      if enabled, format receipt and pass to output routine
      pass EDI data to translator
      Log if successful/unsuccessful
    exit
    if digests do NOT match:
       Log message failure
       save entire message for manual review
       do NOT pass EDI to translator
       do NOT send receipt
       exit
```

7.2.4.6.3 Non-Repudiation

```
30. read next tag (EXPLICIT):     a0

    read next length:             82 .. ..

    read next tag (SEQUENCE):     30

    read next length:             82 .. ..

    read next tag (VERSION):      02

    read next length: (LL)        01

    read (LL) bytes VERSION #:    00

31. read next tag (SET):          31

    read next length:             0b

    read next tag (SEQUENCE):     30

    read next length:             09

    read next tag (OID):          06

    read next length (LL):        05

    read (LL) bytes OID:          2b 0e 03 02 1a      (Digest Algorithm = SHA1)

    read next tag (NULL):         05

    read next length:             00

32. read next tag (SEQUENCE):     30

    read next length:             82 .. ..

    read next tag (OID):          06

    read next length (LL):        09

    read (LL) bytes OID:          2a 86 48 86 f7 0d 01 07 01     (Plain Data)

    read next tag (EXPLICIT):     a0

    read next length (LL):        82 .. ..

    read next tag (OCTETSTRING):  04

    read next length (LL):        82 .. ..

    read next (LL) bytes:         .. .. .. .. .. .. .. .. .. .. .. ..
    (EDI Message Content)

33. read next tag (SET):          31

    read next length:             81 ..

    read next tag (SEQUENCE):     30
```

```
read next length:             81 ..
read next tag (INTEGER):      02
read next length (LL)         01
read (LL) bytes VERSION #:    00
read next tag (SEQUENCE):     30
read next length:             34
read next tag:                30
read next length:             2c
read next tag (SET):          31
read next length:             0b
read next tag (SEQUENCE):     30
read next length:             09
read next tag (OID):          06
read next length (LL):        03
read next (LL) bytes:         55 04 06                (OID for Country)
read next tag(PRINTSTRING):   13
read next length:             02
read next (LL) bytes:         55 53                   (Country)
read next tag (SET):          31
read next length:             1d
read next tag (SEQUENCE):     30
read next length:             1b
read next tag (OID):          06
read next length (LL):        03
read next (LL) bytes:         55 04 0a                (OID for Name)
read next tag(PRINTSTRING):   13
read next length:             14
read next (LL) bytes:         45 78 .. .. .. 69 6f 6e       (Issuer's Name)
read next tag (INTEGER):      02
read next length (LL):        04
read next (LL) bytes:         .. .. .. ..          (Serial Number)
```

```
34. read next tag (SEQUENCE):    30

    read next length:            09

    read next tag (OID):         06

    read next length (LL):       05

    read next (LL) bytes         2b 0e 03 02 1a  (OID of Hash Algorithm =
    SHA1)

    read next tag (NULL):        05

    read next length:            00

35. read next tag (SEQUENCE):    30

    read next length:            0d

    read next tag (OID):         06

    read next length (LL):       09

    read next (LL) bytes:        2a 86 48 86 f7 0d 01 01 01  (OID of Encrypted
    Algorithm = RSA)

    read next tag (NULL):        05

    read next length:            00

36. read next tag (OCTETSTRING): 04

    read next length (LL):       60

    read (LL) bytes:             .. .. .. .. .. .. .. .. .. (Message Digest
    signed w/ Private Key)

37. Calculate Local Message Digest

    using EDI message received in step 32 and algorithm

    specified in step 34

38. Decrypt Remote Message Digest from step 36

    using senders public key in accordance with encryption algorithm in step
    35

39. Compare Local (Step 37) and Remote (Step 38) Message Digests:

    if digests match:

    Log message acceptance

    save received message digest

    pass EDI data to translator

     log if successful
```

```
      exit if unsuccessful

     check if receipting enabled, if not—exit

      if enabled, format receipt and pass to output routine and exit

   if digests do NOT match:

     Log message failure

     save entire message for manual review

     do NOT pass EDI to translator, do NOT send receipt; exit
```

7.2.4.6.4 IAStatus

```
40. read next tag (EXPLICIT):     a0

    read next length:             82 .. ..

    read next tag(BIT STRING):    03

    read next length (LL):        05

    read next (LL) bytes:         .. .. .. .. ..    (Status Code)

41. Log message received

    pass EDI data to translator

    log if successful / unsuccessful

    exit
```

7.2.4.6.5 IAReceipt

```
50. read next tag (EXPLICIT):     a0

    read next length:             81 89

    read next tag (SEQUENCE):     30

    read next length:             81 86

51. read next tag (SEQUENCE):     04

    read next length:             69

    read (LL) bytes ISA:          .. .. .. .. .. .. .. .. .. .. ..
    (105 octet ISA segment)

    read next tag (OID):          17

    read next length (LL):        0f

    read (LL) bytes OID:          .. .. .. .. .. .. .. .. .. .. ..
    (15 Octet UTCtime)
```

```
52. read next tag (OPTIONAL):     a1 or a2          (This section is OPTIONAL

    read next length:             1f or 6f            and may NOT always be

    read next tag (SEQUENCE):     30                present in all receipts.)

    read next length:             1d or 6d

    read next tag (OID):          06

    read next length (LL):        05 or 09

    read (LL) bytes OID:          .. .. .. .. .. (.. .. .. ..)      (Digest or
    Signature Algorithm)

    read next tag (OCTETSTRING):  04

    read next length (LL):        14 or 60

    read (LL) bytes:              .. .. .. .... .. .. .. .. .. (Message Digest
    or Digest Signature)

53. Process receipt according to receipt format in use.

54. Compare Remote (Step 53) and all pending Local Message Digests that have
a status of "Pending Receipt."

    if a match exists:

       Log receipt

       save receipt

         exit

    if digests do NOT match:

       Log receipt mismatch

       save receipt for manual review

       exit
```

7.2.5 Interfaces

This section identifies the external interfaces associated with the Interactive Agent client and server processes.

7.2.5.1 Data Communications Protocol. The Interactive Agent process sends and receives data using the SSL3 libraries. Interactions with the SSL3 libraries will be in the form of SSL3 toolkit functions.

7.2.5.2 EDI Translators. Interfaces to and from EDI translators may be in the form of files, APIs, message queues, and so on, and will pass X12 transactions (e.g., raw EDI, no additional headers, trailers, etc.).

7.2.6 Design Considerations

This section identifies the design considerations/constraints that will make the Interactive Agent efficient and robust.

7.2.6.1 Multiprocessing/Multithreading. Each IA server should provide up to a maximum of 16 concurrent connections per trading partner. This capability may be provided using technical approaches such as multiprocessing or multithreading, or any other comparable technology.

7.2.6.2 Non-Persistent Connections. Each SSL3 connection should support the transfer of a single EDI message. This means that a session should exist only for the duration of transmission of a single EDI message. This capability should avoid the potential of orphaned processes or sockets which may only be cleared by either an interruption in the client/server processes or a restart of the computing platform. This capability is a deliberate tradeoff for reliability over performance.

7.2.6.3 Resumable SSL3 Sessions. For performance reasons, resumable SSL3 sessions are recommended. Full SSL3 establishment is processor intensive; therefore, a mechanism needs to be provided that allows a session to be resumed using an optimal subset of the SSL3 setup processing. The length of time that a session will be maintained as "resumable" is an issue for the pairwise agreement. Implementations should provide for this parameter to be configurable. If sessions are to age out of the resumable session cache, this parameter should be configured to initiate these expirations after the indicated number of minutes has elapsed since the last use of the session ID. Recommended values are in the range of 1 to 30 minutes.

7.2.6.4 Connectivity. Connectivity is established in accordance with TCIF-98-009.

7.2.6.5 Message Priority. The management of priorities (normal and high) is expected to be achieved by TCP/IP port assignments.

7.2.7 Operational Concerns

7.2.7.1 Security. If security services are to be supported, they should be instituted in accordance with TCIF-98-009, which provides for a choice of cipher suites. The Interactive Agent should normally select *RSA_with_DES-CBC_with_SHA1* as its preferred suite. Companies may mutually agree to utilize the *RSA_with_NULL_with_SHA1* suite in those instances where message privacy is not desired.

7.2.7.2 Flow Control. The transport mechanism specified by the EDI/SSL3 Interactive Agent requires an individual SSL3 session to be set up for each EDI Message. The communication is essentially one way, from the client to the server. The IA status message is a mechanism that allows trading partners to exchange errors and other types of flow control information. The specific message codes are to be defined within the JIA. The server also has the option of refusing an offered SSL3 session setup if conditions on its side prevent prompt processing of an incoming message. In this situation, the client is expected to retry the connection after an agreed upon delay.

7.2.7.3 Logging. Logging is provided to support application troubleshooting as well providing an internal tracking mechanism for transactions into and out of the Interactive Agent. The logging process should consist of a standard API that provides (as a minimum) the following:

- Date/time
- ISA segment
- Remote IP address and port number
- Success/fail indicator with interfaces

7.2.7.3.1 Logging Levels. The logging function should have varying degrees of logging from minimal to full transaction logging and can be implemented via a script-based, tunable logging parameter. At a minimum, the data logged should include the data identified above. Other levels of logging may include fixed-length summaries of the record data through full transaction recording. Full transaction logging may have performance impacts and will likely only be used for testing, verification, and diagnostics.

7.2.7.3.2 Log Files. The actual logging mechanism is considered an implementation issue. However, a mechanism must be provided to give persistence to log entries.

As an example, log files may be text-based (for performance and ease of use). Each log entry can take the form of a single-line entry. Entries can be appended to the existing log file as they are created. Log files can be of a fixed length. The maximum number of files can be implemented via a script-based, tunable logging parameter. The log files can be circular (e.g., if log file last is filled, log file first is overwritten) to prevent any interruption of service due to log files being filled. It is the responsibility of the system administrator to rename and archive existing logs in accordance with each company's specific systems administration requirements and processes.

7.2.7.4 Routing. Based on the trading partner identity and other relevant information, the corresponding IP address and port number are determined to route the data to the appropriate Server. After the read is completed, the server decodes the message and transfers the EDI content to the translator.

7.2.7.5 Firewalls. The IA exists internal to the firewall. Corporate firewalls must be configured in such a manner that they are transparent to the connection between Interactive Agents.

7.2.7.6 Digital Certificates. The IA architecture requires that both parties exchange digital certificates during the SSL3 handshake. Upon receiving a certificate from a peer, the certificate information should be passed from the SSL3 transport layer to the Interactive Agent where it is saved and used for enhanced security operations such as non-repudiation. The same key pair/certificate used for authentication to the peer system is also utilized for any digital signing that is required.

Parties to a pairwise agreement should agree to mutually acceptable trusted certificate authorities. Only certificates issued by these trusted authorities should be exchanged or referred to in digital signatures.

7.2.8 Error Handling/Recovery

An Interactive Agent session may be abruptly terminated owing to technical failures. Recovery in this case consists of establishing a new session and resending the transaction in progress. However, there exists a well-defined point at which the client considers a transaction to be successfully sent. This point is when the final *SSL Write* returns with a success indication. After this point, recovery is at the end-to-end application level, which will be triggered by failure to receive an acknowledgment transaction (e.g., X12 997) within the timeout period.

Application level recovery is indicated by the timeout expiring while awaiting the matching acknowledgment transaction (X12 997—Functional Acknowledgment). Separate timeouts may be specified for *Normal* and *High Priority* transaction types for each trading partner.

7.2.9 Implementation Issues

This section identifies implementation issues that directly affect the Interactive Agent's ability to operate with other Interactive Agents built by other vendors and the ability to control port assignments. Also identified are trading partner responsibilities that may affect successful implementation.

7.2.9.1 Interoperability. Adherence to the specifications provided within this document should provide a basis for the independently developed Interactive Agent modules to interoperate successfully. This specification provides the minimum set of requirements to support interoperability. Deviations from these specifications degrade interoperability and may complicate future implementations.

7.2.9.2 Port Assignments. TCP/IP port assignments are to be agreed upon by the trading partners. These ports need not be the same at each end of the connection. The ports are associated with the TCP protocol stack within the system configuration.

The Internet Assigned Number Authority (IANA) has registered the protocol described in the IA specification and assigned it protocol number 117. The protocol is registered as Interactive Agent Transfer Protocol and is abbreviated as **iatp.**

The IANA has assigned port number 6999 to **iatp-normalpri.** Use of this port is recommended for normal priority transactions using this protocol.

The IANA has assigned port number *pending* to **iatp-highpri.** Use of this port is recommended for high-priority transactions using this protocol.

7.2.9.3 Partner Responsibilities. The trading partners are responsible for selecting the message types, level of security, IP address and port numbers, hashing, and encryption algorithms. In addition, *trading partners* must agree to change intervals for IA software changes, IP addresses, port assignments, test and certification criteria, schedules for EDI revision incorporation, certificate issuing authorities/class of certificates, and so on.

8

CORBA-Based TMN Security

Like EDI, CORBA (Common Object Request Broker Architecture) in the TMN is carried over TCP/IP. Therefore, like EDI security, CORBA security uses SSL3 in the transport layer, as well as a different module in the application in order to support non-repudiation.

The richness of CORBA raises new security concerns. The client's ORB (Object Request Broker), using CORBA's trader services, can ask a remote system to perform some activities on behalf of the client. If these activities involve sensitive material, then there is a need to control which server can be trusted for what kind of operations. Managing this kind of information can be cumbersome. In most TMN applications, for example, when an OS reconfigures a digital cross connect, such advanced distributed computing features are not really necessary. Therefore, CORBA security in the context of the TMN can be simplified by specifying that the CORBA trader services be disabled. Similarly, propagation of privileges is to be disabled.

CORBA security modules in the application layer communicate through the Secure Inter ORB Protocol (SECIOP). SECIOP provides for the exchange of a security token. However, SECIOP is invoked after fragmentation of GIOP (General Inter ORB Protocol) messages; therefore, it cannot be used for non-repudiation service. This service must be provided at a higher level, protecting whole PDUs. As carriers were preparing to deploy service order applications using CORBA, there were no commercially available interoperable ORBs that supported non-repudiation. It therefore became necessary to develop a prescription for application-based non-repudiation.

One of the lessons learned from ABS is not to alter any information model in order to accommodate security needs. Thus, there was early agreement that the CORBA solution should have no impact whatsoever on the normal exchange of CORBA messages. All non-repudiation items are to be carried in separate messages. Exchanging separate messages just for non-repudiation carries an obvious transmission penalty. Fortunately, this penalty is quite limited: non-repudiation evidence, consisting essentially of a digital signature, is rather short. Furthermore, its size is independent of the size of the message it protects.

Sending non-repudiation evidence (or proof) separately from the protected message imposes certain constraints on how the evidence is computed and how it is sent.

- **Correlation.** The most obvious requirement is to transmit with the non-repudiation evidence some information that would allow the receiver to correlate that evidence with the message it protects. For non-repudiation of a reply, it must further be possible to correlate the non-repudiation evidence with both the original request and the corresponding reply. Transmitting the digest of the protected message(s) along with the non-repudiation evidence provides that correlation.

- **Replay prevention.** An unscrupulous service provider could drum up some business by reusing non-repudiation evidence: having received a single request for service, such a provider could claim to have received several such requests, and produce the original non-repudiation evidence for each of the fictitious requests. To protect against this attack, the non-repudiation evidence must contain some unique string. In line with earlier security provisions, a time stamp is used to ensure uniqueness.
- **Misplay prevention.** An unscrupulous and (deservedly) unpopular service provider may capture non-repudiation evidence sent to a more deserving competitor and claim to have received the original requests for service. To protect against this attack, the non-repudiation evidence must contain the unique identity of the intended recipient.

The resulting approach, Telecom Non-Repudiation Inter-ORB Protocol (TeNoRIOP), is described in the following section.

8.1 OVERVIEW OF GENERAL INTER ORB PROTOCOL (GIOP)

In order to understand the TeNoRIOP, a basic understanding of GIOP, the message protocol between CORBA ORB domains, is helpful.

When a client initiates a CORBA operation, an application request message is sent from the client to the server. The application request message is a GIOP ***Request*** message that includes the object's key, the "in" and "inout" parameters of the operation, and the name of operation.

When a server responds to the operation or raises an exception to it, an application reply message is sent from the server to the client. The application reply message is a GIOP ***Reply*** message. If the server returns normally from the operation's implementation, the GIOP ***Reply*** message contains a status of NO_EXCEPTION, and the return values and any "inout" and "out" parameters of the interface. If the server raises an exception in the operation's implementation, the GIOP ***Reply*** message contains a status of USER_EXCEPTION and the exception information.

A client can cancel a request before it receives the corresponding reply by sending a cancel-request message. Non-repudiation of origin and of receipt of request and reply messages is discussed below. Non-repudiation of cancel-request messages follows the same principles.

The security mechanism described in the next section provides non-repudiation of origin and of receipt for these two application messages. Whether or not an ORB uses the GIOP ***Fragment*** message to fragment the ***Request*** or ***Reply*** messages is of no consequence to providing non-repudiation for these messages at the application layer.

8.2 TELECOM NON-REPUDIATION INTER-ORB PROTOCOL (TeNoRIOP)

TeNoRIOP is a vendor-independent security mechanism that provides non-repudiation for operations within a single ORB domain or between ORB domains. While in principle TeNoRIOP could provide peer authentication, confidentiality, and integrity, it focuses primarily on non-repudiation since the other security functions are supported by the CORBA/SSL3 Security that is currently available from ORB vendors.

TeNoRIOP provides the following non-repudiation services:

- Non-repudiation of request origin
- Non-repudiation of request receipt

- Non-repudiation of reply origin
- Non-repudiation of reply receipt

Non-repudiation (NR) is provided by the NREvidence object. This object has three types of interfaces:

1. **A local Application Programming Interface (API).** This is an intra-ORB interface between the local client object (in the client domain) or the application object (in the server domain) and the NREvidence object. In the client domain, the client object provides the NREvidence object with a copy of an outgoing request and instructs the NREvidence object to perform one or both of the following:
   - Generate NR of origin evidence and send it to the remote NREvidence object.
   - Wait for NR of receipt from the remote NREvidence object and verify the evidence when it arrives.

   In the server domain, the application object provides the local NREvidence object with a copy of a request received from the client and instructs the NREvidence object to perform one or both of the following:
   - Wait for NR of origin evidence and verify the evidence when it arrives.
   - Generate NR of receipt and send it to the remote NREvidence object.

   Although the preceding discussion addresses NR of requests, it also applies to NR (of origin and of receipt) of replies sent by the server to the client.
2. **An inter-ORB interface between two NREvidence objects in different domains.** It allows the NREvidence object on the server side to recieve NREvidence of origin for a request and of receipt for a reply, and it allows the NREvidence object on the client's side to receive NREvidence of origin for a reply and of receipt for a request.
3. **An administrative interface between the NREvidence object and a local, intra-ORB security administration object.** This interface allows the security administration object to instruct the NREvidence object which signature algorithm to use, which key to use for generating signatures, what to do with the messages and NREvidences it receives (e.g., log all received messages, match requests with NREvidence, . . .), and query any information maintained by the NREvidence object (e.g., obtain a list of all the requests for which the NREvidence has not yet arrived). This interface is not required to ensure interoperable NR service between the two ORB domains. It is not addressed here.

8.2.1 Non-Repudiation for Request

We start with the inter-ORB interface between two NREvidence objects in different domains, first providing a descriptive overview and then proceeding with the details. The subsequent section addresses the local API interface.

Figure 8.1 shows the entities and messages involved in providing non-repudiation of origin and receipt for an application request (GIOP Request message). NR for request is accomplished as follows:

1. The client invokes the operation on the application object that results in an application request message being sent to the server.
2. The client computes a message digest for the request. Using the message digest, the client then computes the digital signature to create the non-repudiation of origin evidence for the request.

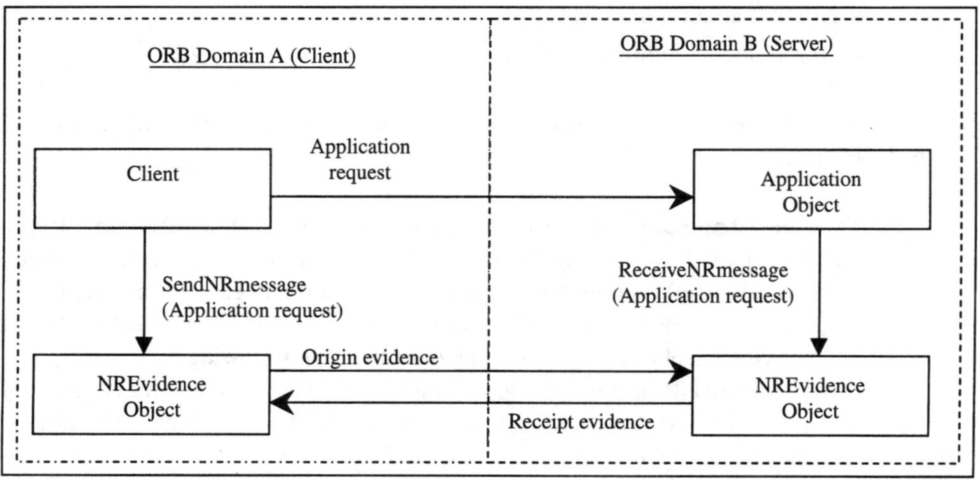

Figure 8.1 Non-Repudiation for request

3. The client sends the non-repudiation of origin evidence to the server by invoking the sendRequestOriginEvidence operation on an ***NREvidence*** instance in the server's domain.
4. The server, upon receipt of the application request, computes the message digest for the request. Using the digest, the server correlates this request with the non-repudiation of origin evidence received by the NREvidence object.
5. The server decrypts the non-repudiation evidence with the client's public key and compares the resulting digest with the locally computed digest. If the two digests match, the request is protected for non-repudiation of origin. If the two digests do not match, the behavior is determined by the security policy of the receiver of the non-repudiation evidence. It could, for instance, notify the sender and continue; notify the sender and stop; ignore the discrepancy; and so on.
6. The server then generates a digital signature to create the non-repudiation of receipt evidence for the request (if non-repudiation of receipt is required).
7. The server sends the non-repudiation of receipt evidence to the server by invoking the ***sendRequestReceiptEvidence*** operation on an ***NREvidence*** instance in the client's domain.
8. The client, upon receiving the receipt evidence, correlates it with the application request using the digest of the request.
9. The client decrypts the non-repudiation evidence with the server's public key and compares the resulting digest with the locally computed digest. If the two digests match, the request is protected for non-repudiation of receipt. If the two digests do not match, the behavior is determined by the security policy of the receiver of the non-repudiation evidence. For instance, it could notify the sender and continue; notify the sender and stop; ignore the discrepancy; and so on.

The following sections describe in detail how the request message digest and evidence for non-repudiation of origin and receipt are generated.

8.2.1.1 Digest for Request. To generate a message digest for a request, the information in a request needs to be converted into a sequence of octets. The request information used to create the secure hash is as follows:

IOR: The Interoperable Object Reference (IOR). The IOR uniquely identifies an object instance and the instance's interface type. The interface type and its semantics should be included in a document agreed upon by the communicating parties (e.g., an ANSI T1 standard or an ECIC GIG).

operation name: The name of the operation being invoked.

in/inout parameters: Parameters of the operation with "in" or "inout" keywords.

message originator: The identity of the initiator of the request. The name should be specified in the Joint Implementation Agreement (JIA) and be chosen carefully to ensure that the identity is unambiguous (e.g., a distinguished name as defined in X.500).

message recipient: The identity of the intended recipient of the message. The name should be specified in the JIA and be chosen carefully to ensure that the identity is unambiguous (e.g., a distinguished name as defined in X.500).

Each item is encoded into a sequence of octets using the following rules:

IOR: The IOR is converted into an ASCII string using the CORBA::object_to_string() operation. The resulting ASCII string consists of ASCII characters 0–9 and a–f, representing hexadecimal digits. It is converted in CDR using the following rule: Starting with each pair of ASCII characters after the "IOR:" prefix, the first character is encoded into the four high bits of an octet, and the second character is encoded into the four low bits.

operation name: The name of the operation is placed in a CORBA string and encoded into a sequence of octets using CDR. The operation name is the IDL-scoped name (i.e., <module name>::<interface name>::<operation name>, for example, **SampleModule::SampleInterface::op**).

in/inout parameters: Each parameter with the **in** or **inout** keyword in the operation is encoded into a sequence of octets using CDR. The resulting octets are then concatenated in the order in which the parameters are specified in the operation's IDL definition, from left to right.

message originator: The identity of the message originator is placed in a CORBA string and encoded into a sequence of octets using CDR.

message recipient: The identity of the message recipient is placed in a CORBA string and encoded into a sequence of octets using CDR.

The octets for each of the items above are concatenated together in the order in which they appear above. For example, given the following IDL:

```
module SampleModule {
interface Sample {
    exception SampleException {
      string   info;
    };
    boolean sampleOp(in string   param1,
             inout string param2,
             out string  param3
```

```
            )
            raises(SampleException);
        };
    };
```

The resulting sequence of octets would look like this:

IOR	"SampleModule::Sample:: sampleOp"	param1	param2	message originator	message recipient

The SHA1 hashing algorithm is then applied to this sequence of octets, which results in the request's message digest. The object identifier for the SHA1 hashing algorithm is "1:3:14:3:2:26."

8.2.1.2 Digital Signature for Request. The message digest for the request is combined with the time when the digital signature is created. The time is encoded into a sequence of octets using the following rule:

> **time stamp:** The format of the time stamp is the same as the ASN.1 GeneralizedTime type. A typical example is "1998011715322020.100 + 0000." The time stamp is placed in a CORBA string and encoded into a sequence of octets using CDR.

The time stamp is combined with the request digest by appending the CDR-encoded time stamp to the request digest as follows:

Request Message Digest	Time Stamp

The SHA1 algorithm is then applied to the resulting octets. The output of the SHA-1 algorithm is then encrypted using RSA public key encryption to create a digital signature. For non-repudiation of origin, the client's private key is used. For non-repudiation of receipt, the server's private key is used. The object identifier for the RSA digital signature is "iso(1):member-body(2):US(840):rsadsi(113549)."

8.2.1.3 Strict Correlation. A client may generate several identical requests to the same server; that is, the IOR, operation name, and parameters all have the same values. This may happen, for example, if the association is used for ordering communication facilities and the client issues several orders for identical transport circuits between the same end points. In this case, the digests for all those requests will be identical and the receiver will not be able to associate any of the corresponding NREvidences with any specific request. (Notice, however, that the server cannot claim to have received more requests that were actually sent, since each NREvidence includes a unique time stamp.) More generally: the server may receive n identical requests and $m < n$ NREvidences pertaining to those requests. The server can then prove only to have received m requests, though the server does not know which of the n identical requests are the ones that have been corroborated. In general, such lack of strict correlation is immaterial. However, if there are some business reasons that mandate a strict correlation, this may be provided by designing the interface to the referenced object in such a way that each request will always include a combination of parameters whose values are always unique. For instance, one of the parameters may be a unique time stamp or a unique sequence number.

8.2.2 Non-Repudiation for Reply

Figure 8.2 shows the entities and messages involved in providing non-repudiation of origin and receipt for an application reply (GIOP Reply message). The following steps describe how non-repudiation for reply is accomplished:

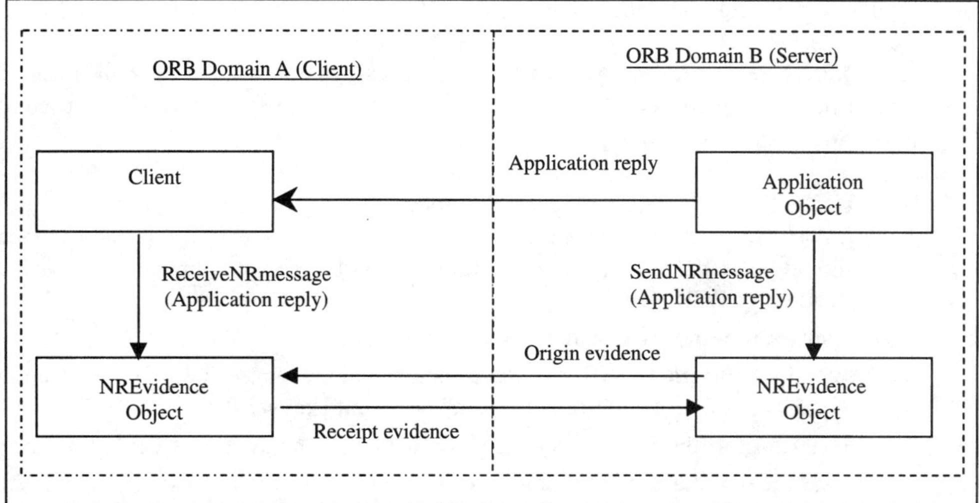

Figure 8.2 Non-Repudiation for reply

1. The server computes a message digest for the reply. Using the message digest for the reply and the message digest of the request, the server then computes the digital signature to create the non-repudiation of origin evidence for the reply.
2. The server sends the non-repudiation of origin evidence to the client by invoking the **sendReplyOriginEvidence** operation on a **NREvidence** instance in the client's domain.
3. The server sends the application reply to the client.
4. Upon receipt of the application reply, the client computes the message digest for the reply. Using the digest, the client correlates the reply with the non-repudiation of origin evidence received by the NREvidence object.
5. The client decrypts the non-repudiation evidence with the server's public key and compares the resulting digest with the locally computed digest. If the two digests match, the reply is protected for non-repudiation of origin. If the two digests do not match, the behavior is determined by the receiver of the non-repudiation evidence based on the local security policy. For instance, notify the sender and continue, notify the sender and stop, ignore the difference, and so on.
6. The client then generates a digital signature to create the non-repudiation of receipt evidence for the reply.
7. The client sends the non-repudiation of receipt evidence to the server by invoking the **sendReplyReceiptEvidence** operation on an **NREvidence** instance in the client's domain.

8. Upon receiving the receipt evidence, the server correlates it to the application reply using the digest of the reply.
9. The server decrypts the non-repudiation evidence with the client's public key and compares the resulting digest with the locally computed digest. If the two digests match, the reply is protected for non-repudiation of receipt. If the two digests do not match, the behavior is determined by the security policy by the receiver of the non-repudiation evidence. For instance, notify the sender and continue, notify the sender and stop, ignore the difference, and so on.

The following sections describe in detail how the reply message digest and evidence for non-repudiation of origin and receipt are generated.

8.2.2.1 Digest for Reply. To generate a message digest for a reply, the information in a reply needs to be converted into a sequence of octets. The reply information used to create the secure hash is as follows:

IOR: The Interoperable Object Reference (IOR). The IOR uniquely identifies an object instance and the instance's interface type. The interface type and its semantics should be included in a document agreed upon by the communicating parties (e.g., T1 standard or ECIC GIG).

operation name: The name of the operation being invoked.

out/inout parameters: If the Reply is a normal response (i.e., not an exception), the parameters of the operation with "out" or "inout" keywords.

exception: If the Reply is an exception response, the exception being raised.

message originator: The identity of the initiator of the request. The name should be specified in the JIA and be chosen carefully to ensure that the identity is unambiguous (e.g., a distinguished name as defined in X.500).

message recipient: The identity of the intended recipient of the message. The name should be specified in the JIA and be chosen carefully to ensure that the identity is unambiguous (e.g., a distinguished name as defined in X.500).

Each item is encoded into a sequence of octets using the following rules:

IOR: The IOR is converted into an ASCII string using the CORBA::object_to_string() operation. The resulting ASCII string consists of ASCII characters 0–9 and a–f, representing hexadecimal digits. It is converted in CDR using the following rule: Starting with each pair of ASCII characters after the "IOR:" prefix, the first character is encoded into the four high bits of an octet, and the second character is encoded into the four low bits.

operation name: The name of the operation is placed in a CORBA string and encoded into a sequence of octets using CDR. The operation name is the IDL scoped name (i.e., <module name>::<interface name>::<operation name>, for example, **SampleModule::SampleInterface::op**).

out/inout parameters: Each parameter with the **out** or **inout** keyword in the operation is encoded into a sequence of octets using CDR. The resulting octets are then concatenated in the order in which the parameters are specified in the operation's IDL definition, from left to right.

exception: The exception is encoded into a sequence of octets using CDR. (CDR encodes both the exception name and the exception data.)

message originator: The identity of the message originator is placed in a CORBA string and encoded into a sequence of octets using CDR.

message recipient: The identity of the message recipient is placed in a CORBA string and encoded into a sequence of octets using CDR.

The octets for each of the items above are concatenated together in the order in which they appear above. For example, given the following IDL:

```
module SampleModule {

 interface Sample {

   exception SampleException {

     string   info;

   };

   boolean sampleOp(in string    param1,

           inout string param2,

           out string  param3

   )

   raises(SampleException);

 };
};
```

The resulting sequence of octets for a normal response would look like this:

IOR	"SampleModule::Sample:: sampleOp"	param2	param3	message originator	message recipient

The resulting sequence of octets for an exception response would look like this:

IOR	"SampleModule::Sample:: sampleOp"	SampleException	message originator	message recipient

The SHA1 hashing algorithm is then applied to the sequence of octets that results in a message digest for the reply. The object identifier for the SHA1 hashing algorithm is "1:3:14:3:2:26."

8.2.2.2 Digital Signature for Reply. The message digest of the reply is combined with the message digest of the request and the time that the digital signature is created. The time is encoded into a sequence of octets using the following rule:

time stamp: The format of the time stamp is the same as the ASN.1 GeneralizedTime type. A typical example is "1998011715322020.100 + 0000." The time stamp is placed in a CORBA string and encoded into a sequence of octets using CDR.

The message digest of the request and the time stamp are combined with the message digest of the reply as follows:

Request Message Digest	Reply Message Digest	Time Stamp

The SHA1 algorithm is then applied to the resulting octets. Next, the output of the SHA1 algorithm is encrypted using RSA public key encryption to create a digital signature as specified in PKCS#7. For non-repudiation of origin, the server's private key is used. For non-repudiation of receipt, the client's private key is used. The object identifier for the RSA digital signature is "1:2:840:113549."

Only the reply message digest, time stamp, and digital signature are sent as evidence of NR of origin of the reply. The request message digest, which is used in the computation of the digital signature, is not sent as part of the NREvidence. Indeed, the reply message digest (which is sent) allows the recipient to correlate the NREvidence with the corresponding reply. The reply is correlated with its corresponding request through the usual CORBA mechanisms, independently of TeNoRIOP.

8.3 IDL SYNTAX FOR NON-REPUDIATION EVIDENCE

TeNoRIOP provides for the delivery of non-repudiation evidence through the interface of a specially defined object that is separate from the objects used for the application. This allows the application to be completely independent of the non-repudiation function. The interface for delivering non-repudiation evidence is the ***TeNoRIOP::NREvidence*** interface. This section details the interface's operations.

sendRequestOriginEvidence

```
void sendRequestOriginEvidence(
            in Digest_t                requestDigest,
            in GeneralizedTime_t       timeStamp,
            in DigitalSignature_t      signature
        )
        raises(DeliveryError);
```

This allows non-repudiation of origin evidence for an application request message to be sent from the application client to the application server. The NREvidence instance resides in the server's ORB domain.

Parameters

requestDigest	The message digest for the request message.
timeStamp	The time that the application client created the signature.
signature	The digital signature of the requestDigest and timeStamp created by the application client.

Return Values

None

Exceptions

DeliveryError	An error occurred when delivering the evidence.

sendRequestReceiptEvidence

```
void sendRequestReceiptEvidence(
        in Digest_t              requestDigest,
        in GeneralizedTime_t     timeStamp,
        in DigitalSignature_t    signature
    )
    raises(DeliveryError);
```

This allows non-repudiation of receipt evidence for an application request message to be sent from the application server to the application client. The NREvidence instance resides in the application client's ORB domain.

Parameters

requestDigest	The message digest for the request message.
timeStamp	The time that the application server created the signature.
signature	The digital signature of the requestDigest and timeStamp created by the application client.

Return Values

None

Exceptions

DeliveryError	An error occurred when delivering the evidence.

sendReplyOriginEvidence

```
void sendReplyOriginEvidence(
        in Digest_t              requestDigest,
        in GeneralizedTime_t     timeStamp,
        in DigitalSignature_t    signature
    )
    raises(DeliveryError);
```

This allows non-repudiation of origin evidence for an application reply message to be sent from the application server to the application client. The NREvidence instance resides in the application client's ORB domain.

Parameters

replyDigest	The message digest for the reply message.

timeStamp — The time that the application server created the signature.

signature — The digital signature of the replyDigest, requestDigest, and timeStamp created by the application server.

Return Values

None

Exceptions

DeliveryError — An error occurred when delivering the evidence.

sendReplyReceiptEvidence

```
void sendReplyReceiptEvidence(
            in Digest_t                replyDigest,
            in GeneralizedTime_t       timeStamp,
            in DigitalSignature_t      signature
        )
        raises(DeliveryError);
```

This allows non-repudiation of receipt evidence for an application reply message to be sent from the application client to the application server. The NREvidence instance resides in the application server's ORB domain.

Parameters

replyDigest — The message digest for the reply message.

timeStamp — The time that the application client created the signature.

signature — The digital signature of the replyDigest, requestDigest, and timeStamp created by the application client.

Return Values

None

Exceptions

DeliveryError — An error occurred when delivering the evidence.

If the application messages and non-repudiation messages are not transmitted over the same transport (e.g., the application messages use the IIOP/SSL3 transport while the non-repudiation messages use the IIOP/TCP transport), the following should be considered:

- The non-repudiation transport need not provide confidentiality since the messages do not reveal any information meaningful to an intruder.

- Integrity protection for the non-repudiation transport may provide some efficiency since the message can be determined to be invalid before doing the public key encryption. Integrity protection for the non-repudiation transport does not add any security value.

8.4 LOCAL API INTERFACE SPECIFICATION

This section defines the IDL interface between the client (on the client side) or application object (on the server side) and the local NREvidence object. In both cases messages flow only toward the NREvidence object, with no reply.

The interface is designed to be as simple as possible, foregoing richer functionality.

sendNRMessage

This call is used by the client (application) object on the client (server) side to provide the NREvidence object with a copy of an outgoing request (reply) and specify which features are required.

It can optionally specify the destination of the NREvidence message. Typically, this destination specification will be the IOR of the receiving object. This destination specification can be omitted if:

- The sending NREvidence object always sends the evidence to the same receiving NREvidence object.
- The sending NREvidence object can autonomously deduce the IOR of the receiving NREvidence object from the IOR, which is part of the request (reply) message contained in the input_buffer.

The IDL defined in this clause imports stringOp_t (an optional string used here to convey the destination) and the ProcessingFailureError exception from the Base IDL module defined in ANSI Standard T1.256.

```
void sendNRMessage (
       in sequence<octet>   input_buffer,
// either request or reply message
       in boolean    input_buffer_complete,
// TRUE if the buffer contains the end of the message
       in boolean    nr_origin,
// if TRUE, then generate and send NR of origin evidence
       in boolean    nr_receipt,
// if TRUE, then expect to receive NR of receipt evidence
       inBase::stringOP_t destination
       )
       raises (Base::ProcessingFailureError);
```

Parameters

input_buffer — The data sent from this ORB that requires NR service. If the data for which NR is required is larger than can conveniently fit into a single buffer, it is possible to issue multiple calls, passing a portion of the data in each call.

input_buffer_complete	This parameter has the value FALSE only if the message sent from this ORB that requires NR service is not yet complete.
nr_origin	This indicates that NR of origin evidence needs to be generated and sent to the object identified by destination
nr_receipt	This indicates that NR of receipt evidence is expected from the object identified by destination.
destination	This parameter identifies the remote NREvidence object. It is the destination of NR of origin evidence to be sent (if any) and the source of NR of receipt evidence to be received (if any).

Return Values

None

Exceptions

Base::ProcessingFailureError	An error occurred when delivering the call.

receiveNRMessage

This call is used by the client (application) object on the client (server) side to provide the NREvidence object with a copy of an incoming reply (request) and specify which features are required.

It can optionally specify the destination of the NR evidence message. Typically, this destination specification will be the IOR of the receiving object. This destination specification can be omitted if:

- The sending NREvidence object always sends the evidence to the same receiving NREvidence object.
- The sending NREvidence object can autonomously deduce the IOR of the receiving NREvidence object from the IOR which is part of the request (reply) message contained in the input_buffer.

The IDL defined in this clause imports stringOp_t (an optional string used here to convey the destination) and the ProcessingFailureError exception from the Base IDL module defined in ANSI Standard T1.256.

```
void receiveNRMessage (
       in sequence<octet>  input_buffer,

// either request or reply message
       in boolean   input_buffer_complete,
// TRUE if the buffer contains the end of the message
       in boolean   n-r_origin,
// if TRUE, then expect to receive NR of origin evidence
       in boolean   n-r_receipt,
// if TRUE, then generate and send NR of receipt evidence
       inBase::stringOP_t destination
       )
       raises (Base::ProcessingFailureError);
```

Parameters

input_buffer	The data sent from this ORB that requires NR service. If the data for which NR is required is larger than can conveniently fit into a single buffer, it is possible to issue multiple calls, passing a portion of the data in each call.
input_buffer_complete	This parameter has the value FALSE only if the message sent from this ORB that requires NR service is not yet complete.
nr_origin	This indicates that NR of origin evidence is expected from the object identified by destination.
nr_receipt	This indicates that NR of receipt evidence needs to be generated and sent to the object identified by destination.
destination	This parameter identifies the remote NREvidence object. It is the destination of NR of receipt evidence to be sent (if any) and the source of NR of origin evidence to be received (if any).

Return Values

None

Exceptions

Base::ProcessingFailureError	An error occurred when delivering the call.

8.5 TIMING OF NON-REPUDIATION EVIDENCE

This section addresses the timing constraints for providing non-repudiation evidence as well as the behavior of the communicating parties if those timing constraints are violated. This section also provides several default timing parameters. The communicating parties can agree by means outside the scope of this document (e.g., through a Joint Implementation Agreement or acceptance of Generic Implementation Guidelines) on any other values for the timing parameters.

8.5.1 Timing Agreements

8.5.1.1 Non-Repudiation of Origin. If non-repudiation of origin is expected, then upon sending a message (either a request or a response), the originator of that message will send non-repudiation evidence for that message at most one minute after sending the message.

Allowing for variability in network delay of up to two minutes, the recipient of the message will expect to receive the non-repudiation evidence at most three minutes after receiving the message.

8.5.1.2 Non-Repudiation of Receipt. If non-repudiation of receipt but not of origin is expected, then the recipient of the message will generate and send evidence of non-repudiation of receipt at most one minute after receiving the message. Allowing four minutes for one-way network delay, the originator of the message will expect to receive evidence of non-repudiation of receipt at most nine minutes after sending the original message. If non-repudiation of both origin and receipt are expected, then the recipient of the message will generate and send evidence of non-repudiation of receipt at most one minute after receiving proof of non-repudiation of origin for that message. Allowing four minutes for network delay, the originator of the message will expect to receive evidence of non-repudiation of receipt at most nine minutes after sending the proof of non-repudiation of the origin of original message.

8.5.1.3 Behavior While Waiting. The behavior of a communicating party while it is waiting for evidence of non-repudiation is a local matter that depends on local security policy. It is governed by the fact that non-receipt of evidence within the expected time frame may be due to any of several causes:

- The message (to be validated) was generated by an intruder and not by the valid remote entity.
- An intruder has intercepted the evidence of non-repudiation.
- The party responsible for providing the evidence of non-repudiation does not want to be accountable for the message and therefore refuses to provide evidence of non-repudiation.
- The evidence of non-repudiation got lost in transit.
- The system of the (valid) originator is temporarily down or overloaded.
- The evidence of repudiation is being delayed by network congestion.

The first bullet above is applicable only to non-repudiation of origin; the remaining bullets are applicable for non-repudiation of both origin and response.

Differences in local security policy may lead to different behaviors, as illustrated in the following examples.

8.5.1.3.1 TRUSTING BEHAVIOR. If a party has a high level of trust in the remote entity as well as in the security of the communication network, it can process and act upon every message it receives without waiting for evidence of non-repudiation. It may even issue evidence of non-repudiation of receipt without waiting for evidence of non-repudiation of origin. Meanwhile, it will log the time of receipt of each message and its corresponding evidence of non-repudiation. The log can be analyzed at regular time intervals (e.g., every hour or day). If expected evidence of non-repudiation is missing from the log, then the party responsible for issuing that evidence should be contacted promptly to verify that indeed it originated (or received) the message and that it will promptly supply the missing evidence. If all the expected evidence items are present, but some of them have been received with excessive delay, then more detailed security analysis may be called for.

8.5.1.3.2 CAUTIOUS BEHAVIOR. If a party has a fair level of trust in the remote entity as well as in the security of the communication network, it can process and act upon every message it receives without waiting for any evidence of non-repudiation. It may even issue evidence of non-repudiation of receipt without waiting for evidence of non-repudiation of origin. However, it will carefully time the arrival of expected evidence of non-repudiation. Every time the evidence is not received within the expected time window, it will promptly alert the remote entity and suspend any further actions stemming from the questionable message.

8.5.1.3.3 SUSPICIOUS BEHAVIOR. If a party has a low level of trust in the remote entity or in the security of the communication network, it will not process or act upon any message until it has received the expected evidence of non-repudiation. If the evidence has not been received within the expected time window, it will promptly notify the remote entity.

Furthermore, in such cases it may suspend all activities related to the remote entity even if other messages have been properly confirmed.

It is suggested that the two communicating parties reach a prior agreement on their respective behavior since such behavior will impact the interface's performance. Regardless of any such agreement, each party will reserve the right to dynamically switch to more cautious behavior whenever it suspects that it is under attack.

8.5.1.4 Notifications. When one of the communicating parties notices the absence or delay of non-repudiation evidence, and it believes this to be the result of mechanical difficulties (e.g., network congestion), it can notify its correspondent over the same connection. If the party that notices the absence or delay of non-repudiation evidence suspects deliberate, malicious action, it may choose to notify its correspondent using a different channel or undertake its own investigation before notifying its correspondent.

8.6 NON-REPUDIATION PROTOCOL MACHINE

The protocol machine for non-repudiation has two elements: one for the sender of a message and one for the receiver of a message. The non-repudiation protocol machine is the same for a request as for a response. It is based on the timing considerations in the previous section.

8.6.1 Message Sender

The protocol machine for the message sender is depicted in Figure 8.3. Notice that the

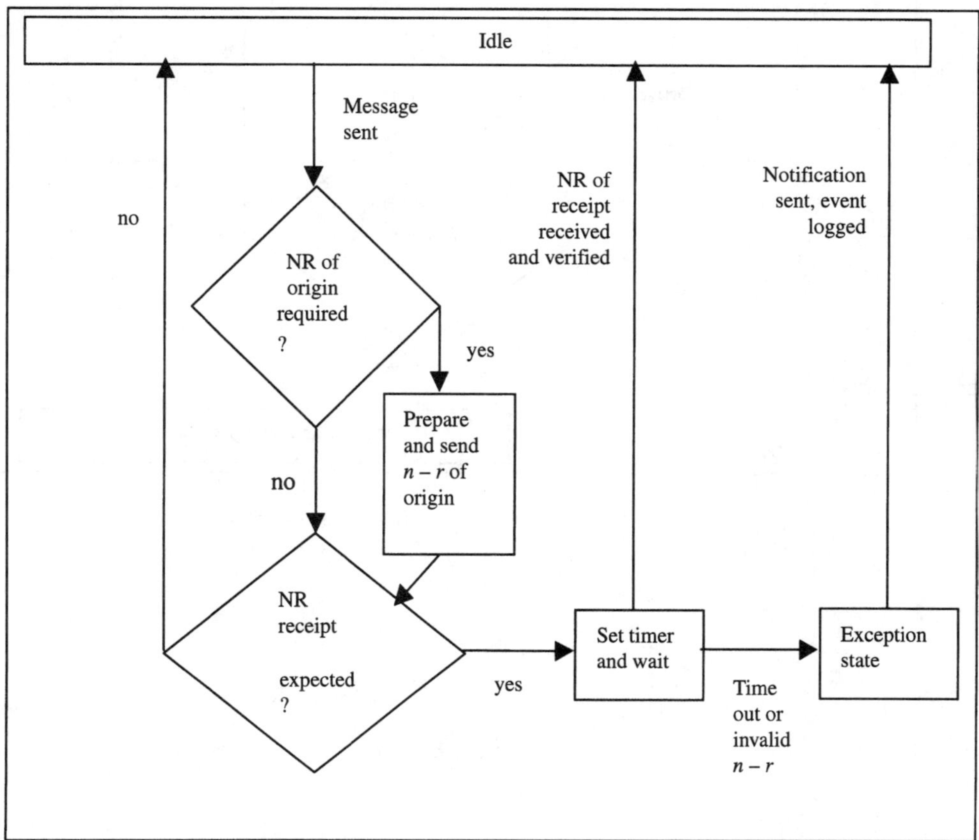

Figure 8.3 Message sender protocol machine

decision points (diamonds) do not represent states of the machine; rather, they are used as an expedient way of labeling the state transition arcs.

8.6.2 Message Receiver

The non-repudiation protocol machine of the message receiver is represented in Figure 8.4.

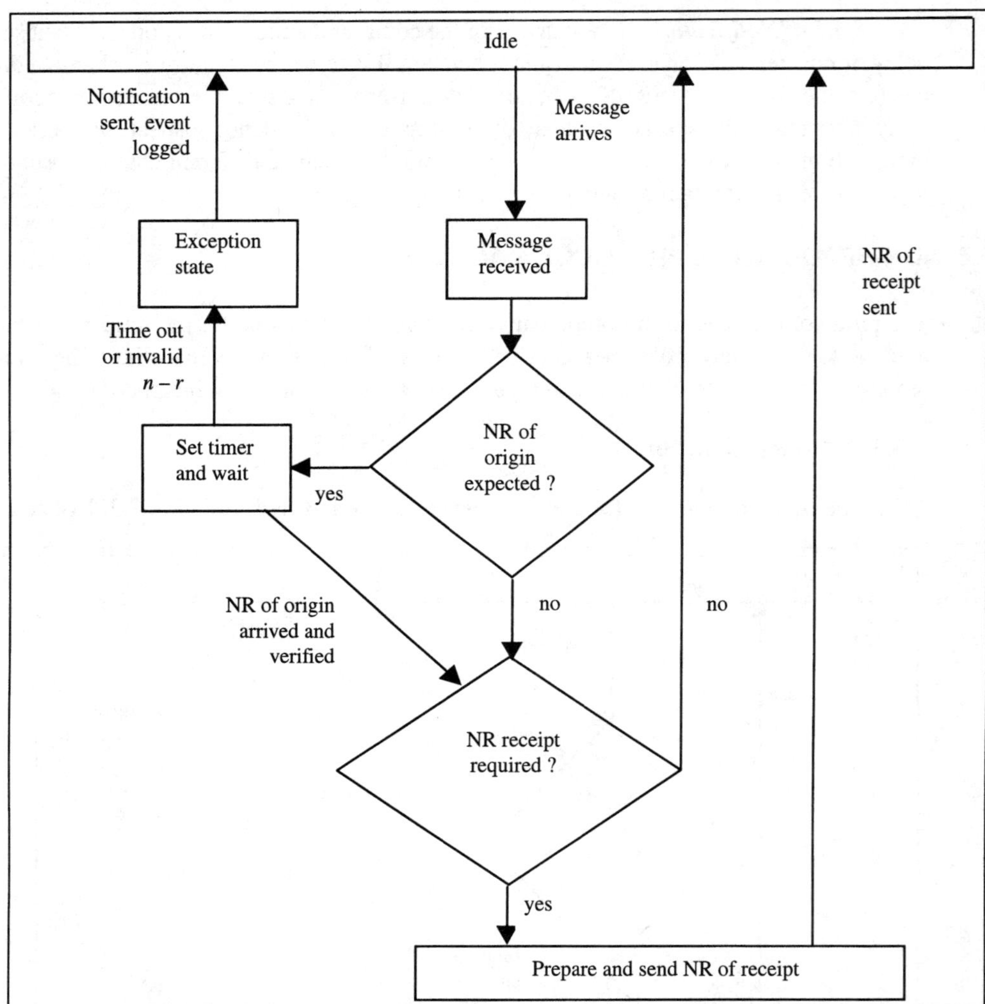

Figure 8.4 Message receiver protocol machine

9

SNMP-Based TMN Security

SNMP comes in three versions. Their security features are so different that they warrant separate discussions.

9.1 SNMPv1 SECURITY

Talking about SNMPv1 security is simple: there isn't any. There isn't even any way of identifying individual users for auditing purposes. Yet, there is a glimmer of hope. SNMP transactions could be protected with IPsec. This would require some manual initialization of IPsec security parameters within each entity managed through SNMPv1—a tedious but feasible task. IPsec's minimum protocols include DES for encryption and MD5 and SHA1 for hashing, in perfect harmony with (default) security specifications for non-SNMP TMN transactions.

IPsec protection for SNMPv1 can be somewhat simplified if several TMN entities share a common Security Gateway (SG), as illustrated in Figure 9.1.

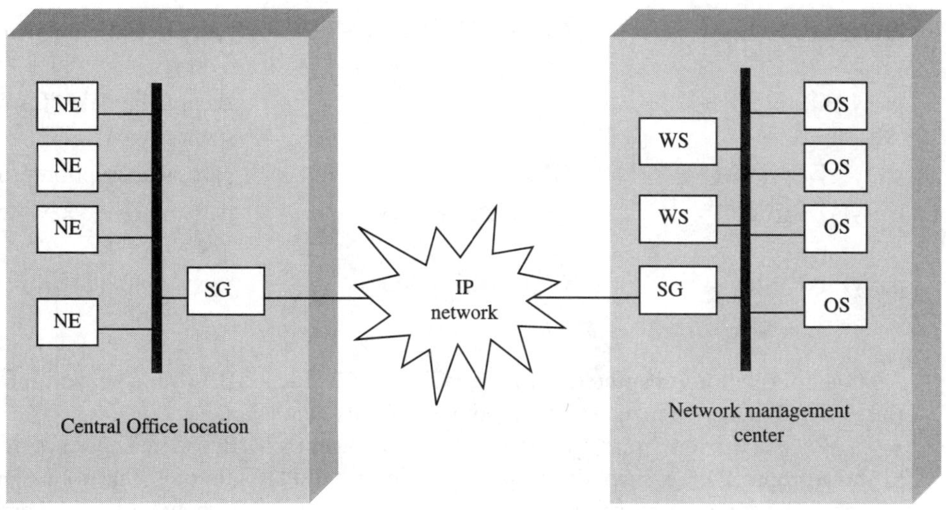

Figure 9.1 Secure Gateways (SG) protecting TMN entities

In Figure 9.1 an SG controls access to several Network Elements (NEs) in a Central Office building, while another SG controls access to several Operation Systems (OSs) and Work Stations (WSs). If each of these locations is secure, IPsec can be used to provide SG to SG (tunneling) security to protect all SNMP messages exchanged between entities inside the two locations. In most cases strong authentication will be sufficient. In such cases only the Authentication Header (AH) protocol of IPsec will be needed. The SG will act as a firewall that does its screening based on authenticated sender identification rather than on an unprotected (and easily spoofed) IP address.

IPsec does not offer non-repudiation service. However, SNMP is being used essentially for interfaces between NEs and element management systems. Such interfaces do not require non-repudiation protection.

9.2 SNMPv2 SECURITY

SNMPv2 messages can be protected with IPsec just like SNMPv1 exchanges. However, a more attractive alternative is to use the security features built into SNMPv2.

9.2.1 Proper ID Required

The most rudimentary element of security is mutual authentication of communicating entities. Even this basic ingredient is missing from SNMPv1, where several entities share the same "community" name. Version 2 endows each communicating entity with a unique **SnmpParty** persona. It is actually a lot more than just a unique ID; it is a SEQUENCE of the following elements:

partyIdentity	OBJECT IDENTIFIER	party's unique ID
partyTDomain	OBJECT IDENTIFIER	party's transport service
partyTAddr	OCTET STRING	party's transport address
partyMaxMessageSize	INTEGER	max acceptable message size
partyAuthProtocol	OBJECT IDENTIFIER	party's authentication protocol
partyAuthClock	INTEGER	party's local time
partyAuthPrivate	OCTET STRING	party's private authentication key
partyAuthPublic	OCTET STRING	party's public authentication key
partyAuthLifetime	INTEGER	acceptable delay for received messages
partyPrivProtocol	OBJECT IDENTIFIER	party's encryption protocol
partyPrivPrivate	OCTET STRING	party's private key for encryption
partyPrivPublic	OCTET STRING	party's public key for encryption

All this information is captured in the partyTable in the local MIB (Management Information Base). Thus, it can be managed by a remote (security) manager.

When a Version 2 manager tries to access an agent's MIB, the manager not only needs to show proper ID (i.e., include the partyIdentity in the PDU), but also has to declare its intentions. Those intentions are captured in the **"context"** or **MIB view** that the manager wants to access. An MIB view consists of a list of all the subtrees of the MIB that are included in the view as well as a list of the subtrees that are explicitly excluded from the MIB view. A context or MIB view is assigned an OBJECT IDENTIFIER. This simplifies access control. When an agent receives a message from its manager, it verifies that all the objects within that message variable-bindings list are within the context specified in the message. It further verifies in its local access control list (another table in its MIB) that the requesting manager does have the required access privileges for the specified context.

9.2.2 On the Relativity of Time

There is one more concept we need to clarify before peering into the security portion of the version 2 protocol, and it is about time. Clocks come in handy for security. Time stamps can be used to thwart replay attacks through their uniqueness (with respect to a single clock); they help establish the proper order of messages; and they help determine the timeliness (or lack thereof) of a message. The last feature requires clocks to be synchronized within some manageable tolerance. In the absolutist OSI universe, this is accomplished with GeneralizedTime or UTCTime pegged to universal time. In the freewheeling IP world each entity has its own, highly individualistic time. It is an INTEGER counter that counts the number of seconds elapsed since its own creation (or last reboot). Thus, comparing the time stamp in an incoming message to the local time is completely useless for evaluating how old that message is. In order to determine the freshness of a message, the receiver must keep track of the sender's clock. If Alice and Bob want to communicate securely through SNMPv2, each must have the other's current time in its MIB. Initially, that time is set to 0. Alice sends a setRequest to Bob, setting the value of her clock in his MIB to her actual time. This is an unprotected message. Indeed, this message is sent only when Alice suspects that Bob has lost track of her clock and therefore using Version 2 security mechanisms would fail. Bob duly reciprocates. From now on, every time (i.e., every second) Alice and Bob increment their own clocks each will increment their version of the other's.

Every time Alice receives a protected message from Bob it will include two time stamps: Bob's time, as well as Bob's version of Alice's time. Since Alice keeps track of Bob's time, she can check the freshness of the message. If her version of Bob's time is higher then Bob's time in the received message, plus the allowed lifetime for the message (this is another manageable parameter), then Alice will discard the message.

If Alice and Bob have been out of touch for a while, their clocks may have drifted apart. The resynchronization procedure is simple. If Alice did accept Bob's message in the preceding paragraph, she would use both time stamps in his message for the procedure. If Bob's time in Bob's message is higher then Bob's time in Alice's MIB, then Alice will set Bob's time in her MIB to Bob's time in the message. Similarly, if Alice's time in Bob's message is higher then Alice's own time, she will set her own time to what Bob thinks it should be. If any of the times Alice keeps for Bob is higher then the corresponding time in Bob's message, Alice need not do a thing. However, if this happens often with many of her correspondents, she may want to recalibrate her system clock. So Alice needs to keep two clocks for each of her potential correspondents: her version of theirs and their version of hers. If Alice is very popular, she will keep lots of clocks. Bob, of course, reciprocates. Notice that with this procedure all clocks are monotonically increasing, as they should in order to be useful for security applications.

9.2.3 Secure PDUs

Having elucidated the lofty matters of individual identities and nonidentical clocks, we can now tackle the version 2 secure exchanges.

The payload of secure v2 message, called SnmpMgmtCom, consists of:

dstParty	OBJECT IDENTIFIER	the intended recipient's partyIdentity
srcParty	OBJECT IDENTIFIER	the sender's partyIdentity
context	OBJECT IDENTIFIER	the target MIB view identifier
pdu	PDUs	unprotected SNMPv2 PDU

SMNPv2 authentication is based on secure hashing. If it is required, the authentication message digest is the hash of:

partyAuthPrivate	OCTET STRING	sender's private authentication key
dstTimestamp	INTEGER	sender's value of the intended recipient's time
srcTimestamp	INTEGER	sender's time
SnmpMgmtCom	payload as defined above	

The actual authentication information (authInfo) consists of:

digest	OCTET STRING	message digest computed as described above
dstTimestamp	INTEGER	sender's value of the intended recipient's time
srcTimestamp	INTEGER	sender's time

If authentication is not required, then authInfo is an OCTET STRING of zero length.

If confidentiality is required, then the concatenation of authInfo and SnmpMgmtCom is encrypted; otherwise it is sent in the clear. In either case it is preceded by

PrivDst	OBJECT IDENTIFIER	the intended recipient's partyIdentity

This field, which has the same value as dstParty, is sent unprotected at the beginning of the message. Thus, SNMPv2 can provide authentication with or without privacy protection as needed.

Since SNMP is connectionless, there is not much of an opportunity to negotiate a security context. Instead, every SNMPv2 entity has a permanent security context with each of its potential correspondents. These security contexts are stored in the local MIB, and they can be remotely managed.

9.3 SNMPv3 SECURITY

Like SNMP v1 and v2, v3 messages can in principle be protected with IPsec. However, SNMPv3 provides its own security features that are easier to manage (with SNMPv3 of course). SNMPv3 security has two major components: User-based Security Model (USM) for version 3 of SNMP to protect SNMPv3 transactions [RFC2274], and View-based Access Control Model (VACM) for the SNMP to implement and manage access control [RFC2275]. Both are discussed below.

9.3.1 User-Based Security Model

USM [RFC2274] provides integrity protection with or without delay detection and privacy protection. Before discussing how USM supports these security services, we examine how it achieves synchronization and how it manages keys.

9.3.1.1 Simple Times. When two SNMPv2 entities interact, each must keep a local notion of the other's time. SNMPv3 simplifies matters. The manager still has to keep a local notion of the agent's time, but the agent does not care how the manager's clock is doing. More precisely, SNMPv3 defines an **authoritative SNMP entity** for each message as follows:

- For a message that requires a response (e.g., a get message), the intended receiver of the message is the authoritative entity.
- For a message that does not require a response (e.g., a trap), the sender of the message is the authoritative entity.

Every secure SNMPv3 message must include the user name (on whose behalf the message is sent), the authoritative SNMP entity for that message, and, if timing information is needed, the authoritative SNMP entity's time. Time consists of two INTEGER indicators: the number of times the SNMP entity has been rebooted since it was created **(snmpEngineBoots),** and the number of seconds since the last reboot **(snmpEngineTime).** For the authoritative SNMP engine these indicators are called **msgAuthoritativeEngineBoots** and **msgAuthoritativeEngineTime.** The maximum value of these parameters is 2,147,483,647 (2^{32}-1), which generously allows for very long-lived SNMP entities.

To ask the agent for (its) time of day, the manager can send an unprotected empty get (i.e., with an empty variable bindings list), with both msgAuthoritativeEngineBoots and msgAuthoritativeEngineTime set to zero. The response will include the correct values of msgAuthoritativeEngineBoots and msgAuthoritativeEngineTime.

Every time a nonauthoritative SNMP engine receives a message from an authoritative SNMP engine, it compares the values of msgAuthoritativeEngineBoots and msgAuthoritativeEngineTime in the received message to the ones it keeps for the corresponding authoritative SNMP engine. If

- the received msgAuthoritativeEngineBoots is larger than the locally kept value, or
- the received msgAuthoritativeEngineBoots is the same as the locally kept value and the received msgAuthoritativeEngineTime is larger than the locally kept value

then the nonauthoritative SNMP engine updates its notion of the authoritative SNMP engine's msgAuthoritativeEngineBoots and msgAuthoritativeEngineTime to the received values. The latter is further incremented by one unit every second. At the same time, the nonauthoritative SNMP engine also updates another value it keeps, the **latestReceivedEngineTime** for the authoritative SNMP engine to the received value of msgAuthoritativeEngineTime.

9.3.1.2 Key Items. SNMPv3 supports both single login to many systems and key management.

Single login is achieved by computing the shared secret between a user (typically a manager) and a target system (typically an agent) from the user's single, secret (nonshared) password and the agent's unique identifier. First, the user's password is repeated as many times as necessary, and then truncated, to obtain a string of 1,048,576 octets. The resulting string is hashed with the secure hashing function used for authentication to obtain digest1.

The agent's snmpEngineID is sandwiched between two copies of digest1, and the resulting string is hashed with the secure hashing function. Again, it is the same hashing function that will be used for message authentication. In 1999 SNMPv3 offers a choice between MD5 and SHA1. (Other choices may be added in the future.) The result of this hash is the shared secret. Initially, the resulting shared secret must be manually loaded into the agent SNMP engine. It need not be kept in the user's system; it can be easily computed for each session. Thus, the user needs to remember only a single password in order to access many different systems.

In general, the procedure described above is used to derive both authentication secret and privacy secret. When it is used to derive an authentication secret, the authentication key is the result of the computation described above without further ado. When it is used to derive a privacy secret, further ado is needed to derive a secret DES key and an IV (Initialization Vector). The DES key is simply the first 8 octets of the privacy secret, with the least significant bit of each octet set to ensure the correct parity for each octet. The IV consists of the next 8 octets of the privacy secret XORed with an 8-octet "salt." The value of the salt is semirandom; it should be different for every message, and it is transmitted as an octet string with every encrypted message.

Every time the user selects a new password, all the secret keys it shares with other systems must be updated. Fortunately, this update can be done through a simple SNMP transaction. That transaction does not even require confidentiality protection. Instead, the hashing function used for authentication is used here to encrypt the new key, so there is no need to further encrypt the whole PDU containing the new key.

To update a shared secret key with one target system, the user's system selects a (pseudo) random octet string as long as the shared secret key. The random string is appended to the old (i.e., current) key, and the result is hashed with the hashing function used for authentication. The random string is then appended to the digest, and the resulting string is hashed. The process goes on until the combined length of all these digests is at least as long as the new key. At this point all the digests are concatenated, and the result is truncated to match the length of the new key. The trimmed string is EXCLUSIVE-ORed with the new key to produce a string called delta. The random string concatenated with the delta string is sent to the remote SNMP engine as the value of an object called KeyChange. Upon receiving the value of KeyChange, the agent SNMP engine performs the same set of hashing rounds using the old key and the random string (extracted from KeyChange) as was done by the user's system. After trimming the result as done on the sender's side, the recipient EXCLUSIVE-ORs that string with the delta string portion of the KeyChange to obtain the new key. Figure 9.2 illustrates the main steps in the use of hashing for encryption and decryption.

One reason for changing encryption keys is forward protection: if an intruder has broken an encryption key, changing that key will protect any messages secured with the new key. In our case, since the old key is used for encrypting the new key, this advantage is lost. Indeed, if an intruder succeeds in obtaining one key and can listen to all subsequent messages, that intruder will be able to compute all subsequent new keys. Nevertheless, changing keys as described here limits the amount of ciphertext available to the intruder for any single key. This can thwart differential cryptographic analysis, which requires vast amounts of ciphertext.

9.3.1.3 USM PDUs. Having established all the preliminaries, we are ready for the USM PDUs. Section 2.3.7.11.4.1 provides the structure of SNMPv3 PDUs. One component of these PDUs is msgSecurityParameters, which is transported as an octet string. For USM, that octet string is the BER (Basic Encoding Rules) encoding of the following SEQUENCE:

- **msgAuthoritativeEngineID,** an OCTET STRING identifying the authoritative SNMP engine for the message.

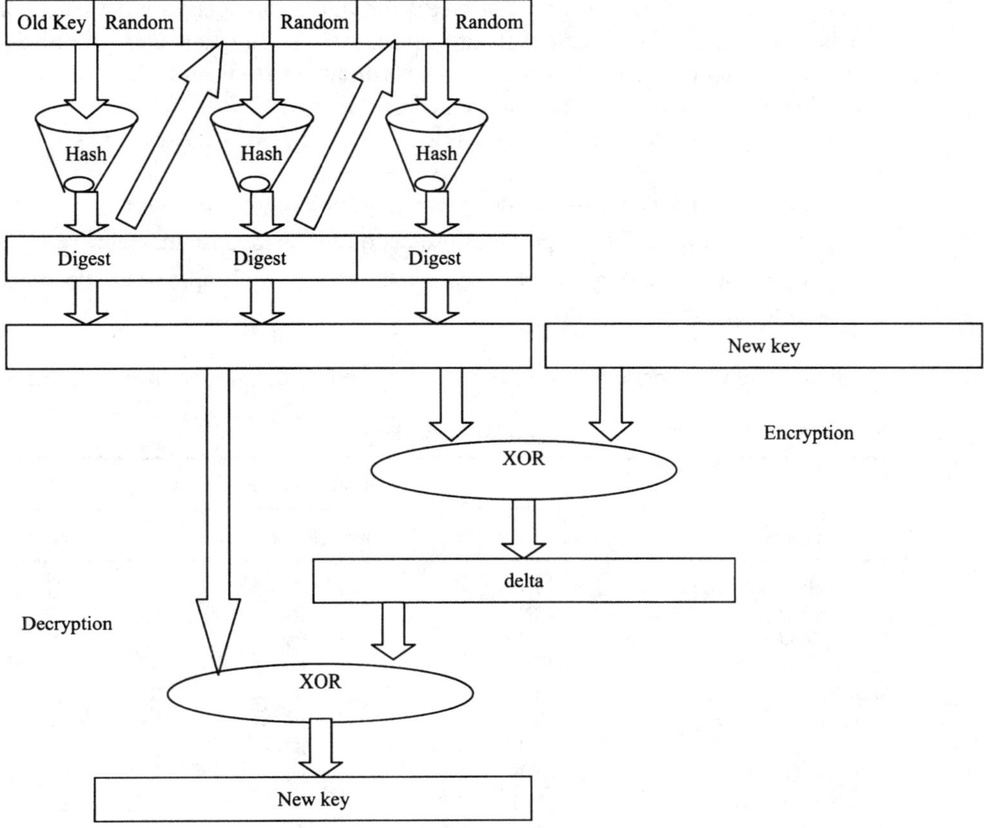

Figure 9.2 Using hashing for encryption/decryption

- **msgAuthoritativeEngineBoots,** an INTEGER with values from 0 to 2147483647 counting the number of times the authoritative SNMP engine has been re-booted since its creation.
- **msgAuthoritativeEngineTime,** an INTEGER with values from 0 to 2147483647 counting the number of seconds since the last re-boot of the authoritative SNMP engine.
- **msgUserName,** an OCTET STRING from 1 to 32 octets identifying the principal on whose behalf the message is sent.
- **msgAuthenticationParameters,** an OCTET STRING containing information that is meaningful to the authentication mechanism, if any, used for protecting the message.
- **msgPrivacyParameters,** an OCTET STRING containing information that is meaningful to the privacy mechanism, if any, used for protecting the message.

USM specifies two authentication mechanisms: HMAC-MD5-96 and HMAC-SHA-96. They correspond to HMAC with MD5 and SHA1, respectively, with the result truncated to the first 96 bits in each case. In each case, these 96 bits, represented as an OCTET STRING, make up the value of the msgAuthenticationParameters. The HMAC function is computed over the whole SNMPv3 message, with the msgAuthenticationParameters consisting of 12 zero octets. Those 12 octets are then replaced with the result of the HMAC computation. To promote interoperability, support for HMAC-MD5-96 is mandatory.

USM specifies one mandatory mechanism for privacy: DES with padding, in the CBC mode. The value of the salt (used to define the IV) is carried as an OCTET STRING of 8 octets in the msgPrivacyParameters.

9.3.1.4 USM APIs. USM provides APIs for all the components it defines, promoting a high degree of modularity and software reuse. All the function call primitives return statusInformation, providing indication of success or an errorIndication. The API also specifies the IN and OUT parameters for each function call.

USM offers three services to the Message Processing (MP) subsystem:

- **generateRequestMsg**—provides protection for a request message
- **generateResponseMsg**—provides protection for a response message
- **processIncomingMsg**—verifies the authenticity and, if appropriate, decrypts an incoming message

The IN and OUT parameters for these three services are summarized in Table 9.1.

TABLE 9.1 IN and OUT Parameters of USM Service Primitives

	Function Call Primitives		
Parameters	generateRequestMsg	generateResponseMsg	processIncomingMsg
messageProcessingModel	IN	IN	IN
maxMsgSize	IN	IN	IN
securityModel	IN	IN	IN
securityEngineID	IN	IN	IN
securityName	IN	IN	IN
securityLevel	IN	IN	IN
globalData	IN	IN	IN
scopedPDU	IN	IN	OUT
securityParameters	OUT	OUT	IN
wholeMsgLength	OUT	OUT	IN
wholeMsg	OUT	OUT	IN
maxSizeResponseScoped PDU			OUT
securityStateReference		IN	OUT

Four of the parameters listed in Table 9.1 are passed through the USM without any processing:

- **messageProcessingModel**—The SNMP version number for the message to be generated.
- **globalData**—The message header (i.e., its administrative information).
- **maxMessageSize**—The maximum message size as included in the message.
- **securityModel**—The securityModel in use. Should be User-Based Security Model.

The other parameters in Table 9.1 are used by the USM to do its job:

- **securityParameters**—The security parameters as received in an incoming message. They are filled in by the User-Based Security module for an outgoing message.
- **securityName**—Together with the snmpEngineID it identifies a row in the usmUserTable that is to be used for securing the message. The securityName has a format that is independent of the Security Model.
- **securityLevel**—The Level of Security from which the User-based Security module determines if the message needs to be protected from disclosure and if the message needs to be authenticated. In the case of a response, this parameter is ignored and the value from the cache is used.

- **securityEngineID**—The snmpEngineID of the authoritative SNMP engine for the message.
- **scopedPDU**—The message payload. The data is opaque to the User-based Security Model.
- **securityStateReference**—For an incoming message this is a handle/reference to **cachedSecurityData** which is to be used when securing the corresponding outgoing Response message. When the Message Processing Subsystem calls the User-Based Security module to generate a response to an incoming message, it must pass this handle/reference. For an outgoing response message, this is the same handle/reference generated by the User-Based Security module when processing the incoming Request message to which the Response message corresponds.
- **wholeMsg**—The fully encoded and secured message ready for sending on the wire.
- **wholeMsgLength**—The length of the encoded and secured message (wholeMsg).
- **maxSizeResponseScopedPDU**—The maximum size of a scopedPDU to be included in a possible Response message.

Every invocation of a USM module returns statusInformation that indicates the success or failure of the invocation. In the case of failure it may provide an error code specifying, for example, authentication failure, decryption failure, failure to parse the security parameters, message not in time window, unknown SNMP engine ID, and unknown security name. USM maintains counters that keep track of how many such mishaps have occurred. Every failure causes an update of the corresponding counter. In such cases, the statusInformation includes the OID and value of the counter that has been incremented.

If authentication is used, then USM invokes its authentication module. It offers two services: **authenticateOutgoingMsg** and **authenticateIncomingMsg** are invoked on the originating and receiving side, respectively. On the receiving side, authenticateIncomingMsg verifies the validity of message integrity code (in authParameters) and the validity of the timing indicators. The IN and OUT parameters of the authentication module services are summarized in Table 9.2.

TABLE 9.2 IN and OUT Parameters for USM Authentication Primitives

	Function Call Primitives	
Parameters	authenticateOutgoingMsg	authenticateIncomingMsg
authKey	IN	IN
wholeMsg	IN	IN
authenticatedWholeMsg	OUT	OUT
authParameters		IN

The parameters listed in Table 9.2 have the following semantics:

- **authKey**—The user's localized private authKey is the secret key that can be used by the authentication algorithm.
- **wholeMsg**—The complete serialized message to be authenticated.
- **authenticatedWholeMsg**—The same as the input given to the authenticate-IncomingMsg service but after authentication has been checked.

If privacy is provided, then the privacy module of USM is invoked. Its two services, **encryptData** and **decryptData,** are invoked on the originating and receiving side, respectively. Their IN and OUT parameters are summarized in Table 9.3.

TABLE 9.3 IN and OUT Parameters for Privacy Primitives

Parameters	Function Call Primitives	
	encryptData	decryptData
encryptKey	IN	
decryptKey		IN
dataToEncrypt	IN	
encryptedData	OUT	IN
privParameters	OUT	IN
decryptedData		OUT

The parameters listed in Table 9.3 have the following semantics:

- **encryptKey**—The user's localized private privKey is the secret key that can be used by the encryption algorithm.
- **dataToEncrypt**—The serialized scopedPDU is the data to be encrypted.
- **decryptKey**—The user's localized private privKey is the secret key that can be used by the decryption algorithm. It is the same as the key used for encryption.
- **privParameters**—The msgPrivacyParameters (the "salt" for the IV), encoded as an OCTET STRING.
- **encryptedData**—The encryptedPDU represents the encrypted scopedPDU, encoded as an OCTET STRING.
- **decryptedData**—The serialized scopedPDU if decryption is successful.

Secure SNMPv3 messages allow strong authentication of the message sender. The receiving system uses this information to decide if any request from the sender should be granted or denied. This function of SNMPv3 is described in the next section.

9.3.2 View-Based Access Control Model (VACM)

The SNMPv3 security module uses VACM to decide the disposition of incoming messages and outgoing notifications. In particular, it is used to determine if an incoming request should be granted or denied. It makes this decision based on the (hopefully authenticated) identity of the message originator and locally stored access control information. In principle, access control information can be very cumbersome. It could explicitly specify the access privileges for every user for every information item. Such a simplistic approach can result in access control information that is more voluminous than the information it is intended to protect. Clever designs, however, can dramatically condense the representation of access control information. VACM addresses the representation of both potential users and protected information.

9.3.2.1 Who's Calling. In network management several users may have the same access privileges. For instance, all the repair technicians for routers may have identical access privileges. Rather than list these privileges again and again for every technician, it is more economical to have one list of access privileges for router repair technicians and a corresponding list of the actual router repair technicians.

Individual users are identified through their corresponding securityModel and their unique securityName within the securityModel. A group of users, such as the router repair technicians from the previous paragraph, is given a unique groupName which represents zero or more individual users.

9.3.2.2 Domain of Discourse. A large MIB may contain hundreds of thousands of managed object instances. Explicit enumeration of all the object instances that are accessible to each group would be horrendously inefficient. VACM does much better with successive refinements of the allowed access modes. This approach saves memory, and more importantly, improves performance by allowing speedy rejections of requests that are way off the mark.

The payload of an SNMPv3 message is a scoped PDU that starts by identifying the context for the PDU. A context typically corresponds to a physical or logical device within an SNMP entity. A context is defined through a set of MIB subtrees that are included in the context and a set of subtrees that are explicitly excluded from the context. The PDU can only refer to managed objects that are within the included subtrees but not within the excluded subtrees. The first step of access control is to verify that the message sender is a member of a group that has some legitimate business with the device represented by the context. At the same time, the access control function must verify that all the elements of the variable bindings list in the PDU are within the stated context. If either test fails or if the context is unknown, the incoming request will be denied.

Different (groups of) users may have different access privileges to any given device within an SNMP entity. To accommodate this level of granularity, VACM provides for the definition of several MIB views for the different (groups of) users with legitimate needs to access that device. An MIB view is defined through a set of MIB subtrees that are included in the view and a set of subtrees that are explicitly excluded from the view, neatly reusing the mechanism for defining contexts.

Sometimes even enumerating all the subtrees in an MIB view may be cumbersome. VACM offers an effective way of specifying a whole family of related subtrees. A simple example illustrates this concept:

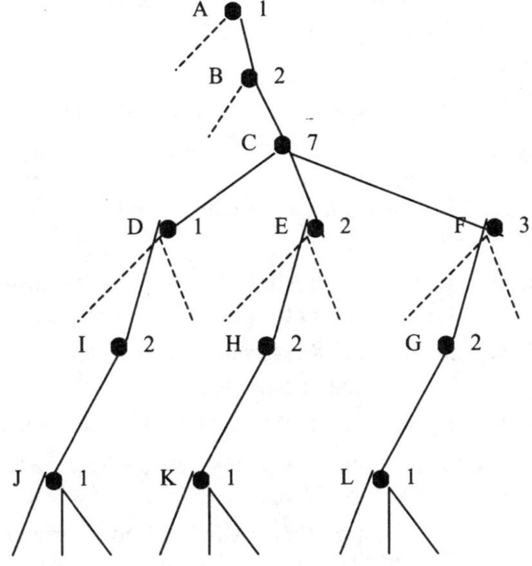

Figure 9.3 A view tree family

Figure 9.3 represents a portion of an MIB. Every node in the MIB is labeled with an integer that distinguishes it from all its siblings. The sequence of integers along the arc from the root of the tree to any node in the tree is that node's OBJECT IDENTIFIER (OID). For ease of reference in this discussion, each node in the figure is further identified with a capital letter. Thus, the OID (1,2,7,1,2,1) uniquely identifies node J. In an MIB view definition, this OID refers to the subtree rooted at J. Suppose the MIB view is to further include the subtrees rooted in nodes K and L. Rather than list their OIDs in addition to J's OID, VACM offers a convenient shorthand. It is based on the observation that J, K, and L have strong family ties: their OIDs differ only in the fourth subidentifier. VACM allows specifying the family of the three subtrees by specifying the OID for J: (1,2,7,1,2,1), followed by the mask (111011). The mask indicates that all the OIDs that are identical to J's OID, except for the fourth subidentifier (which can thus have any value), are members of the same view tree family. Notice that J's OID together with the mask (111011) corresponds to J, K, and L exactly only if nodes D, E, and F are C's only children.

In general, a mask can have several zeros, meaning that the corresponding subidentifiers can have any values. Furthermore, the nonzero entries in the mask can have any values; they need not be one. The semantics of all the nonzero values is the same (i.e., the corresponding subidentifier must be equal to the corresponding subidentifier in the OID).

Now that we have some fairly efficient tools for specifying chunks of information that are accessible to (a group of) users, we are ready for the next level of detail: specifying the access privileges. VACM supports three MIB views for every group of users: read-view, write-view, and notify-view. Their meaning is exactly as expected.

9.3.2.3 VACM Services. VACM offers a single service: it determines whether an incoming request or an outgoing notification should be granted or denied. The service is called **isAccessAllowed,** and it is invoked with the following parameters:

- **securityModel,** which is being used for the current message (e.g., USM)
- **securityName** of the principal who wants access
- **securityLevel,** which is being used for the current message (e.g., for USM: no security, authentication only, authentication + privacy)
- **viewType,** that is, read, write, or notify
- **contextName** for the scoped PDU
- **variableName** OID of the managed object, within the specified context, that is being accessed

isAccessAllowed returns stausInformation which can be one of the following:

- **accessAllowed,** indicating that the request is granted (all other values of statusInformation correspond to denial of the request)
- **notInView,** indicating that the variable name is not in the MIB view for the specified view type for the specified principal
- **noSuchContext,** indicating that VACM does not recognize the specified contextName
- **noGroupName,** indicating that no groupName was found for the specified securityModel and securityName
- **noAccessEntry**, indicating that VACM did not find any entry corresponding to the specified securityModel, securityName, contextName, securityModel, and securityLevel
- **otherError,** indicating an undefined error

VACM bases its decision on the values of the input parameters as well as on locally stored access control information. This information is captured in an MIB that can be managed remotely.

10

Portraits Gallery

This part of the book ends with a family portrait of the security communications protocols used in TMN, followed by an individual snapshot of each member.

The backdrop of the family portrait is the dichotomy of in-stack and out-of-stack security communications protocols introduced in Section 5.5. Distinguishing between in-band and out-of-band security signaling further refines it. In the first case, the security-related PDUs are carried along with other PDUs. In the second case, security-related PDUs may be carried over separate channels. The resulting picture is shown in Figure 10.1.

	In-band	Out-of-band
In-stack	GULS SSL3, TLS, IPsec TR40, ABS, STASE-ROSE IA, SNMPv2, SNMPv3	Not applicable
Out-of-stack	GSS-API	TeNorIOP

Figure 10.1 TMN security communications protocols

The backdrop for the individual pictures, in Figures 10.2 to 10.11, is rather more elaborate. It captures the functionality of each security communications protocol. Shaded boxes in the applicable cells in the table represent the individual protocols. The width of a shaded box is a quantitative representation of the functionality covered by the protocol within the cell. The height of a box depicts the efficiency of the protocol in providing the corresponding functionality.

Layer	Handshaking		Secure transfer phase						
	Security context negotiation	Peer entity authentication	Dynamic security context update	Whole PDU protection				Data flow integrity	Selective field protection
				Data origin authentication	Integrity	Privacy	Non-repudiation		
Application									
Application layer									
Presentation layer									
Session layer									
Transport layer									
Network layer									

Figure 10.2 GULS—for all OSI-compliant protocols

Layer	Handshaking		Secure transfer phase						
	Security context negotiation	Peer entity authentication	Dynamic security context update	Whole PDU protection				Data flow integrity	Selective field protection
				Data origin authentication	Integrity	Privacy	Non-repudiation		
Application									
Application layer									
Presentation layer									
Session layer									
Transport layer									
Network layer									

Figure 10.3 SSL3/TLS1 for TCP/IP only

Layer	Handshaking		Secure transfer phase						
	Security context negotiation	Peer entity authentication	Dynamic security context update	Whole PDU protection				Data flow integrity	Selective field protection
				Data origin authentication	Integrity	Privacy	Non-repudiation		
Application									
Application layer									
Presentation layer									
Session layer									
Transport layer									
Network layer									

Figure 10.4 IPsec—for IP only

Layer	Handshaking		Secure transfer phase						
	Security context negotiation	Peer entity authentication	Dynamic security context update	Whole PDU protection				Data flow integrity	Selective field protection
				Data origin authentication	Integrity	Privacy	Non-repudiation		
Application									
Application layer									
Presentation layer									
Session layer									
Transport layer									
Network layer									

Figure 10.5 TR40—for ACSE-using protocols

Layer	Handshaking		Secure transfer phase						
	Security context negotiation	Peer entity authentication	Dynamic security context update	Whole PDU protection				Data flow integrity	Selective field protection
				Data origin authentication	Integrity	Privacy	Non-repudiation		
Application									
Application layer									
Presentation layer									
Session layer									
Transport layer									
Network layer									

Figure 10.6 Application-Based Security—for CMIP only

Layer	Handshaking		Secure transfer phase						
	Security context negotiation	Peer entity authentication	Dynamic security context update	Whole PDU protection				Data flow integrity	Selective field protection
				Data origin authentication	Integrity	Privacy	Non-repudiation		
Application									
Application layer									
Presentation layer									
Session layer									
Transport layer									
Network layer									

Figure 10.7 STASE-ROSE for ROSE-based protocols

Layer	Handshaking		Secure transfer phase						
	Security context negotiation	Peer entity authentication	Dynamic security context update	Whole PDU protection				Data flow integrity	Selective field protection
				Data origin authentication	Integrity	Privacy	Non-repudiation		
Out-of-stack									
Application layer									
Presentation layer									
Session layer									
Transport layer									
Network layer									

Figure 10.8 GSS-API

Layer	Handshaking		Secure transfer phase						
	Security context negotiation	Peer entity authentication	Dynamic security context update	Whole PDU protection				Data flow integrity	Selective field protection
				Data origin authentication	Integrity	Privacy	Non-repudiation		
Application									
Application layer									
Presentation layer									
Session layer									
Transport layer									
Network layer									

Figure 10.9 Interactive Agent—for EDI only

Layer	Handshaking		Secure transfer phase						
	Security context negotiation	Peer entity authentication	Dynamic security context update	Whole PDU protection				Data flow integrity	Selective field protection
				Data origin authentication	Integrity	Privacy	Non-repudiation		
Application									
Application layer									
Presentation layer									
Session layer									
Transport layer									
Network layer									

Figure 10.10 TeNorIOP—for CORBA only

Layer	Handshaking		Secure transfer phase						
	Security context negotiation	Peer entity authentication	Dynamic security context update	Whole PDU protection				Data flow integrity	Selective field protection
				Data origin authentication	Integrity	Privacy	Non-repudiation		
Application									
Application layer									
Presentation layer									
Session layer									
Transport layer									
Network layer									

Figure 10.11 SNMPv2 and SNMPv3 security

PART IV SECURITY MANAGEMENT

This part addresses the management of security information—both information related to the security of the TMN and information related to the security of the telecommunication network. It covers in some detail the standards and information models available for security management. That section is especially relevant to system designers and developers.

The use of CORBA or EDI for security management has not yet been addressed within the context of TMN. The use of SNMP for managing SNMP-related security information is discussed in Chapter 9 within the context of SNMP security.

This part focuses on the extensive information models for CMIP-based security management.

11

Management of Security Information

This chapter shows how CMIP can be used for the management of security information. This is possible when the interface between the OS performing security administration and the OS or NE that is being managed is based on CMIP. This approach applies whether or not the systems that are managed use OSI system management-based interfaces to communicate with any other systems. This is illustrated in Figure 11.1 where A manages security information in B and/or C over CMIP-based interface(s), while B and C may communicate using any type interface.

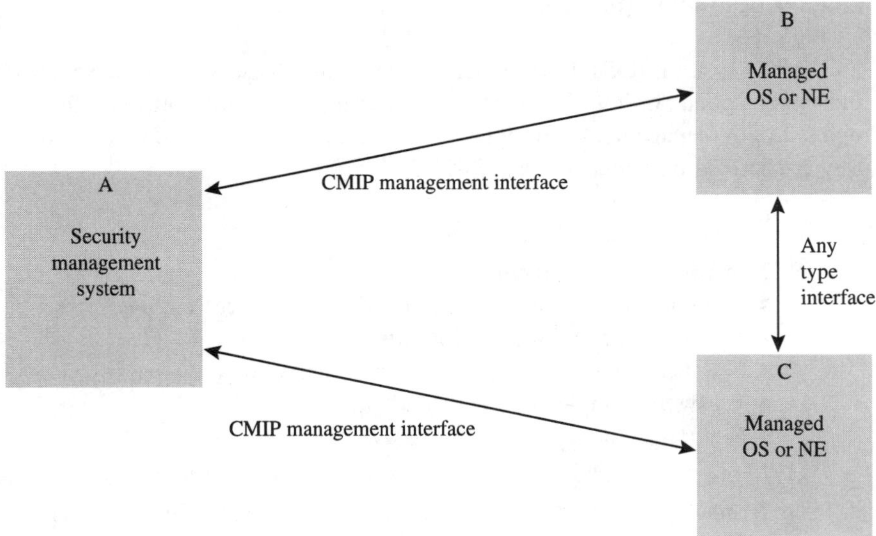

Figure 11.1 Example of security manager and managed systems

Chapters 4 and 5 specify services and mechanisms for securing TMN interfaces. Implementation of such security mechanisms typically requires the storage of security-related information in all the participating TMN elements. Such information may include encryption keys, access control information, and guidelines about which security-related events shall be recorded and/or communicated to other entities. In a large TMN, including thousands of NEs and OSs, it is not practical to manage such information locally. It is far more efficient to manage it from one site or a few central locations. In order to manage such information over OSI-based interfaces, it is necessary to define an information model for this task. This chapter presents the object classes, notifications, actions, and attributes that make up the information models for security administration.

11.1 SECURITY ADMINISTRATION FUNCTIONS

For the purpose of enhancing the clarity of presentation, security administration functions can be divided into four fairly distinct areas:

- Login management, which supports security management of users logging into servers
- Notifications management, which supports the management of security alarms and audit trails
- Access control management
- Key management

The information models that support the security management activities listed in this chapter use numerous object classes and attribute types defined in national and international standards. This section explains the role of such objects and attributes but refers to the original standards for formal definitions rather then repeating their details. This section also includes definitions of additional object classes and associated properties to address requirements that are not currently available in standards but are defined in Bellcore's GR-1253.

11.1.1 Login Management

Many systems (OSs, NEs, servers) allow (human) users to log in. Security administrators spend a good portion of their efforts managing the security-related information for such logins. Login management is designed to support such activities. The functions in this category are derived from Bellcore's [TR815].

A. Activate, suspend, revoke, reactivate users.
B. Manage passwords, including:
   - Assignment of initial user passwords and reset current passwords
   - Setting of the password aging interval
   - Number of passwords used and time interval before a password can be reused
   - Password complexity algorithm
   - Method for notifying user of password expiration
C. Manage pre-login warning message and post-login information that users see.
D. Manage parameters associated with a communication link, such as:
   - Number of unsuccessful login attempts before an alarm is generated
   - Channel lockout period
   - Channel idle time out period
E. Retrieve information associated with a current user session, such as location, session starting time, list process IDs, and session idle time.
F. Terminate a user session (e.g., when a security violation is suspected).

11.1.2 Notification Management

Many managed objects, as well as non-OSI systems, can issue notifications that are relevant for overall security. Some notifications, such as a security alarm announcing an unauthorized intrusion, may require immediate action. Such notifications should be promptly routed to the appropriate destination. Other notifications may result from ordinary events, such as associations being set and terminated. Although such events do not individually raise any suspicion, they can be analyzed together to ferret out security violations. The purpose of notification management is to control which notifications are routed to which destination and under what circumstances, and which notifications are recorded in a security log, as well as to manage the security log (e.g., adjust the size of the log). In order to use the OSI management information model for this management activity, notifications from non-OSI systems may have to be converted to notifications specified in the appropriate OSI standards.

The functions of notification management are as follows.

A. Define security events for audit and alarm purposes.
B. Manage security audit trails.
C. Manage security alarms.
D. Manage the log for security events.

11.1.3 Access Control Management

Access control management consists of managing all the information used to determine what should be done when a request to access a resource is received. Such information includes the rules used to make access control decisions, information related to the initiator of the request, and information related to the target of the request (the resource being accessed and the operation[s] requested).

Access control rules can be either positive or negative. That is, if the conditions in the rule are satisfied, the result may be either to grant or to deny the requested access. An example of a deny rule is that if an initiator presents an access control certificate that is listed in a revocation list, then the access request is denied.

A request to establish an association (or a communication channel in a non-OSI environment) is also a request to access a resource. Therefore, access control management overlaps with some of the login management functions listed earlier. If OSI's general access control management is available, it may assume some of the login management functions, or it may work with the MOs defined in support of login management.

11.1.4 Management of Encryption Keys

Any security mechanism is based on secret information (keys or passwords) that are shared only by the communicating entities, or secret information (e.g., the private key in a public key crypto system) that is known to only one entity and public information that is securely available to other entities.

When a new entity is introduced into a security domain, it must receive, securely, some encryption keys. For instance, it may need to share a secret key with a Kerberos server, or it may need to securely receive (i.e., without any possibility of receiving false information) the public key of the Certification Authority of its security domain. No existing standards support the distribution of the very first encryption information that the entity receives. Furthermore, such distribution typically needs to happen only once in the life of a system. Such distribution is therefore outside the scope of this section.

Several mechanisms are involved in the subsequent distribution of encryption keys. Some of the most prominent ones are:

- Use of the Diffie–Hellman algorithm by two communicating entities to establish a secret key
- Use of the IEEE standard for key management over a LAN/MAN
- Use of a third party, such as a Kerberos server to securely distribute a secret key to two entities
- Use of the TMN Directory, based on X.500, for the secure distribution of certified public keys

Use of such mechanisms does not require any special security administration activity, and it is therefore outside the scope of this section.

When two communicating entities share several secret keys and both include STASE-ROSE with the capability to negotiate and update security parameters, they can negotiate which key to use without intervention of a security administrator.

In the absence of a security infrastructure that includes Kerberos service, a secure TMN Directory, or STASE-ROSE, some other means are needed for key distribution. This section discusses an MO class defined in [GR1253] that supports the key distribution mechanism described in ANSI TR 40 for Electronic Bonding [TR40].

Security management may also include the management of encryption algorithms—the downloading and deletion of specific algorithms in managed systems. However, this activity is rather infrequent, and therefore it is not included in this section.

11.2 INFORMATION MODEL DESCRIPTION

This section introduces the information models that support the security management functions listed in Section 11.1. It provides a brief (and incomplete) description of the Managed Object (MO) classes that make up the information model. The complete, detailed definitions of the MOs are given in referenced standards and Bellcore GRs. The present section explains the information model and the usage of the MOs.

11.2.1 Login Management

Login management covers security management functions associated with users logging to servers such as UNIX systems. All the managed object classes in this section are defined in Bellcore GR-1253. The possible name bindings among the MOs introduced in this subsection are depicted in Figure 11.2. All those MO classes are derived from the Top managed object class. Their use is explained in the following subsections.

11.2.1.1 Pre- and Post-Login Messages. When users attempt to log into a system, a pre-login message may be displayed warning the would-be users of any restrictions regarding use of the system; for example, the system is strictly for business-related use by authorized employees. After a user has successfully logged in, a post-login message may provide information about the last session. The **securitySystem** managed object (MO) class has attributes that represent such messages, thereby allowing a remote security administrator to manage such messages. Typically, there is one securitySystem MO instance per system (OS or NE).

The security system object class is a class of managed support objects that represent security information related to the managed system. There is only one instance of this object

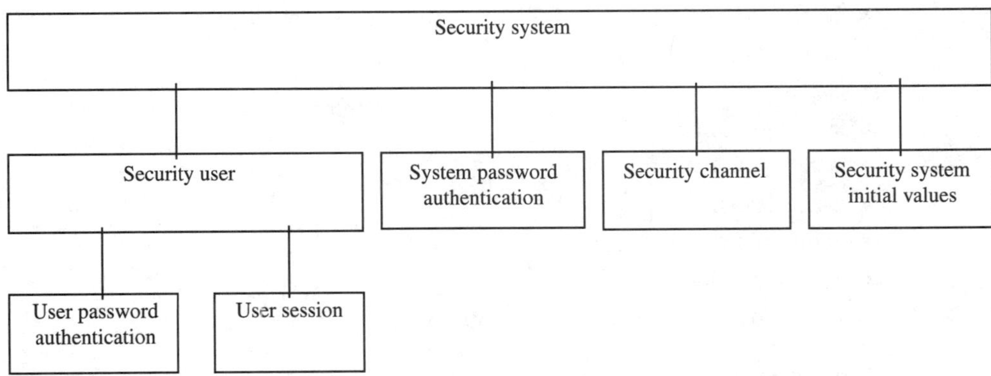

Figure 11.2 Possible name bindings (containment) for login management MOs

class per managed system, and this instance should exist for the duration of the existence of the managed system. Attributes in this object class are applicable to all users accessing the managed system through any channels. Those attributes are:

- preLoginMessage
- postLoginMessage
- securityEventList
- securitySystemId

If a user attempts to log in using a non OSI-based protocol, the information specified by the preLoginMessage attribute may be displayed to the user before full identification and authentication information has been received. The information specified by the postLoginMessage attribute will be displayed only after the user has logged in successfully.

If the user attempts to log in using an OSI-based protocol, then the postLoginMessage may contain information based on values of parameters in the reply information to the user login request action.

The securityEventList attribute is used to define conditions under which a security alarm report shall be generated by the managed system.

The securitySystemId attribute is used strictly for naming. Since there can be only one securitySystem object per system, this attribute carries no information, and therefore it always has the value NULL.

11.2.1.2 User Management. Access control includes control over which entity is allowed to establish communications with any given system. In many current cases, however, login is seen as a separate function from access control. In order to accommodate such views, the **securityUser** MO class is defined. Typically, there is one securityUser MO instance per user per system. The attributes of this object represent user-specific information such as the user's ID, last login and logout times, and the user's status. These attributes allow, among other activities, activating, suspending, revoking, and reactivating a user.

In general, if the access control discussed below is deployed, then the securityUser MO is not required.

The security user object class is a class of managed support objects that represent security information related to a user who has (currently active) or had (currently suspended) access right to the managed system. An instance of this object class represents security information related to a user such as a person, process, or remote system in the role of a person, who accesses the managed system. It has the following attributes, the last one being optional:

- userId
- userIdStatus
- lastLoginTime
- lastLogoutTime
- lastLoginLocationId
- lastUserIdDisabled
- numLastUnsuccessLogin
- nonUsedPeriod
- systemAccessControlContextualInformation

If the attribute userIdStatus is equal to new, this indicates a new user, and the attributes lastLoginTime, lastLogoutTime, lastLoginLocationId, lastUserIdDisabled, and numLast-UnsuccessLogin may not contain meaningful values. The nonUsedPeriod attribute is used to decide when a user ID should be put in a suspended status due to its being nonused. The nonused period is calculated from the difference between current date and the date of last login by the user ID (stored in the attribute lastLoginTime). If no initial value is specified for the nonUsedPeriod attribute at instantiation time, the initial value should be obtained from the corresponding attribute in the securitySystemInitialValues managed object.

systemAccessControlContextualInformation contains information that pertains to system access control. System access control is used to limit user access to the managed system based on conditions such as location of entry, method of access, time of day, day of week, and/or calendar date. If any of these conditions is specified, then all conditions must be met before user access is allowed.

A securityUser MO can respond to two actions: userLoginRequest and userLogout-Request. Those actions can be used to establish and terminate, respectively, sessions between the user and the system.

An instance of securityUser object class is created when a user is authorized to access the managed system. The CMIS M-CREATE and M-DELETE services can be used to create and delete an instance of this object class. A temporarily suspended user ID should not be deleted by using M-DELETE. Instead, the userIDstatus attribute should be used to indicate such a status. The maximum number of instances of securityUser object class that can be contained in an instance of securitySystem object class is dependent on the maximum number of users allowed by the managed system.

11.2.1.3 Password Management. Two MO classes are used to support password management: **systemPasswordAuthentication** and **userPasswordAuthentication.**

There is one **systemPasswordAuthentication** MO instance per system. Its attributes represent information such as a password reuse period and how many different passwords must be used before an old password may be reused. It has the following attributes, of which the last two are not mandatory:

- passwordReusePeriod
- passwordReuseNum
- systemPasswordAuthenticationId
- passwordComplexityAlgorithm
- passwordExpirationNotification

The system password authentication object class is a class of managed support objects that represent password authentication information associated with a managed system. Only one instance of this object class should exist for a managed system. However, a password authentication is only one example of authentication mechanisms. Should a different type of authentication mechanism be used, a similar object class should be defined and registered. The password authentication information contained in this managed object is applicable to all user IDs that use a password-based authentication mechanism. The passwordReusePeriod and passwordReuseNum attributes together decide how soon a previous password can be used again by the same user. The password structure is restricted by the passwordComplexityAlgorithmPkg package. If this package does not exist, there may not be any restriction on the complexity of the password, or the complexity cannot be managed through the interface.

The systemPasswordAuthenticationId attribute is used strictly for naming; since there can be only one systemPasswordAuthentication object per securitySystem, this attribute carries no information and therefore it always has the value NULL.

The passwordComplexityAlgorithm attribute specifies the complexity of the user password. In particular, it specifies the minimum length of a valid password, and the number of alphabetic, numeric and special characters it should have. When the userPassword attribute in userPasswordAuthentication object class is modified, the new value has to conform to the complexity specified by this attribute. Otherwise, any changes should be denied by the managed system.

The passwordExpirationNotification attribute specifies how a user should be notified of its password expiration. Upon the expiration of a user password, the managed system has to notify the user to change the password. Such notifications can be sent either a certain period of time before the actual expiration date of the password or at the time the password has expired. In the first case, a notification will start to be sent to the user a certain period of time before the actual expiration date of the password each time the user logs in to the managed system in order to advise of the need to change of the password. If the user decides not to change the password at this time or any time before the actual expiration date of the password, login will still be allowed by the managed system. However, the change of password will be mandated at the actual expiration date of the password. Otherwise, login will not be allowed. In the second case, a notification will start to be sent to the user at the actual expiration date of the password. The managed system will still allow logins during a certain period of time after the actual expiration date of the password, at which time no login will be allowed without changing the password. In either case, the period of time is specified either by number of dates or number of logins.

An instance of systemPasswordAuthentication object class is created at system initialization time if password authentication is (one of) the chosen authentication mechanism(s). After system initialization, an instance of this object class can be created using CMIS M-CREATE service. An instance of this object class can be deleted using CMIS M-DELETE service if and only if another authentication mechanism exists or can be created to replace it. In other words, a managed system is required to have at least one type of authentication mechanism available at any instance of time. When a different authentication mechanism is used, it should be ensured that this mechanism will be applied to the appropriate users. Only one instance of this object class may be contained in any one instance of the superior object class (securitySystem).

There is one **userPasswordAuthentication** MO instance per user per system. Its attributes represent information such as a password, when it was last changed and by whom, and when the user shall be notified that the password is about to expire. It has the following attributes:

- userPassword
- lastPasswordChange

- lastPasswordChangedBy
- passwordAgingInterval
- userPasswordAuthenticationId

The user password authentication object class is a class of managed support objects that represent password authentication information associated with a user. A password authentication is only one example of authentication mechanisms. The userPassword attribute may only be changed; it cannot be retrieved by anyone in order to ensure its protection. The passwordAgingInterval attribute is used to decide when a user will be notified of the password expiration and when such a user will be required to change the password. If no value is specified for the passwordAgingInterval attribute at instantiation time, the value should be obtained from the corresponding attribute in the securitySystemInitialValues managed object.

The userPasswordAuthenticationId attribute is used strictly for naming. Its sole purpose is to distinguish one instance of userPasswordAuthentication from all other instances of userPasswordAuthentication contained within the same securityUser object. However, a securityUser object contains exactly one userPasswordAuthentication object instance; therefore, this attribute carries no information, and therefore it always has the value NULL.

An instance of userPasswordAuthentication object class is created by using CMIS M-CREATE when an instance of its superior object class is created if and only if password authentication is the choice of authentication mechanism for the user. The CMIS M-DELETE service can be used to delete an instance of this object class if and only if the deletion is replaced by another type of authentication mechanism. A maximum of one instance of userPasswordAuthentication object class can be contained in any one instance of securityUser object class.

11.2.1.4 Channel Management. There is one **securityChannel** MO instance per channel such as an NE port or a logical communications channel. Its attributes represent information such as how many unsuccessful login attempts are allowed before the channel is locked and for how long it remains locked; how long a channel is allowed to remain idle (i.e., no activity takes place on the channel) before the channel is terminated. It has the following attributes:

- unsuccessLoginAttempt
- channelLockoutPeriod
- idleTimeOutPeriod
- channelId

The security channel object class is a class of managed support objects that represent security information related to a communication channel such as an NE port or a logical communication channel. These security information items are used for controlling user access to the managed system through a communication channel. One and only one instance of this class should exist for each channel. If a userLoginRequest action (in securityUser object class) is performed unsuccessfully at this channel for a number of times equal to the value specified by the unsuccessLoginAttempt attribute, this channel is locked out by the managed system from any further access for a period of time equal to the value of the attribute channelLockoutPeriod.

During a user session, if the user remains idle for a period of time (the attribute sessionIdleTime in the userSession object class) equal to the value of the attribute idleTimeOutPeriod, this user session will be terminated involuntarily by the managed system. Idle is defined as no activity taking place at a specific channel. This idle period is dependent on the channel that the user session is on and may vary from channel to channel. In order to accommodate certain user activities, which may take longer than the idleTimeOutPeriod to be executed by the managed system, and if during this time no other activity is occurring in the communication channel, the results of these activities should be properly handled by the managed system so that they will be properly delivered to the originated user. If no attribute values (except for channelId) are specified at the instantiation time, the values of these attributes should be obtained from the corresponding attributes in the securitySystemInitialValue managed object.

One instance of securityChannel object class is created at system initialization time for each existing communication channel. The CMIS M-CREATE and M-DELETE services can be used to create and delete an instance of this object class when a new channel is activated or when an existing channel is eliminated. One and only one instance of securityChannel object class exists for each communication channel at any one instant of time. The number of instances of securityChannel object class that can be contained in any one instance of its superior object class (securitySystem) is dependent on the number of communication channels that the managed system can support.

11.2.1.5 Session Management. There is one **userSession** MO instance per session. It allows a security adminstrator to gather information about the sessions that are currently taking place. It has the following attributes, which can only be read:

- userSessionId
- sessionLocationId
- sessionStartTime
- processIdList
- sessionIdleTime

The user session object class is a class of managed support objects that represent a user login session. The existence of an instance of this object class indicates that a user is currently logged into the managed system or its associated support entity. One instance of this object class is created by the managed system when a user login request is accepted. This instance exists at all times during the user session. This instance is deleted by the managed system as a result of either user session idle time out or a successful user logout request action. At the creation of an instance of this object class, the sessionLocationId attribute reflects the current session location (the current channel or port), the sessionStartTime attribute reflects the starting time of this session, the processIdList attribute reflects the initial processes, if any, and the sessionIdleTime attribute is set to 0. When a CMIS M-GET service (i.e., read) is used to retrieve the value of the attribute processIdList, such a value should reflect the list of processes at the time that the M-GET service is executed. When the sessionIdleTime attribute value is equal to the idleTimeOutPeriod attribute value in securityChannel object class, the sessionIdleTimeout notification is emitted. The emitting of this notification can result in the immediate termination of the user session or a warning to the user followed by the termination of the session after a specified period of time if the user session still remains idle. Information contained in this object class is used to support session control functions.

An instance of userSession object class is created automatically when a user is logged into the managed system (as a result of userLoginRequest action, for example). An instance of this object class cannot be created using CMIS M-CREATE service. However, an instance of this object class can be deleted using CMIS M-DELETE service (by a system/security administrator, for example) to force out (logout) a user. Only one instance of this object class may be contained in any one instance of the superior object class securityUser. This is to prevent multiple login sessions by a single user. However, should this requirement be modified to allow multiple login sessions by the same user, no changes need be made to the managed object as the user session is identified by a session ID. At the creation of an instance of the userSession object class, the sessionLocationId attribute shall be set to reflect the current session location, the sessionStartTime attribute shall be set to reflect the starting time of this session, the processIdList attribute shall be set to reflect initial processes, if any, and the sessionIdleTime attribute shall be set to 0.

11.2.1.6 Security MOs Management. Some attributes in different instances of the login management MOs are likely to have identical values. For instance, the number of unsuccessful login attempts before a channel is locked is likely to be the same for most channels for a given system. To facilitate the management of such values, each system has one instance of the **securitySystemInitialValues** MO. Its attributes contain the default initial values of attributes in other objects. It has the following attributes:

- nonUsedPeriod
- passwordAgingInterval
- unsuccessLoginAttempt
- chanelLockoutPeriod
- idleTimeOutPeriod
- securitySystemInitialValuesId

The securitySystemInitialValues object class is a class of managed support objects that contain initial values for attributes in other managed objects. If values for those attributes are not specified during instantiation time, the corresponding values of those attributes in this managed object should be used instead. The securitySystemInitialValuesId attribute is used strictly for naming. Since there can be only one securitySystemInitialValues object per securitySystem, this attribute carries no information and therefore it always has the value NULL.

The values of the attributes of securitySystemInitialValues can be changed to reflect changes in local security policy. If, for example, the security manager decides that passwords should be changed at least every 60 days, rather then every 90 days, the value of the passwordAgingInterval attribute in securitySystemInitialValues will be changed from 90 to 60 days. Thereafter, every time an instance of userPasswordAuthentication is created without specifying the value of its passwordAgingInterval attribute, the value of that attribute will be automatically set to 60 days.

Only one instance of security system initial value managed object may exist for an instance of security system managed object.

11.2.2 Notification Management

Several objects are capable of issuing notifications that contain information relevant to security. Some notifications are security alarms that require immediate action. Some event notifications need to be analyzed together to determine if a security alarm is warranted. Some event notifications need to be logged for possible security audit. The security administrator needs to control which security events are sent to whom and under what circumstances, which event notifications should be recorded in a security log, and manage the security log (e.g., archive its content when it is full). The MO classes and service definitions for notifications management are defined in ISO/IEC 10165-2 | ITU-T X.721 [X.721], ISO/IEC 10164 parts 5, 6, 7 and 8 | ITU-T Recs. X.734, X.735, X.736 [X.734, X.735, X.736, X.740], X.740, and

TMF (Tele Management Forum, formerly the NMF or Network Management Forum) OMNIPoint 1, Forum 006 [NMF006]. To keep matters lively, there are some differences between the ITU-T and TMF models for the same functions.

The cast of characters that participate in security notification management are introduced in order of appearance.

MOs in a local open system issue notifications; in particular, any application in the system can use an instance of the forumSecurityNotifier MO class to issue security notifications. The notifications issued by MOs, as well as PDUs (other than system management event reports)[1] that are to be logged are sent to a conceptual **log preprocessing function** that forms potential log records. If the local open system contains a log, then the conceptual log preprocessing function sends the security alarm report records to the log MO. If the local open system contains a security audit trail log MO, the conceptual log preprocessing function sends the security audit trail records to this log.

An MO may be primed to issue **security alarms** when especially suspicious events occur. For instance: an unauthorized user attempts to modify access privileges stored in the system or the system receives executable code with invalid digital signature. The information in a security alarm includes:

- Identity of the managed object issuing the alarm
- Event time
- Event type (integrity violation, operational violation, physical violation, security service or mechanism violation, or time domain violation)
- Security alarm cause (different causes apply to different event types)
- Security alarm severity
- Security alarm detector
- Service user
- Service provider

A security alarm is issued by the agent using CMIS M-EVENT-REPORT service, which may or may not require confirmation upon receipt by the manager.

Security alarms need to be forwarded to the appropriate party (typically the security administrator) most expeditiously. The forwarding is done by an **Event Forwarding Discriminator** (EFD). An EFD is an MO that accepts potential event reports and decides on their disposition. The decision is based on a filter, which is a Boolean expression on the type of potential event reports and values of their fields. It is used mostly to separate the urgent from the irrelevant. An EFD contains the address (typically an Application Entity Title) for the destination of event reports that need forwarding. It may also contain a list of backup destinations, just in case the primary destination is not available. An EFD may further contain a schedule specifying when it should be active: which days of the week and which hours of the day, from which starting time until which termination time. The filter, destination addresses, and scheduling information can all be controlled by a remote manager. The manager can also temporarily disable an EFD. A single system may have several EFDs, some of which may be active simultaneously.

Security alarms may be recorded in a **security alarm log** for off-line analysis. In addition to security alarms, there are more mundane events that, in the aggregate, may help spot security breaches. Such events may include all login attempts and all attempts to access security specific information, such as access control lists. Such events are typically recorded in

[1]PDUs must be modeled as notifications before they can be logged or sent to an EFD. No such modeling is currently available in standards documents.

a **security audit trail.** The security administrator may choose to record security alarms in the general-purpose security audit trail rather than in a separate security alarm log.

The security alarm trail is a subclass of the more general log MO class. A log contains an attribute, which represents a discriminator construct, the same as in an EFD. This discriminator construct allows the log to decide which of the notifications it receives shall be recorded. This filter can be managed by a remote system. The manager can also control the size of the log.

The information in a security audit trail is captured in instances of **security audit trail record** MO class. The information they contain for each security-related event includes:

- Identity of the managed object
- Event type (service report—event related to the provision, denial, or recovery of service; or usage report—statistical information related to security)
- Event time
- Service report cause (request for service, denial of service, response from service, service failure, service recovery, others)

This data can be used to analyze any security violations and identify adjustments to be made to the operational procedures. Security audit messages or logs are generated as a result of the occurrence of security-related events. Security-related events are determined based on predefined conditions prescribed by security policies or requirements. These security policies or requirements are often specific to a particular application and to specific administrations. Audit mechanisms process audit logs to determine whether security-related actions need to be taken. Some intrusion attacks may be detected by analyzing audit logs generated by entities in different management domains. The procedures for analyzing audit logs are outside the scope of this book. Even though security audit is not directly involved in preventing security violations, it helps in deterring them as well as in detecting them so that actions can be taken to enhance current security policy.

The log and/or security audit trail log MOs may receive event reports from other open systems. They use their discriminators to decide which event report to store.

The security administrator, through the managing process, controls the security audit trail log, log, and eventForwadingDiscriminator MOs and receives responses from them.

Figure 11.3 illustrates security notification management.

The MOs used in this section for notification management are defined in the following standards:

system	ISO/IEC 10165-2 I ITU-T X.721 and ISO/IEC 10164-1 I ITU-T X.730
managedElement	ITU-T M.3100
log	ISO/IEC 10165-2 I ITU-T X.721 and ISO/IEC 10164-6 I ITU-T X.735
securityAuditTrailLog	ISO/IEC 10164-8 I ITU-T X.740
logRecord	ISO/IEC 10165-2 I ITU-T X.721 and ISO/IEC 10164-6 I ITU-T X.735
eventLogRecord	ISO/IEC 10165-2 I ITU-T X.721
securityAlarmReportRecord	ISO/IEC 10165-2 I ITU-T X.721 and ISO/IEC 10164-7 I ITU-T X.736
securityAuditTrailRecord	ISO/IEC 10164-8 I ITU-T X.740

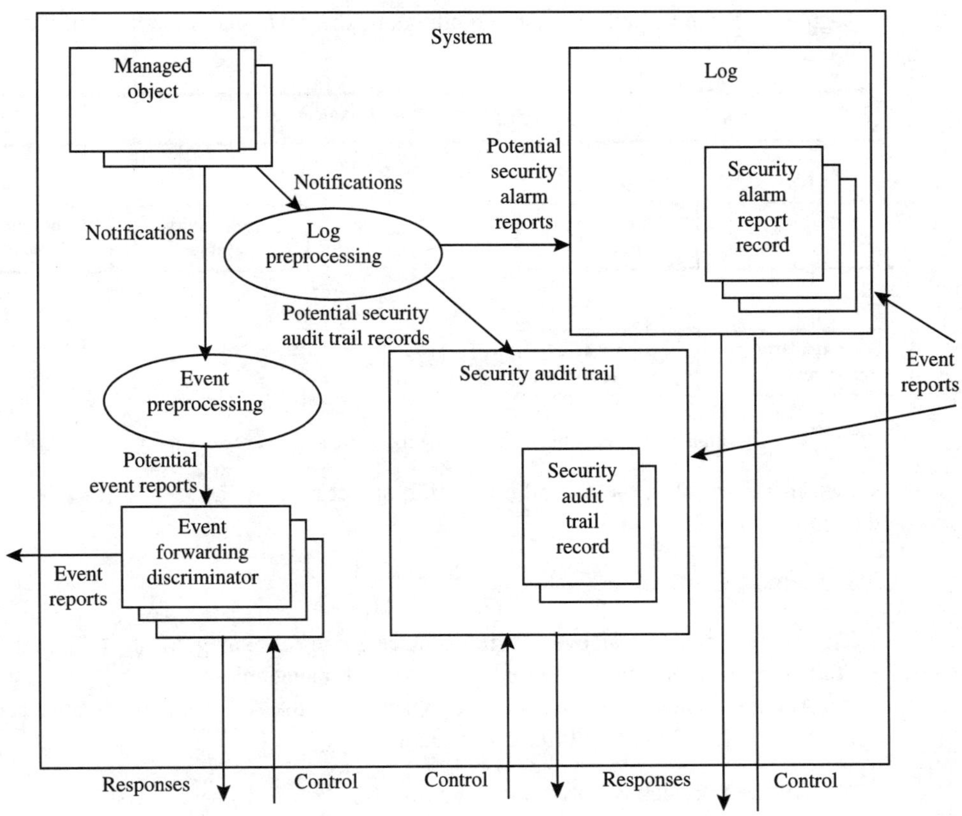

Figure 11.3 Notification management interactions

eventForwardingDiscriminator	ISO/IEC 10165-2 ǀ ITU-T X.721 and ISO/IEC 10164-5 ǀ ITU-T X.734
forumSecurityNotifier	NMF OMNI*Point 1,* Forum 006 [NMF006]

The inheritance of MO classes that support notification management is depicted in Figure 11.4.

Although the security audit trail log is shown explicitly in Figure 11.4, it is defined in ISO/IEC 10164-8 ǀ ITU-T X.740 to be a log as defined in ISO/IEC 10164-6 ǀ ITU-T X.735.

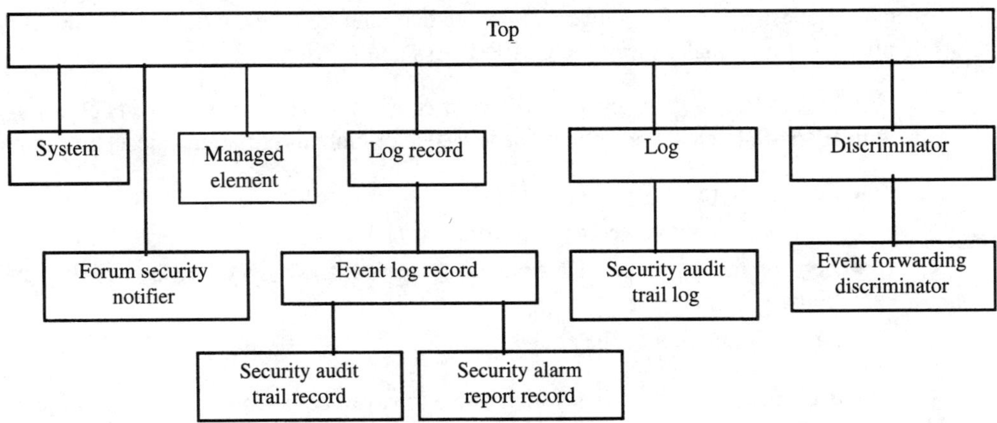

Figure 11.4 Inheritance of notification management MO classes

Figure 11.5 shows possible name bindings of the MOs that support notification management.

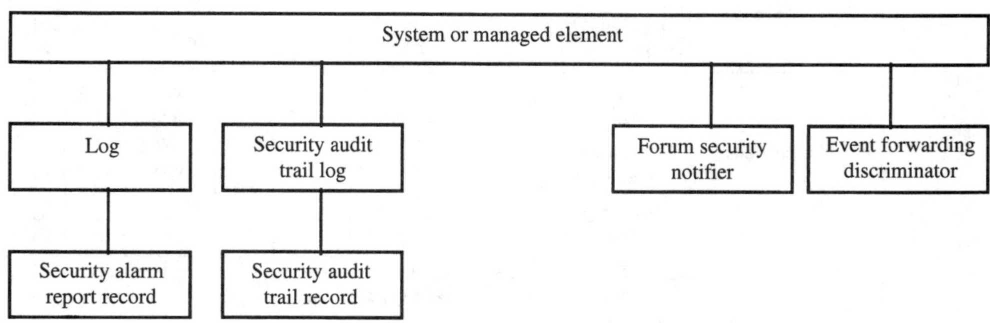

Figure 11.5 Possible name binding for notification management MOs

As shown in Figure 11.5, the overall containing object may be the system for OSs, or managed element for NEs.

11.2.3 Access Control

This section provides an overall view of access control management; it then introduces (in Section 11.2.3.5) the MOs used for access control management.

An Agent's managed resources are represented by the Agent's MIB. Secure access to specific MIB views is based on three components:

- Definition of each MIB view or Target.
- A set of Rules that specify which (group of) Initiator(s) has access to any particular view.
- Initiator bound ACI (Access Control Information); it may, for example, be an authenticator or an access control certificate presented by the Initiator. ACI is used in conjunction with the set of Rules to grant or deny access.

The first two items above are covered by ISO/IEC 10164-9 | CCITT Rec X.741 (Information Technology—Open System Interconnection—System Management—Part 9: Objects and Attributes for Access Control) [X.741]. 10164-9 supports the management of access control information that an Agent uses to determine whether access to a Target (i.e., a specific operation on a specific information element) by an Initiator is to be granted or rejected. The last bullet item above can be handled, among other options, with the ACSE authentication functional unit, or the CMIP access control field.

11.2.3.1 Targets. A distinct MIB view consists of one or more Targets object(s) as defined in 10164-9. Each Targets object specifies one or more information objects by enumerating

- Zero or more managed object classes
- Zero or more managed object instance(s)
- Zero or more groups of managed objects instances specified by scoping and (optionally) filtering
- Any combination of the above

A Targets object further specifies which operations can be performed on those objects or on attributes of those objects. It also specifies any constraints on the values of parameters used in such operations.

The use of Targets objects therefore allows a wide range of access control granularity: from specifying access to all objects in a MIB to allowing access for only some specific operations, with limited values of their parameters, on some specific attributes of some specific object instances.

11.2.3.2 Rules. 10164-9 provides an information model that consists of a set rules that either grant or deny access to a MIB view (i.e., a set of targets).

Each Rule managed object associates one or more Initiators to one or more Targets. A rule also specifies the context within which it applies (e.g., time of day, day of the week, some conditions on the values of some attributes specified by scoping and filtering).

A Rule managed object therefore provides a distinct MIB view to a specific set of Initiators. The Initiators that have access to this view are specified in the Initiators managed object.

11.2.3.3 Initiators. A Rule MO points to an Initiators MO. The Initiators MO instance identifies the initiators (requestors) of management operations to which the rule applies. ISO/IEC10164-9 | ITU-T X.741supports three basic schemes for specifying Initiators:

- Access Control Lists (ACL)
- Capability-based schemes
- Label-based schemes

The three schemes correspond to different ways of representing the same information.

11.2.3.3.1 Access Control Lists. Access Control Lists (ACL) specify the allowed initiators for each (group of) Target(s). Typically, with ACLs, the Initiator-bound ACI consists of an authenticated Initiator ID. This ACI can be provided, for example, in the access control field of CMIP as in the case of Electronic Bonding (EB). When requesting an operation, in this example, the value of the CMIP access control parameter is verified against the values in the ACL.

11.2.3.3.2 Capabilities. Capability-based schemes are similar to ACLs, except that they specify the Targets accessible to a (group of) Initiator(s). The Initiator bound ACI typically consists of a proof (a certificate or a token) that the Initiator is a member of one or more groups.

11.2.3.3.3 Security Labels. In an access control scheme based on security labels, each Initiator and each target is assigned to one or more security groups (e.g., confidential, secret, top secret). Each security group has well-defined access privileges. The Initiator of a request must demonstrate to the Agent that it belongs to a security group that has the access privileges that match the security label (e.g., top secret) of the target. To this end, the Initiator may have a certificate, issued and signed by an appropriate Certification Authority (CA), stating that the Initiator belongs to the relevant security group. If an entity belongs to several security groups, it may have several such certificates; each one can attest to membership in one or more security groups. Each certificate contains the entity's identity (such as an AETitle) and a list of one or more security group identifiers. It may also contain the date of issue of the certificate, time interval (from, until) during which the certificate is valid, and other information.

In order to prevent unauthorized replay of a security label certificate, it must be bound to a specific request message. To achieve such binding, the security label certificate may be contained in a message signed by the Initiator. That message must include a nonrepeating string such as a time stamp. Alternatively, the security label certificate may be sent to the Agent outside any message signed by the Initiator. In this case, the Initiator must provide authentication protection for the message that actually requests the operation. Again, the authenticated message must include a nonrepeating string such as a time stamp. In the second case, if the Agent keeps a copy of the security label certificate, then it may not be necessary to submit this certificate again for subsequent requests to the Agent. In either case, the Agent must have received (previously or in the same message) the Initiator's original certificate (containing the Initiator's identifier and public key).

11.2.3.4 Authentication for Access Control. Initiators must provide some Initiator-bound ACI (Access Control Information) that is used in conjunction with a Rule to grant or deny access. Typically, this ACI provides the authenticated ID of the Initiator. The format of such information may be, for example, authenticated user id or access control certificate.

The ACI can be supplied at association setup time as well as with any management operation (i.e., m-Create, m-Delete, m-Get, m-Set, and m-Action). In the case of Electronic Bonding (EB) for Trouble Administration (TA), which is presented here for illustration purposes, the CMIP access control field is used both at association setup and during subsequent CMIP operations to supply the authenticated Initiator ID. In the TA EB, the access control field is defined to consist of the following components:

- The identity of the Initiator
- The identity of the secret key, known only to the two communicating parties, which is used for the current PDU (the key being listed in a secret list that both parties share)
- An Initialization Vector (IV), which is an arbitrary string of 64 bits that is used for the current PDU.
- The GeneralizedTime corresponding to the issuance of the PDU, encrypted using the Digital Encryption Standard (DES) in the Cipher Block Chaining (CBC) mode, with the key and IV specified above.

When the Agent receives the Initiator-bound ACI, it verifies the validity of that information (e.g., in the TA EB example, it verifies that the encrypted portion of the access control field decrypts into valid GeneralizedTime). The Agent then uses its set of Rules to determine if the Initiator should be granted or denied access to the request in the PDU.

11.2.3.5 MOs for Access Control. This section provides a brief description of the MO classes defined to support access control management. The MOs are defined in ISO/IEC 10164-9 |ITU-T X.741.

Figure 11-6 illustrates the inheritance hierarchy of the MO classes. Most of the access control MO classes are derived from the access control MO class. The access control MO class is not instantiated; it is used simply to provide a single point of specialization for other object classes.

Figure 11-7 illustrates the relationships between the access control MOs. (This figure is somewhat different from the corresponding figure in ISO 10164-9. In particular, the figure in this book does not have any unlabeled MOs, and it does show the assigned labels MO and its three subordinate MOs.) Both in Figure 11-7 and in ISO 10164-9, the initiators MO represent all three subclasses of initiators (i.e., ACL initiators, capability initiators, and label initiators).

When the Agent receives a management operation request, the access control decision function starts by using the access control rules MO instance, along with the ACI provided by the initiator.

The **access control rules** MO contains attributes that represent information such as what to do if a request is denied. It also contains a notification emitter MO and rule MOs.

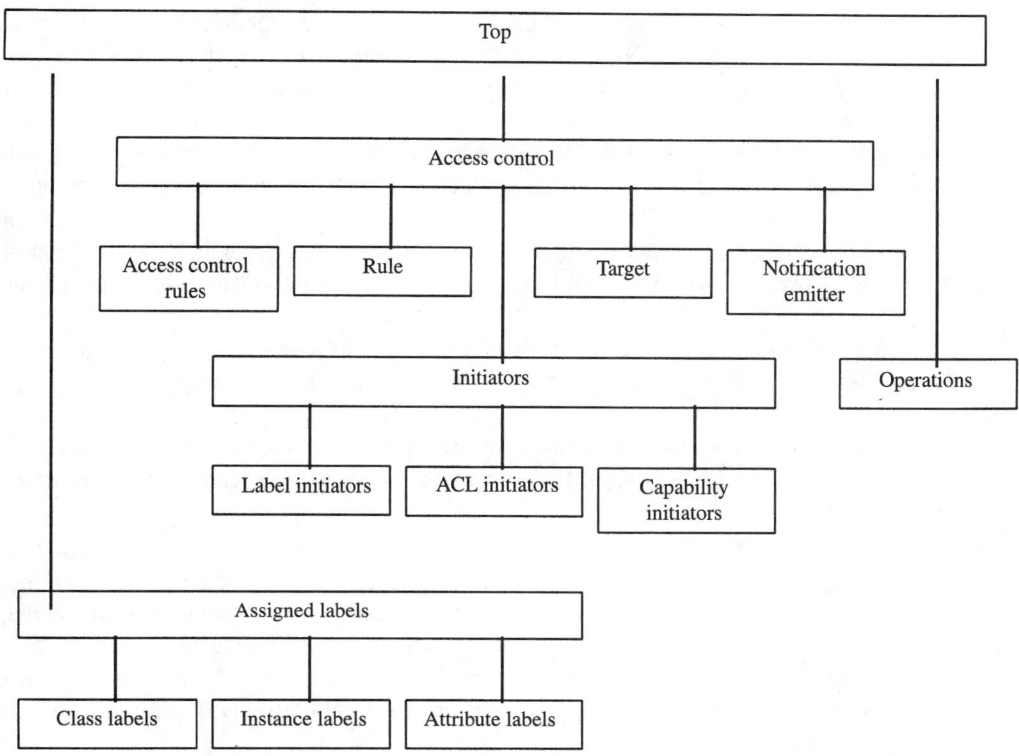

Figure 11-6 Access control MO class inheritance hierarchy

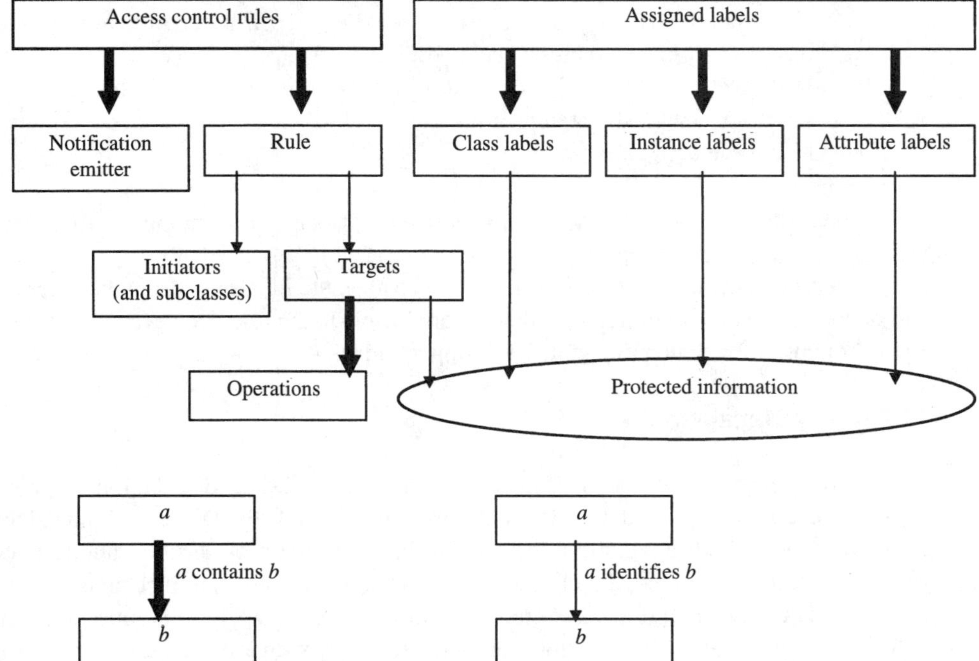

Figure 11-7 Relationship of access control MOs

A **rule** MO contains several packages that allow detailed specification when, and under what circumstances, the rule is applicable. It contains an attribute that specifies what to do if the rule is satisfied (i.e., grant or deny the request). It also contains attributes that point to targets and initiator MOs.

A **targets** MO represents the protected information to which the rule applies. Its attributes can point to MO classes, MO instances (listed explicitly or via scoping and filtering), and specific attributes within MO instances. It also contains operations MO instances. An **operations** MO lists the management operations (including actions) for which the rule is applicable. It also represents constraints on the values of the parameters of the operations by using filters.

An **initiators** MO identifies permissible initiators of management operations. It is specialized into three MOs, depending on the access control scheme used.

- The **ACL initiators** MO contains a list representing the initiators to which the rule applies. This MO is used if the access control decision depends on the identity of the initiator.
- The **capability initiators** MO can be used to specify the security domains and the specific management operations for which these security domains can grant access control privileges. This MO is used if the access control decision depends on the initiator belonging to some specific group (which may include just the initiator) or if the initiator has certain access control certificates. In this case, the initiator must prove it is a member of some group or that it has the right certificates, rather than prove its identity.
- The **label initiator** MO is used if the access control decision is based on security clearance (e.g., secret, top secret). In this case the access control decision function may also use the **assigned labels** MOs, which specify the security clearance needed for accessing specific information. The assigned labels MO contains three types of MOs:
  —**Class label** MO assigns a security label (i.e., a security classification) to all MO instances of one or more MO classes.
  —**Instance label** MO assigns a security label to one or more MO instances.
  —**Attribute label** MO assigns a security label to one or more attributes of a specific MO instance.

The **notification emitter** MO contains several packages for issuing security alarms, as well as usage and service reports.

The rule MO may contain a package whose attribute identifies the authentication policies and authentication requirements that are applicable to the rule. However, use of this package is not currently contemplated in TMN applications.

11.2.4 Key Management

Several mechanisms involved in the secure exchange and distribution of encryption keys do not require any management operations. These include the Diffie–Hellman algorithm, Kerberos servers, and the X.500 directory for the distribution of certified public keys. The present book introduces an MO class that can be used when no such mechanisms are available. This is the case, among other circumstances, when a new entity is introduced into the security domain. Initially, it does not share any secret key with any Kerberos servers (if there are any), nor does it know the public key of the Certification Authority for the security domain (if there is any) that would be needed to verify public key certificates.

The MO presented here is modeled after, though it is not identical to, the key distribution method in Electronic Bonding. The Agent maintains a list of keys in the key list MO. The attributes of this MO represent the following information:

- List identifier (integer or printable string)
- Key type identifier (Object Identifier [OID])
- List size (integer)
- List creation date (optional)
- List validity period (optional)
- List creator (optional)
- List encryption method (OID, optional, default [if not present]: triple DES)
- Key encryption keys (a sequence of key identifiers—optional; if the keys are encrypted, this attribute identifies the key[s] used to encrypt the keys in the list)
- Other parties having the same list (set of OIDs or AETitles)
- A list of keys, where each entry contains the following information:
  —Key identifier (integer)
  —Key (determined by key type, default: octet string)
  —Key status (not used, in use, used, not to be used)

This MO can be created and deleted by the Manager. The Manager and the Agent can change the status of any key in the list to "not to be used." The Agent can change the status of a key from "not used" to "in use" and to "used." All other information in the MO can only be read. When the manager creates an instance of the key list, the values of all the attributes are specified. The Manager has the option of transmitting the actual key in clear text or in ciphertext. If the keys are encrypted, then the encryption algorithm is identified, as well as the key(s) used for encryption. Those keys may be included in another key list MO instance. It is recommended that each key be triple DES-encrypted, using three different DES keys. In this case, the first (second, third) key in the sequence is the first (second, third) key that was used for encryption and therefore the third (second, first) key to be used for decryption.

PART V
SECURITY OPERATIONS

Parts 2 and 3 of this volume covered a bewildering array of security features. A TMN security planner may (rightly) be bewildered when trying to decide which features to deploy where. More worrisome, after making such decisions, our planner may (and should) wonder: "Ok, what have I left out?" It would be nice to have a neat checklist, to serve as a memory jogger, of all the security issues that need to be considered (though not necessarily acted upon). The industry's best efforts to compile such guidelines, for the whole TMN, are captured in ITU-T Rec. M.3400 and Bellcore GR-2869. This part of the book is based on the security-related portions of those two documents.

Parts 2 and 3 also focused on aspects of TMN security that are inherently automated. In addition to machines' contribution to security, humans still have plenty to do. Although machines are in charge of encryption and of logging security significant events, people have to decide which encryption algorithm to use for what and how to tease something meaningful out of endless security logs. This part of the book addresses the (people-oriented) operations needed to secure the TMN and to manage security.

It starts with an overview of the functions and operations that provide TMN security. Such functions are described in ITU-T Rec. M.3400 and Bellcore GR-2869 as part of Common Operations Management (COM) functions that may have to be supported by individual TMN elements. COM functions include securing the TMN. Chapter 12 provides an overview of COM security functions.

This part also further provides an overview of security management functions and operations. Those functions may be automated, though quite a few are manual. This discussion is relevant mostly to security administrators and operation planners.

The functional architecture of the TMN explicitly addresses Security Management (SM). Chapter 13 of this book, based on ITU-T Rec. M.3400 and Bellcore GR-2869, lists the SM functions identified so far. It also portrays the information that needs to be exchanged over (standardized) interfaces in order to perform those functions. In many cases, the SM functions (e.g., key management, management of access control information) can be performed by one or a few centralized systems.

12

Security Functions and Operations

ITU-T Rec. M.3400 and Bellcore GR-2869 list several functions that may have to be supported by individual TMN elements under the banner of Common Operations Management (COM). COM functions include securing the TMN. Security Common Operations Management comprises security features that protect all management operations, that is, covering all the TMN layers and all the TMN Management Functional Areas. They may be exercised in the course of any communication among systems, customers, and internal users, and they are needed in all the TMN components (OSs, NEs, and WSs). They can be divided into two categories: prevention and detection services. Security services that are within Common Operations Management are quite separate from, and complementary to, the Security Management functions covered in Chapter 13. As can be seen from the scenarios in this section, security services interact frequently with the Security Management functions in Chapter 13.

12.1 PREVENTION SERVICES

Prevention services are a set of security services that prevent unauthorized intrusion. These services are described in ISO IS 7498-2/Recommendation X.800; they are:

- Authentication
- Access control
- Data confidentiality
- Non-Repudiation

Every component of the TMN must have authentication and access control capabilities. A centralized authentication and access control server(s) may simplify the task of individual components (such as NEs) but not eliminate them. One possible scenario (and by no means not the only one) is that every time an OS or an NE needs to perform an authentication or access control (authorization) function, it sends the request to a centralized authentication/authorization system. The response from that system indicates whether the authentication is valid and whether the authorization requested should be granted. With this simplified scenario, the OS or NE must be capable of recognizing that a central authentication/authorization system must be consulted; securely send a request for verification/authorization to that central server to authenticate the response (verifying that it comes from the correct central server); and check its integrity (that no information in the message had been altered).

The advantage of a centralized authentication/authorization system is that each NE it serves must securely exchange encryption keys and related material, with only one centralized system, thereby greatly simplifying key management.

12.2 DETECTION SECURITY SERVICES

Detection services allow the detection of suspected security violations and their reporting. Integrity assurance provides for the detection of unauthorized insertion and the deletion or alteration of data in transit or in storage. It is described in ISO IS 7498-2/Recommendation X.800. Software virus detection checks for the presence of a known software virus or any other type of "bad code" (e.g., worm, Trojan horse) when the system receives new software. Security event detection and reporting provide for detecting security-significant events, logging such events, and reporting security alarms.

Analysis of logs of security-significant events may allow the detection of repeated, low-profile intrusions.

12.3 ILLUSTRATIVE SCENARIOS

A couple of examples are used to illustrate the operations involved in TMN security.

12.3.1 Authentication and Access Control Scenario

The scenario shown in Figure 12.1 is an example of functionality that may be required to authenticate and validate a user's request for access to network capabilities. This scenario illustrates authentication and access control for an external customer (at the SML level). Similar security procedures (not shown) will be applied to check the authentication and access control of the internal user (who originates a request caused by the external customer's initial request). This scenario highlights the interactions between Common Operations Management (COM) and Security Management (SM) if necessary.

1. A customer request to perform a function or set of functions is initiated from one of the Management Functional Areas for Performance, Fault, Configuration, or Accounting Management.

   Before the request is carried out, it is necessary to authenticate the requesting customer and verify that the customer is authorized to submit the request. These are authentication and access control COM functions that would be present in every system, though in this illustration they are located in the SML

   Associated Data

   - Customer ID
   - Authentication information (key ID, encrypted authenticator)
     Access Control Information (Access Control Certificate, User Permissions)

2. A customer profile check may be made to determine whether the requested functionality is outside of the normal range of activities for this customer. If so, a warn-

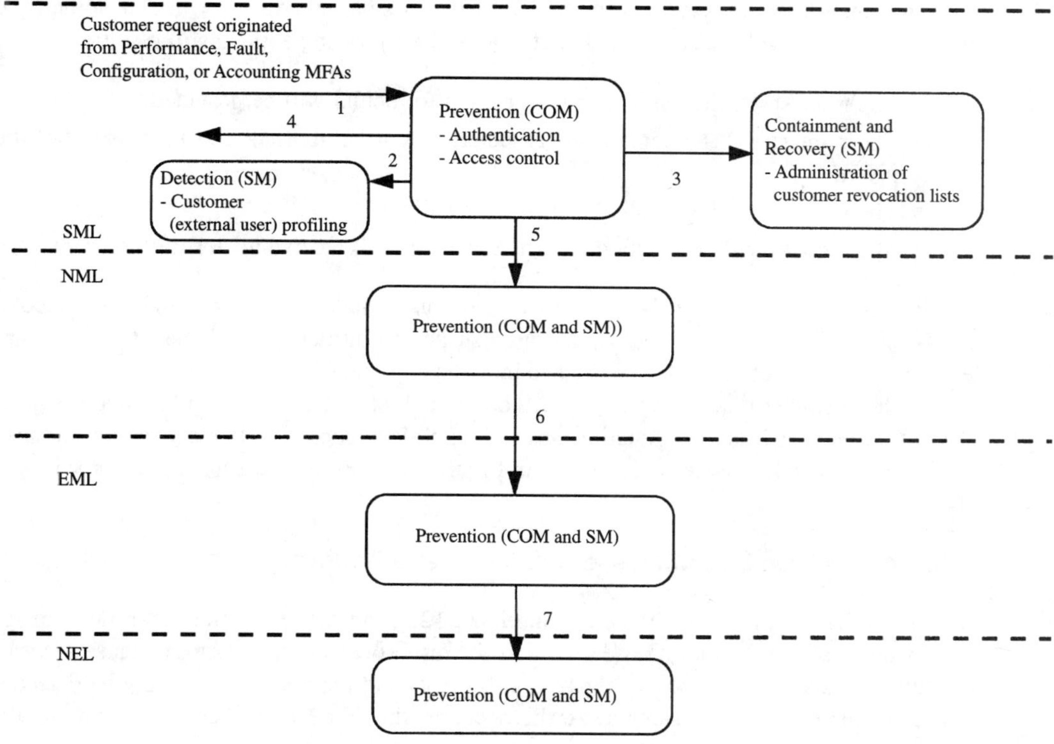

Figure 12.1 Authentication and access control

ing alert should be made to indicate the possibility of an unauthorized intruder using an authorized customer's permissions. Requesting the customer profile check is part of the COM prevention and detection security functions; performing the customer profile checking is SM functionality. Indeed, this check need not be performed by every system; it is usually done by a single, central system. Furthermore, checking a customer's profile typically belongs in the SML.

Associated Data

- Customer ID
- Request type

3. Administration of the Customer Revocation List will be invoked to check if any of the customer's encryption keys, access control certificates, or other security information in the customer's request has been revoked. The initiation of the check is again a COM Prevention function, while the actual checking is an SM function that is typically performed by a central server.

Associated Data

- Customer ID
- Key ID
- Access Control Certificate ID

If the customer's request is within the range of permissible activities for that customer, the customer's request will be accepted. If not, one of the following actions will result:

- The request will be denied, and the reason for denial will be provided.
- The request will be denied, the association will be aborted, and no reason will be given.
- The request will be satisfied as if it were valid.
- The request will be satisfied, but the requestor will receive misleading information.

The first case is typical if it is believed that a valid user made an honest mistake; the second case is intended to protect the network from a suspected intruder; and the last two cases are designed to "hook" the intruder in order to help his apprehension.

If the customer request is valid, it will be carried into the NML and other layers as necessary. Similar COM security functions, as well as SM functions, will be carried out in those layers as appropriate. Those steps (5 to 7 in Figure 12.1) are not described explicitly in this scenario.

12.3.2 NEL Alarm Detection, Containment, and Recovery

The following scenario, illustrated in Figure 12.2, shows the functions that might be invoked when a security violation is detected as the result of a security alarm. Because security violations have a wide variety of impacts on services and the network and may be detected via a wide range of security safeguards, this scenario should be viewed as typical but not all-inclusive.

Some examples of detection may be the result of:

- Detection of a security alarm about a customer's service (e.g., a customer service usage is outside the scope of that customer's service profile) or customer interface (e.g., integrity check failure on a customer's message)
- Detection of a security alarm about the network
- Detection of a security alarm about one or more individual NEs
- Physical detection of an intruder by a guard or building alarm

This scenario is initiated through the detection of a security event in one or more individual NE(s). In this example the COM detection functionality happens to be in a NE. In general, this functionality is present in every TMN system at every layer. If the COM Detection occurs in any layer other than the NEL (those cases are not shown in this illustrative scenario), the alarm will be sent to Security Management (SM) Detection in that same layer.

1. Detection of a security violation may occur as a result of any of multiple security safeguards that are in place in the network. At the NEL an **NE Detection** may detect an alarm or a breach of security and report the event to EML–Detection–NE(s) security alarm management. EML–Detection–Support Element security alarm reporting may also be invoked to provide additional information about the alarm so that the appropriate Containment and Recovery activities may be initiated. Detection of a security threat by an NE is an example of a COM detection functionality. All other security functions in this illustrative scenario are SM functions.

   Associated Data

   - Security Alarm

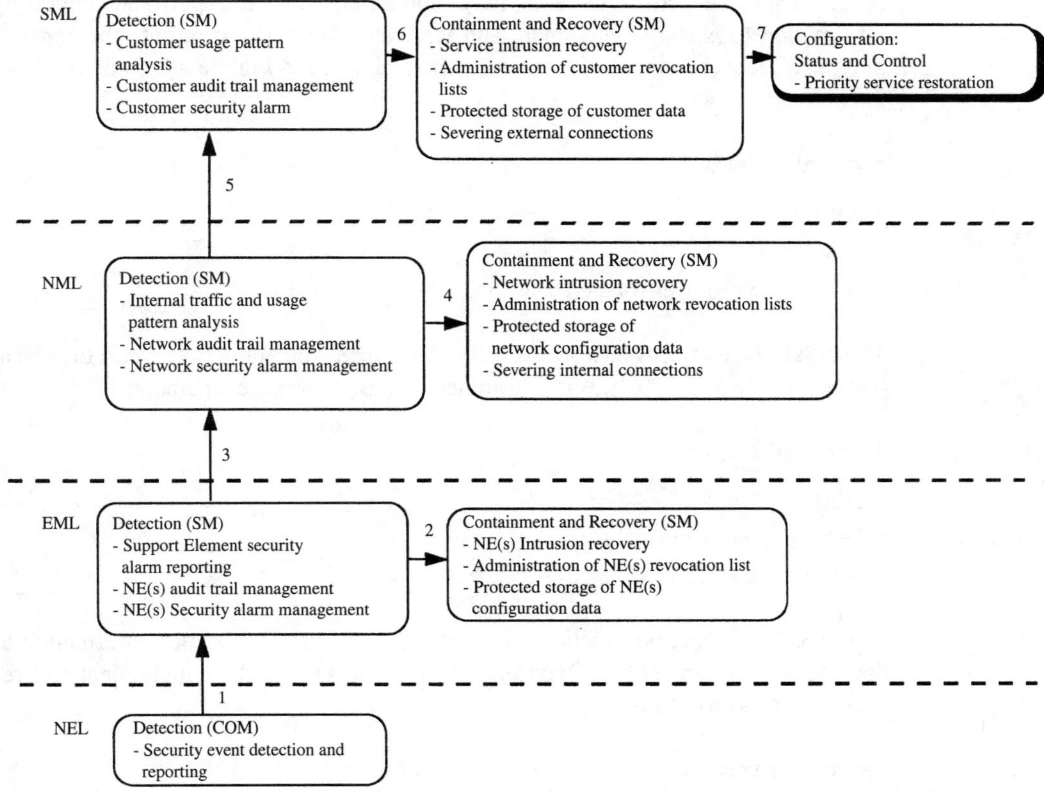

Figure 12.2 NE alarm detection of a security violation, containment, and recovery

2. At the **EML Detection,** functions may send an alarm event report and supporting information to EML–Containment and Recovery functions to initiate the containment and recovery process and, if data is corrupted, to reinstate NE data. The alarm event report sent by the EML Detection function may contain raw information from the security alarm it just received, or it may contain other information based on analysis and event correlation performed at the EML Detection function.

Associated Data

- NE ID
- Key ID
- Access Control Certificate ID
- Security Alarm Information

3. **EML–Detection–**NE(s) security alarm management may forward the alarm or event and the associated information to the NML–Detection–Network security alarm management.

Associated Data

- NE ID
- Security Alarm Information
- Access Control Certificate ID

4. At the **NML, Detection** functions may send an alarm event report and supporting information to NML–Containment and Recovery functions to initiate the containment and recovery process for the network such as isolating the affected NEs from the network.

   Associated Data

   - NE ID
   - Key ID
   - Access Control Certificate ID

5. The **NML–Detection–**Network security alarm management may forward the alarm or event to the SML–Detection–Customer security alarm management.

   Associated Data

   - NE ID
   - Security Alarm Information
   - Access Control Certificate ID

6. At the **SML, Detection** functions may send a trigger and supporting information to the SML–Containment and Recovery functions to initiate the containment and recovery process for services.

   Associated Data

   - NE ID
   - Key ID
   - Access Control Certificate ID

7. If the security breach has corrupted the network, **SML–Containment & Recovery–**Service intrusion recovery will notify SML–Configuration Management–Status & Control–Priority service restoration to initiate the recovery process.

   If customer data has been corrupted as a result of the detected security violation, the Protected storage of customer data functions may be accessed to restore the customer's data to a state prior to the corruption activity. The Configuration Management–Priority service restoration function may be used to establish an order for restoring the network configuration based on the services riding on the network with priority status such as 911 Emergency Service.

   Associated Data

   - Customer ID
   - NE ID
   - Security Alarm Information
   - Customer Backup Data

13

Security Management Functions and Operations

The TMN presents a daunting challenge to security managers: how to secure such a sprawling, evolving behemoth. On the other hand, it also offers a helping hand. It brings to bear the awesome power of the TMN to the management of security. Some of this help is presented in this chapter. It attempts to enumerate and organize the various functions invoked in the management of security and exhibit some of their interactions.

Security Management (SM) is one of the five management functional areas (MFAs) of the TMN. The first challenge of SM is to clearly define its boundaries. To this end we divide telecommunications security functionality into the following categories:

- *Ensuring the security of management transactions.* This functionality applies to all transactions of all the TMN MFAs, at all the TMN layers, and for all the TMN elements. All security-related aspects of the TMN, other than security management, are discussed in detail in Section 13.4.
- *Ensuring the security of the TMN and the Telecommunications Network.* This functionality provides protection for both targets. Unlike the case with the previous category, the functionality of this category is not needed in all TMN elements. Notice that security of the Telecommunications Network does not include any security services provided by the network to the customers (e.g., customer data encryption).
- *Ensuring security administration,* or the management of security information needed for security of the TMN and of the Telecommunications Network.

The partitioning of SM functionality is illustrated in Figure 13.1.

Figure13.1 further illustrates that security of management transactions (which must be provided at each system) and security of the TMN (which can be provided by a few centralized systems) provide for the security of management information. Similarly, security of the TMN (which can be provided by a few centralized systems) and security administration make up security management, which is the focus of this section.

Security of the TMN and the Telecommunications Network is further subdivided into three groups that characterize the life cycle of security activities:

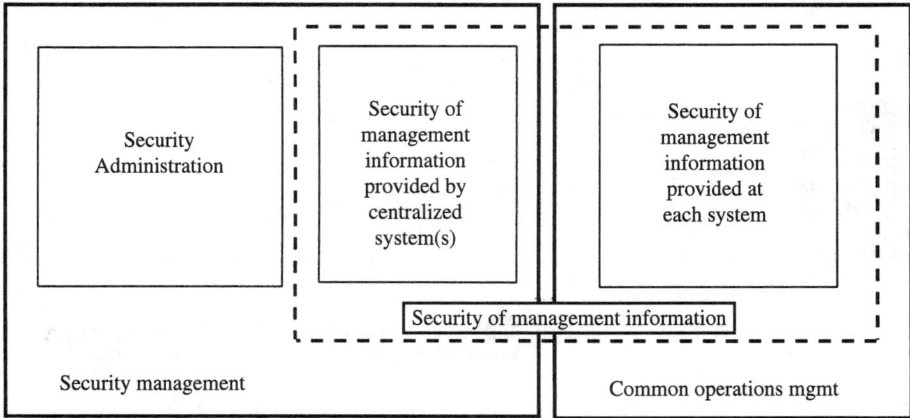

Figure 13.1 Partitioning of TMN security functionality

- Prevention
- Detection
- Containment and Recovery

Category 3 functionality is represented by just one group:

- Security Administration.

In this book a customer is considered to be any external user. "External" refers to another jurisdiction such as a customer or user from a different company or organization. "Internal" refers to an entity within a jurisdiction such as a user from the same company or organization.

The Management Application Functions (MAFs) of Security Management are summarized in Table 13.1. These MAFs are described in Sections 13.1 through 13.4. Generic scenarios that illustrate the use of these MAFs for various business purposes are discussed in Section 13.5. As a result of legislation passed in 1994 in the United States, telecommunications carriers are required to support new requirements for electronic surveillance. Some of the new requirements include additional OAM&P functionality (e.g., real-time monitoring of usage records, tracing and delivery of call data and information content). Security Management must support the requirements that only authorized surveillance shall be carried out and only those with absolute need to know and with proper clearance should know what surveillance is actually carried.

At present, not enough information is publicly available to fully analyze the impact of electronic surveillance on Security Management. Therefore, electronic surveillance is not covered in this section.

13.1 PREVENTION

Prevention is the group of functions that restrict access to management information and capabilities to authorized users. In addition, prevention includes some functions that protect management operations: authentication, access control, data confidentiality, and non-repudiation, as described in ISO IS 7498-2. Since these functions are needed in all the TMN layers and for all the TMN MFAs, they are included in Common Operations Management.

13.1.1 BML

- *Legal review:* involves legal review of corporate documents, service offerings, and policies to allow security activities such as monitoring, establish legally binding "no

TABLE 13.1 Security Management

Layer	Prevention	Detection	Containment & Recovery	Security Administration
BML	Legal review. Physical access security. Guarding. Personnel risk analysis.	Investigation of changes in revenue patterns. Support of element protection.	Protected storage of business data. Exception report action. Theft of service action. Legal action. Apprehending.	Security policy. Disaster recovery planning. Manage guards. Audit trail analysis. Security alarm analysis. Assessment of corporate data integrity.
SML	Security screening.	Customer security alarm. Customer (external user) profiling. Customer usage pattern analysis. Investigation of theft of service.	Service intrusion recovery. Administration of customer revocation list. Protected storage of customer data. Severing external connections.	Administration of external authentication. Administration of external access control. Administration of external certification. Administration of external encryption and keys. Administration of external security protocols. Customer audit trail management. Customer security alarm management.
NML		Internal traffic and activity pattern analysis. Network security alarm. Software intrusion audit.	Network intrusion recovery. Administration of network revocation list. Protected storage of network configuration data. Severing internal connections.	Testing of audit trail mechanism. Administration of internal authentication. Administration of internal access control. Administration of internal certification. Administration of internal encryption and keys. Network audit trail management. Network security alarm management.
EML		Support element security alarm reporting.	NE(s) intrusion recovery. Administration of NE(s) revocation list. Protected storage of NE(s) configuration data.	NE(s) audit trail management. NE(s) security alarm management. Administration of keys for NEs.
NEL				Administration of keys by an NE.

trespassing" signs (e.g., computer screen banners that spell out access restrictions at login time), and protect the enterprise against claims and litigation.

- *Physical access security:* includes the installation, maintenance, and monitoring of all equipment used to restrict or validate access (e.g., electronic access systems requiring a coded badge or other device to validate person seeking access, metal detectors, etc.).
- *Guarding:* includes human monitoring of access such as a guard checking identification badges or checking bags and packages of people entering and leaving a building.
- *Personnel risk analysis:* responds to requests for checking the trustworthiness of current and prospective employees. Similar to security screening (for customers) but usually more thorough.

13.1.2 SML

- *Security screening:* validates that a customer is trustworthy (e.g., has a good credit rating that indicates an acceptable risk for providing service). This function set accesses appropriate databases in other functions.

13.1.3 NML

No NML functionality has been identified for this group.

13.1.4 EML

No EML functionality has been identified for this group.

13.1.5 NEL

No NEL functionality has been identified for this group.

13.2 DETECTION

Detection enables the identification of a security breach such as unauthorized access, corruption of data, customer fraud, and unauthorized actions. The intent of these functions is to determine illegal activities that may be occurring and provide sufficient information about the activities and their source so that corrective action can be taken.

13.2.1 BML

- *Investigation of changes in revenue patterns:* analyzes significant shifts in revenue that might indicate fraud or theft of service.
- *Support element protection:* determines need, monitoring, and analysis of alarm systems that support structures that house network equipment. These alarms may include power, heating, ventilation and air conditioning (HVAC), fire, flood, and open door or cabinet systems.

13.2.2 SML

- *Customer security alarm:* allows a customer access to security alarm information that indicates security attacks on their portion of the network.

- *Customer (external user) profiling:* creates profiles of customer usage data to be used in the analysis of customer activities to identify anomalies and irregularities that may indicate a breach of security, fraud, or theft of service. A customer is considered to be any external user.
- *Customer usage pattern analysis:* provides for the analysis of audit trail information (e.g., originating party, called party, time of day, length of call, etc.) to identify service irregularities or anomalies such as billing and usage abnormalities.
- *Investigation of theft of service:* analyzes customer and internal users whose usage patterns indicate possible fraud or theft of service. This may include possible credit checking, employment record checking, and so on.

13.2.3 NML

- *Internal traffic and activity pattern analysis:* collects audit trail information and determines whether anomalies or abnormalities exist that may indicate a breach of security or theft of customer services.
- *Network security alarm:* allows an internal user access to security alarm information that indicates network security violations.
- *Software intrusion audit:* checks for signs of software intrusion (e.g., the presence of a known virus) in the TMN. Such checks can apply to any TMN system. They can be regularly scheduled, or they may be triggered by security events.

13.2.4 EML

- *Support element security alarm reporting:* identifies the appropriate functions that need information about security alarms so that containment and recovery activities may be initiated. This function may include the reporting of environment alarms (e.g., integrity violation, fire, or moisture) and intrusion detection alarms.

13.2.5 NEL

No NEL functionality has been identified for this group.

13.3 CONTAINMENT AND RECOVERY

Containment is the group of functions that assist in limiting damage to the network such as isolating viruses that cause data corruption or revoking the customer's privileges to perform certain activities that have been identified as unacceptable for that user. Recovery enables the restoration of service and network integrity by providing protected storage of data that can be used to reinstate data that has been corrupted.

13.3.1 BML

- *Protected storage of business data:* provides for mechanisms, such as maintaining back copies of data and monitoring for data corruption, to secure business data.
- *Exception report action:* based on input from an exception report, such as a security alarm, initiates action to limit the security breach (e.g., isolate equipment or data so that corruption is not propagated) and to restore any corrupted data or equipment.
- *Theft of service action:* based on input from an exception report, such as a usage

anomaly, initiates action to limit the security breach (e.g., remove user's access privileges, etc.) and to restore any corrupted data or equipment.

- *Legal action:* initiates litigation against a perpetrator of an illegal action. This may be done in cooperation with law enforcement agencies.
- *Apprehending an intruder:* may be done in cooperation with law enforcement agencies. An intruder may be identified, for example., by analyzing security logs, monitoring targets of intrusion, or feeding misinformation to a suspected intruder.

13.3.2 SML

- *Service intrusion recovery:* includes a set of functions that allow a customer with customer network management capabilities to access backup files in order to restore service after detection of a security violation. If customer network management capabilities are not available, a customer will notify the service provider to initiate intrusion recovery.
- *Administration of customer revocation lists:* maintains a list of all customer public keys and access control certificates for the current time period that are known or suspected of being invalid due to security violation (e.g., stolen secret keys) or administrative procedures (e.g., a customer has moved elsewhere).
- *Protected storage of customer data:* provides for the backup and storage of customer data that can be used in support of intrusion recovery.
- *Severing external connections:* allows connections with a customer to be severed in an attempt to contain data and system corruption as the result of a detected security violation.

13.3.3 NML

- *Network intrusion recovery:* includes a set of functions that allow an internal user access to backup files in order to restore the network configuration after detection of a security violation.
- *Administration of network revocation lists:* maintains a list of all network public keys and access control certificates for the current time period that are known or suspected of being invalid due to security violation (e.g., stolen secret keys) or administrative procedures (e.g., a system has been replaced).
- *Protected storage of network configuration data:* provides for the backup and storage of network configuration data that can be used in support of intrusion recovery.
- *Severing internal connections:* allows internal user connections to be severed in an attempt to contain data and system corruption as the result of a detected security violation.

13.3.4 EML

- *NE(s) intrusion recovery:* includes a set of functions that allow an internal user access to backup files in order to restore Element Management or Element information after detection of a security violation.
- *Administration of NE(s) revocation list:* maintains a list of all keys and access control certificates used for Element Management and Element access that are known or suspected of being invalid due to security violation (e.g., stolen secret keys) or administrative procedures (e.g., an Element Management System or NE has been replaced).

- *Protected storage of NE(s) configuration data:* provides for the backup and storage of NE configuration data that can be used in support of intrusion recovery.

13.3.5 NEL

No NEL functionality has been identified for this group.

13.4 SECURITY ADMINISTRATION

The implementation of a security policy requires the creation and secure dissemination of substantial amount of sensitive information. Such information may include encryption keys, access control rules, access control certificates, directories, security event definitions, and security audit logs. Security administration manages such information by creating, disseminating, and updating security data, and controlling the flow of security-related information (e.g., by defining which security events shall be reported, with what information, under what circumstances and to whom) and analyzing security data.

13.4.1 BML

- *Security policy:* provides company standards for establishing and maintaining a secure environment for personnel, hardware, and software.
- *Disaster recovery planning:* provides for methods and procedures to be used in restoring the network in the event of a security breach and the resulting corruption of data.
- *Manage guards:* provides for the management of physical and mechanized devices used to provide security.
- *Audit trail analysis:* provides methods and procedures for audit trail information to be collected and evaluated to identify possible or potential security violations by individuals or groups of users.
- *Security alarm analysis:* provides guidelines for monitoring, evaluating, and correlating security alarm.
- *Assessment of corporate data integrity:* determines the need for security, monitoring, and analysis of security measures instituted to protect corporate data from unauthorized access, and/or corruption.

13.4.2 SML

- *Administration of external authentication:* provides for verification that customers are who they present themselves to be. It also provides for an authentication path involving external authenticators. If a customer has been authenticated by an authentication agent outside the TMN, this function certifies, if appropriate, that the external authentication agent is a valid entity for providing that kind of authentication. This function also provides for the administration of external authentication for X interfaces at the NML and EML.
- *Administration of external access control:* provides control over what a customer can do with any given resource and includes establishing and validating customer permissions and credentials.[2] This function also provides for the administration of external access control for X interfaces at the NML and EML.

[2]M.3400 distinguishes between vertical and horizontal access control. This function includes both where horizontal access security controls customer access to only those domains belonging to that customer, and vertical access security controls customer permissions to establish and modify the privileges of restricted login user types which are allowed access to only specified subsets of the customer's full capabilities.

- *Administration of external certification:* provides a certification path involving external Access Control Certificates (ACC). A customer may have received from an external Certification Authority (CA) an ACC that permits access to a previously agreed upon set of capabilities. The external certification function certifies, if appropriate, that the external CA is authorized to issue an ACC. This function also provides for the administration of external certification for X interfaces at the NML and EML.
- *Administration of external encryption and keys:* provides for the generation and distribution of encryption keys to be used in communications between an external customer and the TMN. Such keys may be used for authentication, integrity, confidentiality, and non-repudiation. This function also provides for the administration of external encryption keys for X interfaces at the NML and EML.
- *Administration of external security protocols:* provides for the management of Joint Implementation Agreements with other jurisdiction to ensure interoperability of security protocols—for example, ensuring that both communicating parties use the same encryption algorithm with the same set of options and parameters—and agreement on the kind of security information that shall be provided for authentication. This function also provides for the administration of external security protocols for X interfaces at the NML and EML.
- *Customer audit trail management:* allows a customer to establish and configure audit trails to obtain information about service usage. This function allows a customer access to usage and security event information related to their portion of the network.
- *Customer security alarm management:* allows a customer to control which security alarms to receive, with what information, under what circumstances, and at which destination.

13.4.3 NML

- *Testing of audit trail mechanism:* allows for the request of an audit trail mechanism test for data integrity.
- *Administration of internal authentication:* provides for the verification that internal users are who they present themselves to be.
- *Administration of internal access control:* provides control over what an internal user can do with any given resource and includes establishing and validating internal user permissions and credentials.
- *Administration of internal certification:* provides an internal user with an ACC that permits access to a previously agreed upon set of capabilities.
- *Administration of internal encryption and keys:* provides for the generation and distribution of encryption keys to be used in communications between internal users. Such keys may be used for authentication, integrity, and confidentiality.
- *Network audit trail management:* allows for internal users, generally support personnel, to establish and configure audit trails to obtain information about network usage. This function allows an internal user access to network usage and security event information.
- *Network security alarm management:* allows internal users to control which network security alarms to receive, with what information, under what circumstances, and at which destination.

13.4.4 EML

- *NE(s) audit trail management:* allows for internal users, generally support personnel, to establish and configure audit trails to obtain NE usage data. This function allows an internal user access to NE usage and security event information.

- *NE(s) security alarm management:* allows internal users to control which NE(s) security alarms to receive, with what information, under what circumstances, and at which destination.
- *Administration of keys for NEs:* provides for the generation and distribution of local encryption keys to be used in communications between NEs and between an EMS and the NEs it controls. Such keys may be used for authentication, integrity, and confidentiality.

13.4.5 NEL

- *Administration of keys by an NE:* provides for the generation and distribution of local encryption keys to be used in communications between NEs. Such keys may be used for authentication, integrity, and confidentiality.

13.5 SECURITY MANAGEMENT SCENARIOS

This section provides a set of scenarios that may be used to validate the Security Management functional definitions, to provide natural groupings of functionality, and to identify possible interactions that occur as a result of the scenarios. The scenarios for Security Management are:

- Establish/Change Privileges
- Audit Detection of a Security Violation, Containment, and Recovery

Each scenario describes the activities that may occur, identifies interactions with other MFAs as appropriate, and provides a high-level view of the associated data. The scenarios and the associated data are provided only for the purpose of illustrating how MAFs may be linked to form useful management processes. They are neither comprehensive nor exclusive.

13.5.1 Establish/Change Privileges

The scenario shown in Figure 13.2 depicts how customer (external user) permissions are established or changed.

1. A request to establish or change customer or external user permissions is received by the SML Security Administration group of functions. The appropriate function or set of functions for authentication, access control, certification and encryption, and keys are accessed.

   Associated Data

Customer ID	Password
User Permission(s)	Change Request Data
Access Control Certificates	Certification Authority Name
Public Key	

2. BML Security policy information may be accessed to determine if the request is acceptable based on company policy.

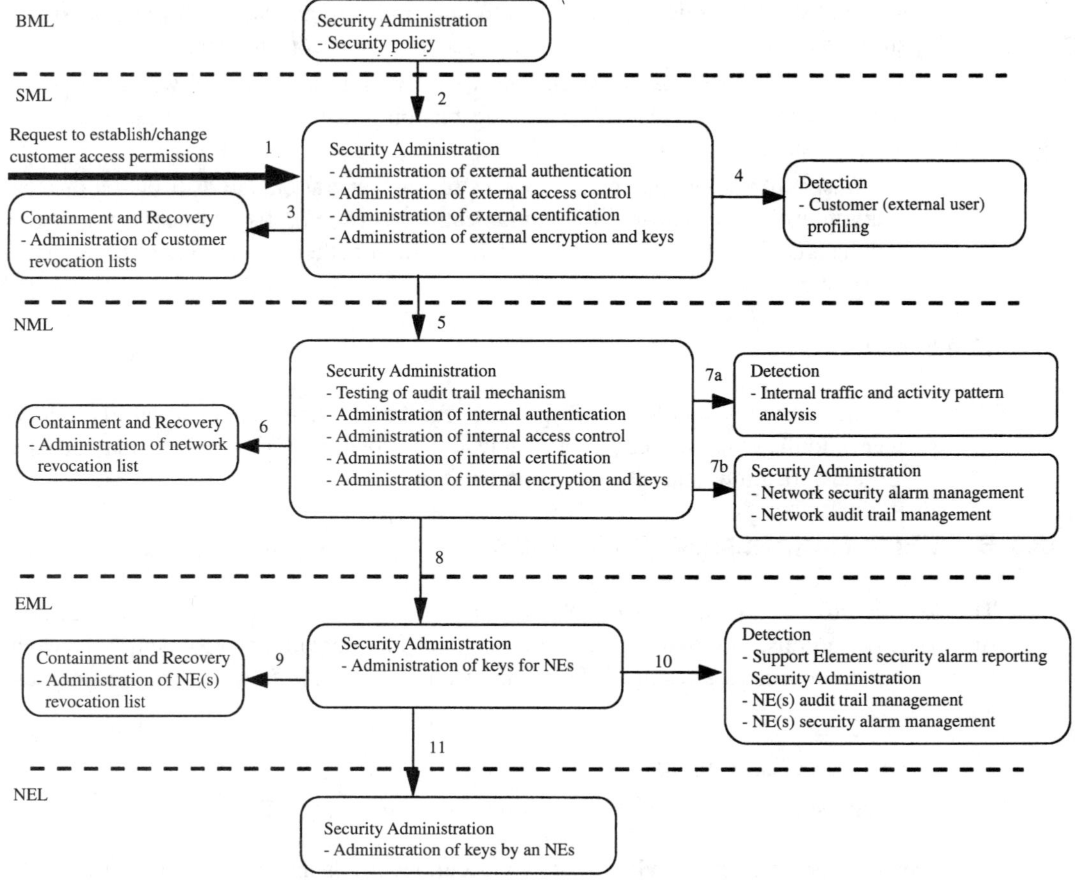

Figure 13.2 Establish/change privileges

Associated Data

Security Guidelines

3. The SML–Containment and Recovery–Management of customer revocation lists function is accessed to ensure that the customer's permissions have not been revoked.

Associated Data

Customer ID
Public Key
Access Control Certificates
Certificate revocation list

4. The SML–Detection–Customer (external user) profiling function is also accessed to establish or change the expected range of customer functions.

Associated Data

Customer ID

User Profile Data
Time & Date Stamp

5. Security Administration at the SML will update the appropriate NML security administration functions with the new or changed permissions. If no changes to lower layer functions are required, the flow may terminate at this point.

Associated Data

Customer ID
Password (if password authentication for user login is implemented)
User Permission(s)
Activation Time / Date

6. The NML–Containment and Recovery–Management of network revocation lists function may be accessed if the external permissions affect internal user permissions. For example, if an internal user is changed to an external user, that user's internal permissions may be placed on the revocation list.

Associated Data

Customer ID
Customer public key
Access Control Certificates
Time & Date Stamp

7. The NML–Detection functions are established or modified to monitor traffic and activity patterns and to establish access to audit trails and security alarms as appropriate.

Associated Data

Customer ID
Alarm Setting(s)
Traffic Pattern Profiles
Audit trail setting
Alarm ID(s)
Activity Pattern Profiles
Activation Time / Date

8. Security Administration at the NML will update the appropriate EML security administration functions with new or changed permissions. If no changes to lower layer functions are required, the flow may terminate at this point.

Associated Data

Customer ID
User Permission(s)
Activation Time / Date
Password (if password authentication for user login is implemented)

9. The EML–Containment and Recovery–Management of NE(s) revocation lists function may be accessed if the external permissions affect NEs. For example, if a customer has had access to certain NEs via a customer network management ca-

pability and is no longer permitted to access one or more of these NEs, the NE revocation list may have to be updated.

Associated Data

Customer ID
Customer public key
Time & Date Stamp

10. The EML–Detection functions are established or modified for security alarms as appropriate.

Associated Data

Customer ID
Alarm ID(s)
Alarm Setting(s)
Activation Time / Date
Audit trail setting

11. Security Administration at the EML will update the appropriate NEL security administration functions with new or changed permissions, as appropriate.

Associated Data

Customer ID
User Permission(s)
Audit trail setting
Password (if password authentication for user login is implemented)
Activation Time / Date

13.5.2 Audit Detection of a Security Violation, Containment, and Recovery

Figure 13.3 shows the functions that might be invoked when a security violation is detected as the result of an audit. Because security violations have a wide variety of impacts on services and the network and may be detected via a wide range of security safeguards, the scenario depicted in the figure should be viewed as illustrative rather then all-inclusive.

Detection may be the result of:

Customer usage pattern analysis indicating a significant variation from normal usage patterns

Analysis of a customer audit trail placed on a customer suspected of security violations such as credit card fraud

Internal traffic and activity pattern analysis that results in the detection of a customer or user (external or internal) security violation

Analysis of a network audit trail placed on the network to detect a suspected security violation

Analysis of a NE audit trail placed on a NE to detect a suspected security violation

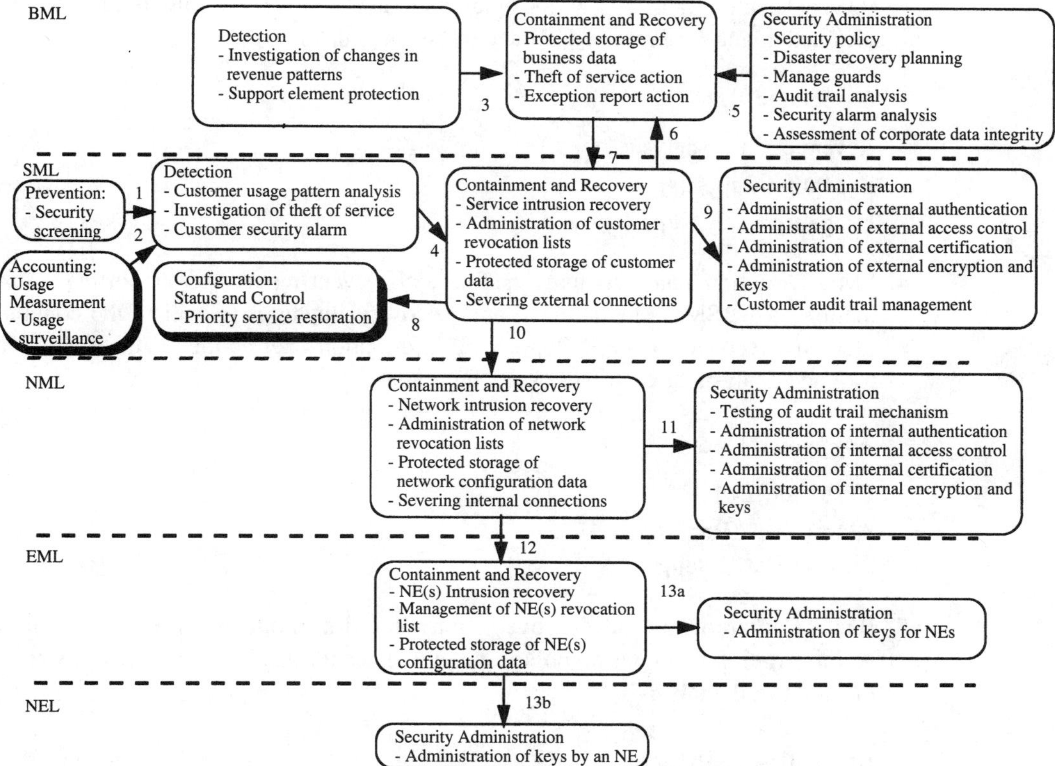

Figure 13.3 Audit detection of a security violation, containment, and recovery

1. A security violation may be detected as a result of the SML Prevention–Security screening function providing information to the SML–Detection functions indicating that a customer does not have sufficient credit.

   Associated Data

   Customer ID
   Credit Limit
   Time & Date Stamp

2. A security violation may be detected as a result of analysis of usage measurement information being forwarded to SML Detection–Customer usage pattern analysis from Accounting Management–Usage Surveillance indicating usage anomalies.

   Associated Data

   Customer ID
   Deviation Usage Data
   Measurement Rules
   Usage Data Records

3. A security violation may be detected as a result of BML–Detection–Investigation of changes in revenue patterns and/or Investigation of theft of service.

BML–Detection notifies BML–Containment and Recovery to initiate theft of service action and /or Protected storage of business data.

Associated Data

Revenue Shift Analysis
Exception Report(s)
Usage Anomaly Report

4. SML–Detection functions may send an alarm event report and supporting information to the SML–Containment and Recovery functions to initiate the containment and recovery process. Supporting information may be the result of an audit trail and analysis of customer usage anomalies.

Associated Data

Customer ID
Alarm Event Report
Time & Date Stamp

5. BML–Containment and Recovery may require input from BML Security Administration functions on policy and guidelines for handling a security breach or the audit trail analysis.

Associated Data

Audit Trail Guidelines
Security Alarm Guidelines
Disaster Recovery Guidelines
Corporate Security Guidelines

6. SML–Containment and Recovery may trigger BML–Containment and Recovery Theft of service action.

Associated Data

Customer ID
Exception Report
Time & Date Stamp

7. BML–Containment and Recovery may provide the SML Containment and Recovery policy information on how to handle a security breach based on the information provided by the Detection functions. For example, if the security breach can be tied to a customer and if the customer committing the security violation is still connected to the network, the Severing external connections function may be used to remove the violator from the network. In addition, that customer's permissions may be revoked and placed on a permission revocation list to ensure that no further activity by that customer will be permitted.

Associated Data

Customer ID
Audit Trail Guidelines
Time & Date Stamp
Customer Backup Data
Security Alarm Setting

8. If customer data has been corrupted as a result of the detected security violation, the Protected storage of customer data function may be accessed to restore the customer's data to a state prior to the corruption activity.

   If the security breach has corrupted the network, SML–Containment & Recovery–Service intrusion recovery will notify SML–Configuration Management–Status & Control–Priority service restoration to initiate the recovery process.

   Associated Data

Customer ID	Circuit ID
Customer Backup Data	Recovery Data
Time & Date Stamp	

9. If customer data has been corrupted as a result of the detected security violation, the Protected storage of customer data function may be accessed to restore the customer's data to a state prior to the corruption activity. The Configuration Management–Status and Control–Priority service restoration function may be used to establish an order for restoring the network configuration based on the services riding on the network with priority status such as 911 Emergency Service. SML Security Administration will be notified if a customer's permissions are being revoked.

   Associated Data

Customer ID	User Permissions
Access Control Certificate Data	Time & Date Stamp

10. The corresponding NML Containment and Recovery functions may also be invoked to contain and recover system and data integrity.

    Associated Data

Customer ID	Circuit ID
Customer Backup Data	Customer Public Key
Access Control Certificate Data	Activation Time / Date

11. If the customer has access to internal user permissions, Security Administration at the NML may need to be updated with the revocation of permissions and certifications.

    Associated Data

Customer ID	User Permissions
Access Control Certificate Data	Time & Date Stamp

12. NML Containment and Recovery notifies EML Containment and Recovery of the security breach to initiate NE intrusion recovery. In addition, if a user's permissions have been revoked, the NE(s) revocation list must be updated. Protected storage of NE(s) configuration data is invoked if data has been corrupted as a result of the security breach.

Associated Data

Customer ID
Customer Backup Data
Access Control Certificate Data
NE ID
Customer Public Key
Activation Time / Date

13. EML Containment and Recovery will update Security Administration–Administration of local keys at both the EML (13a) and the NEL (13b) if a user's permissions have been revoked.

Associated Data

Customer ID
User Permission
Activation Time / Date

14

Future Enhancements to the TMN Security

Although much has been accomplished in securing the TMN, plenty remains to be done. Some of the glaring gaps in the current state of TMN security are enumerated here.

14.1 SECURE INTERWORKING

When two TMN systems interface through a communication protocol (e.g., X.25) they form a (small) domain. As more systems are added to the domain, it grows like a crystal in a supercooled fluid. Meanwhile, other domains (e.g., TCP/IP) may be similarly growing within the same TMN. When domain boundaries clash, they form fault lines like disoriented crystals in a solid. More specifically, straightforward interaction between systems in different domains is not possible. Much effort has been invested in bridging across such boundaries. The Joint Inter Domain Management (JIDM) in particular has provided specifications for mediating between CORBA and CMIP-based systems. Such mediation is far from trivial. Doing it securely adds to the challenge.

The most pedestrian way to provide secure mediation is to provide the desired security between the mediation device and each of the two (indirectly) communicating systems. This requires the two systems to trust a third party. Furthermore, the mediation device may have to share secret information (e.g., symmetric keys) with the other systems. Although this scenario adds risk and complexity, it is often the only option available. This section explores a few alternatives for secure communications through an untrusted mediation device.

14.1.1 Application-Based Security

If the two communicating systems use different application layer protocols (e.g., CORBA and CMIP), then the only choice is to secure individual fields within the communicating applications. Specific fields (e.g., a customer's name) may be encoded in a predefined way (e.g., as ASCII characters) and then encrypted or otherwise protected. The mediation device need not, and cannot, read the encrypted field(s). Such fields are passed as opaque octet strings.

Detailed specifications for application-based security in a CMIP environment are provided in T1.254. If this approach is to be extended to allow secure communications between

CORBA and CMIP-based systems, then similar specifications must be provided for CORBA. In particular, much of the ASN.1 syntax in T1.254 must be translated into CORBA IDL.

14.1.2 CMIP/CORBA

One hypothetical scenario for the TMN posits that almost all TMN systems will interoperate through CORBA. If any systems (during a transition phase) use CMIP, their CMIP messages will be carried over CORBA. In this scenario, encoded CMIP PDUs will be transported inside CORBA GIOP messages as opaque octet strings. CORBA will be able to provide PDU-level security, for example, by encrypting the enveloping GIOP messages in SECIOP.

During an earlier transition phase some of the CMIP-based systems may not have been converted to CORBA. Such systems may transport encoded CMIP PDUs directly over, for example, TCP. It may be possible for CMIP-based systems in the two domains to communicate securely through an untrusted mediation device. CORBA 2.2 provides for interceptors that allow an application to intercept a CORBA message.

14.2 PUBLIC KEY INFRASTRUCTURE

TMN security is based on the use of public key certificates for the secure distribution of public keys. A common format for those certificates needs to be defined. It is generally agreed that the certificates will conform to X.509, version 3. This standard allows the definition of any number of extensions to the certificate structure. The IETF has produced RFCs (Request for Comments) that define numerous useful extensions. A crucial item in TMN security is the specification of similar X.509 extensions that are needed for network management. Most likely, IETF defined extensions that are relevant to network management will be adopted in the TMN context. However, it is unlikely that all the extensions defined by the IETF need to be supported for TMN applications. It is also possible that additional extensions are needed for the TMN. It is expected that those issues will be examined in 1999 and 2000.

14.3 INTERNAL CERTIFICATION OF EXTERNAL ENTITIES

Eventually, the TMN may interface with a large number of small users (any customer). Many such customers may not know what a public key certificate is, let alone where and how to get one. To this end, a TMN Certification Authority (CA) may offer limited certification service. For instance, it may authenticate users as customers, without checking their true identity; it may restrict use of Certificate Revocation Lists it maintains to internal TMN entities only. This approach needs to be examined more closely. If it appears viable, then there may be a need to define standard guidelines for the services offered by such CA for external entities.

14.4 EXTERNAL CERTIFICATION AUTHORITIES

In principle, an external entity dealing with a TMN can freely choose the CA from which it gets its public key certificate. Although most CAs are expected to be thorough and ethical, we cannot assume that this will always be the case. In order to reject a certificate from a questionable CA, it is useful to lean on standardized guidelines for CAs that are recognized and accepted by a TMN.

14.5 SECURITY ALARM MANAGEMENT

Security alarm management is needed to control the disposition of a security alarm: what information it should contain, where it should be logged, where it should be routed (depending on alarm type, day of week, hour of the day, origin of the alarm). ISO | ITU-T standards support this function. However, it still needs to be specialized for the TMN.

14.6 SECURITY AUDIT TRAIL MANAGEMENT

Security audit trail management is needed to:

- Control which events are to be recorded
- Ascertain what information shall be kept
- Determine how large the log shall be
- Learn what notifications it should issue, and when, as it nears saturation
- Know what action shall be taken when it is full

ISO | ITU-T standards support this function. However, it still needs to be specialized for the TMN.

14.7 X INTERFACE SECURITY

T1.261 specifies security for management transactions over the Q interface (between TMN systems). A similar standard is needed for the X interface. Although most of T1.261 can be used for the X interface (between a TMN OS and an external entity), the latter presents some additional issues:

- An external entity may have its own set of encryption algorithms that may not match those used internally by the TMN over its Q interfaces.
- An external entity is most likely to have public key certificates from an external CA; therefore, support for certificate chaining is necessary.
- Access control may need to be more flexible since some external entities (e.g., new customers) may not be known to the TMN.

14.8 F INTERFACE SECURITY

F interface (between an OS and a work station in the same TMN) security cannot be specified until the F interface itself is fully defined. (This work is currently in progress at the ITU-T.) It may require security functionality that is not currently supported over the Q interface. For instance:

- It may need to support different logon procedures (e.g., using secure ID cards).
- It may require the management of passwords, logon parameters (e.g., how many failed attempts are allowed), and logon messages.
- It may require access control based both on the identity of the user behind the work station and the OS through which access is attempted.

14.9 UPDATE OF RELATED STANDARDS

Some of the advances to date in TMN security require updates to other, existing standards as illustrated below.

X.741 provides a powerful information model for managing access control for a system. However, managing access control for a network of thousands of systems requires a small addition, which is specified in T1.261. X.741 needs to be updated to reflect the added functionality.

Abbreviations

3DES	Triple DES
AARE	ACSE Association Response
AARQ	ACSE Association Request
ABS	Application-Based Security
ACC	Access Control Certificate
ACI	Access Control Information
ACSE	Association Control Service Element
AE	Application Entity
AH	Authentication Header
AM	Accounting Management
ANSI	American National Standards Institute
AP	Application Process
APDU	Application Protocol Data Unit
API	Application Programming Interface
ASCII	American Standard Code for Information Interchange
ASE	Application Service Element
ASN.1	Abstract Syntax Notation 1
ATIS	Alliance for Telecommunications Industry Solutions
ATM	Asynchronous Transfer Mode
BER	Basic Encoding Rules
BML	Business Management Layer
CBC	Cipher Block Chaining
CF	Coordination Function
CLEC	Competitive Local Exchange Carrier

CM	Configuration Management
CMIP	Common Management Information Protocol
CMIS	Common Management Information Service
CMISE	Common Management Information Service Element
CO	Central Office
COM	Common Operations Management
CORBA	Common Object Request Broker Architecture
CUG	Closed User Groups
DCN	Data Communications Network
DEA	Data Encryption Algorithm
DER	Distinguished Encoding Rules
DES	Digital Encryption Standard
DIB	Directory Information Base
DSA	Digital Signature Algorithm
DSS	Digital Signature Standard
EB	Electronic Bonding
EBCDIC	Extended Binary Coded Decimal Interchange Code
ECB	Electronic Code Book
ECIC	Electronic Communications Implementers Committee
EDE	Encrypt-Decrypt-Encrypt
EDI	Electronic Data Interchange
EML	Element Management Layer
EPROM	Electronically Programmable Read Only Memory
ESP	Encapsulating Security Payload
ETSI	European Telecommunications Standards Institute
FIPS PUB	Federal Information Processing Standards Publication
FM	Fault Management
FTAM	File Transfer Administration and Management
FU	Functional Unit
FW	Firewall
GDMO	Guidelines for Definition of Managed Objects
GIOP	Generic Internet Inter ORB Protocol
GMT	Greenwich Mean Time
GR	Generic Requirements
GSS-API	Generic Security Service-Application Programming Interface

GULS	Generic Upper Layers Security
IA	Interactive Agent
IC	Inter exchange Carrier
ICI	Interface Control Information
ICMP	Internet Control Message Protocol
ICV	Integrity Check Value
IDL	Interface Definition Language
IDU	Interface Data Units
IEC	International Electrotechnical Commission
IEEE	Institute of Electrical and Electronics Engineers
IETF	Internet Engineering Task Force
IIOP	Internet Inter ORB Protocol
ILEC	Incumbent Local Exchange Carrier
IOP	Inter ORB Protocol
IP	Internet Protocol
Ipsec	Internet Protocol Security
IS	International Standard
ISO	International Organization for Standardization
ITU-T	International Telecommunication Union – Telecommunication Standardization Sector
IV	Initialization Vector
LAN	Local Area Network
LEC	Local Exchange Carrier
MAC	Message Authentication Code
MAF	Management Application Function
MD	Message Digest
MFA	Management Functional Area
MIB	Management Information Base
MIC	Message Integrity Code
MIM	Management Information Model
MIT	Management Information Tree
MO	Managed Object
NBS	National Bureau of Standards
NE	Network Element
NEF	Network Element Function
NEL	Network Element Layer
NIST	National Institute for Standards and Technology
NMF	Network Management Forum
NML	Network Management Layer

N-R	Non-repudiation
OID	Object Identifier
OIW	Open Systems Environment Implementers Workshop
OMG	Object Management Group
ORB	Object Request Broker
OS	Operations System
OSF	Operations System Function
OSI	Open System Interconnection
PC	Personal Computer
PCI	Protocol Control Information
PDU	Protocol Data Unit
PIC	Primary Interexchange Carrier
PKCS	Public Key Cryptography Standard
PM	Performance Management
PROM	Programmable Read Only Memory
QoP	Quality of Protection
RFC	Request for Comments
ROSE	Remote Operations Service Element
RSA	Rivest Shamir Adelman
SA	Security Association
SDU	Service Data Unit
SECIOP	Secure Inter ORB Protocol
SG	Security Gateway
SHA	Secure Hashing Algorithm
SHS	Secure Hashing Standard
SIF	SONET Interoperability Forum
SM	Security Management
SMASE	System Management Application Service Element
SML	Service Management Layer
SNMP	Simple Network Management Protocol
SONET	Synchronous Optical Network
SR	STASE-ROSE
SSL3	Secure Socket Layer version 3
ST	Security Transformation
STASE-ROSE	Security Transformations Application Service Element-ROSE
TA	Trouble Administration
TCP	Transport Control Protocol
TeNoRIOP	Telecommunication Non-Repudiation Inter ORB Protocol
TFF	Technology Frontier Foundation

TL1	Transaction Language 1
TLS	Transport Layer Security
TMF	TeleManagement Forum
TR	Technical Report
UDP	User Datagram protocol
WS	Work Station

Suggested Reading

ON TMN

Aidarous, Salah, and Plevyak, Thomas (Eds.). *Telecommunications Network Management: Technologies and Implementations.* Piscataway, NJ: IEEE Press, 1997.

Aidarous, Salah, and Plevyak, Thomas (Eds.). *Telecommunications Network Management into the 21st Century: Techniques, Standards, Technologies, and Applications.* Piscataway, NJ: IEEE Press, 1996.

Raman, Lakshmi. *Fundamentals of Telecommunications Network Management.* Piscataway, NJ: IEEE Press, 1999.

ON OSI

Rose, Marshall. *The Open Book.* Englewood Cliffs, NJ: Prentice Hall, 1990.

ON SNMP

Stallings, William. *SNMP, SNMPV2, SNMPV3, and RMON 1 and 2.* Reading, Mass.: Addison Wesley, 1999.

ON CRYPTOGRAPHY

Schneier, Bruce. *Applied Cryptography.* 2nd ed. New York: John Wiley & Sons, 1996.

References

[FIPS81] NBS FIPS PUB 81, "DES Modes of Operation," National Bureau of Standards, U.S. Department of Commerce, December 1980.

[FIPS74] NBS FIPS PUB 74, "Guidelines for Implementing and Using the NBS Data Encryption Standard," National Bureau of Standards, U.S. Department of Commerce, April 1981.

[FIPS46-1] NBS FIPS PUB 46-1, "Data Encryption Standard," National Bureau of Standards, U.S. Department of Commerce, January 1988.

[FIPS46-2] NIST FIPS PUB 46-2, "Data Encryption Standard," National Institute of Standards and Technology, U.S. Department of Commerce, December 1993.

[FIPS180-1] NIST FIPS PUB 180-1, "Secure Hash Standard," National Institute of Standards and Technology, U.S. Department of Commerce, April 17, 1995.

[FIPS186] NIST FIPS PUB 186, "Digital Signature Standard," National Institute of Standards and Technology, U.S. Department of Commerce, May 1995.

[GR1253] GR-1253-CORE, *TMN Security Administration,* Issue 1 (Bellcore, June 1995).

[GR1332] GR-1332-CORE, *Generic Requirements for Data Communications Network Security,* Issue 1 (Bellcore, February 1994).

[GR1469] GR-1469-CORE, OTGR Section 15.5: *Generic Requirements on Security for OSI-Based Telecommunications Management Network Interfaces (a module of OTGR FR-439),* Issue 1 (Bellcore, September 1994).

[GR2869] GR-2869-CORE, *Generic Requirements for Operations Based on the TMN Architecture,* Issue 1 (Bellcore, October 1995).

[IEEE802.10C] ISO/IEC JTC 1/SC 21 N6998, IEEE 802.10c/D1, IEEE Standard for Interoperable LAN/MAN Security, Part 3—Key Management Protocol, March 21, 1993.

[IEEEP802.10B] IEEE P802.10B/D7, IEEE Standard for Interoperable LAN Security, (SILS) Part B—Secure Data Exchange, November 9, 1991.

[ISO10164-1] ISO/IEC IS 10164-1, Information Technology—Open Systems Interconnection—Systems Management—Part 1: Object Management Function, 1992.

[ISO10164-5] ISO/IEC IS 10164-5, Information Technology—Open Systems Interconnection—Systems Management—Part 5: Event Report Management Function, 1992.

[ISO10181-3] ISO/IEC 10181-3, Committee Draft—Information Technology—Security Frameworks for Open Systems—Part 3: Access Control Framework, 1994.

[ISO11770-1] ISO/IEC 11770-1, Information Technology—Security Techniques—Key Management—Part 1: Key Management Framework.

[ISO7498-2] ISO/IEC 7498-2, *Information Processing Systems—OSI Reference Model—Part 2: Security Architecture,* 1989.

[ISO8072] ISO/IEC 8072, Information Processing Systems—Open Systems Interconnection—Transport Service Definition, June 1986.

[ISO8073] ISO/IEC 8073, Information Processing Systems—Open Systems Interconnection—Connection Oriented Transport Protocol Specification, July 1986.

[ISO8571] ISO/IEC 8571, Information Processing Systems—Open System Interconnection—File Transfer, Access and Management—Parts 1–4, 1988.

[ISO8602] ISO/IEC 8602, Information Processing Systems—Open Systems Interconnection—Protocol for Providing the Connectionless—Mode Transport Service, June 1987.

[ISO8649-1]ISO 8649-1, *Information Processing Systems—Open Systems Interconnection—Authentication Addendum to ACSE Service Definition,* 1992.

[ISO8650] ISO/IEC 8650, Information Processing Systems—Open Systems Interconnection—Protocol Specification for the Association Control Service Element, 1988.

[ISO8650-1] ISO 8650-1, *Information Processing Systems—Open Systems Interconnection—Authentication Addendum to ACSE Protocol Definition,* 1992.

[ISO9798-3] ISO/IEC 9798-3:1993, Information Technology—Security Techniques—Entity Authentication Mechanisms—Part 3: Entity Authentication Using a Public Key Algorithm.

[KOB87] N. Koblitz, "Elliptic Curve Cryptosystems," *Mathematics of Computation,*vol. 48, 1987, pp. 203–209.

[M.3010] ITU-T, Draft M.3010, Principles for a Telecommunications Management Network, April 1995.

[M.3100] ITU-T Draft Rec. M.3100, Generic Network Information Model, January 1995.

[M.3400] ITU-T Rec. M.3400, *TMN Management Functions,* 1996.

[MIL85] V. Miller, "Uses of Elliptic Curves in Cryptography," Advances in Cryptology, CRYPTO '85, *Lecture Notes in Computer Science* 218, Springer-Verlag, pp. 417–426.

[MILL87] S. P. Miller, B. C. Neuman, J. I. Schiller, and J. H. Saltzer, "Project Athena Technical Plan Section E.2.1: Kerberos Authentication and Authorization System," Project Athena, MIT, December 1987.

[NMF006] Network Management Forum, OMNI *Point 1 Definitions,* Forum 006, Issue 1.0, August 1992.

[NMF016] Application Services: Security of Management, *Network Management Forum,* Forum 016, Issue 1.0, August 1992.

[OIW12] Working Implementation Agreement for Open System Interconnection Protocols: Part 12—OS Security, Open Systems Environment Implementors Workshop, December 1992.

[OMG-COS2.2] Object Management Group, *CORBAservices, Common Object Services,* Version 2.2, November 1997.

[OMG-ORB2.2] Object Management Group, *The Common Object Request Broker: Architecture and Specification,* Version 2.2, February 1998.

[OMG-SSL] Object Management Group, OMG Document orbos/97-02-04, *Secure Socket Layer / CORBA Security,* adopted June 24, 1997.

[PKCS1] RSA Data Security Inc., PKCS#1: RSA Encryption Standard, Version 1.5, November 1993.

[PKCS7] RSA Laboratories, a division of RSA Data Security, Inc., *Public-Key Cryptography Standards #7—Cryptographic Message Syntax.* (The Network Working Group of the IETF as an Informational Request has recently published this standard for Comments, RFC 2315.)

[Q.811] ITU-T Rec. Q.811, Lower Layer Protocol Profiles for the Q3 and X Interfaces.

[Q.812] ITU-T Rec. Q.812, Upper Layer Protocol Profiles for the Q3 and X Interfaces.

[Q.813] ITU-T Rec. Q.813, Security Transformations Application Service Element for Remote Operations Service Element (STASE-ROSE).

[RAB79] M. O. Rabin, "Digital Signatures and Public Key Functions as Intractable as Factorization," MIT Laboratory for Computer Science, Technical Report, MIT/LCS/TR-212, January 1979.

[RC4] R. L. Rivest, "The RC4 Encryption Algorithm," RSA Data Security Inc., March 1993.

[RC5] R. L. Rivest, "The RC5 Encryption Algorithm," *Dr. Dobb's Journal,* 20, no. 1 (January 1995), pp. 146–148.

[RFC1319] R. Rivest, IETF RFC 1319: "The MD5 Message Digest Algorithm, April 1992.

[RFC1320] R. L. Rivest, "The MD4 Message Digest Algorithm," RFC 1320, April 1992.

[RFC2078] IETF RFC 2078, "Generic Security Service Application Program Interface," Version 2, Internet Engineering Task Force, January 1997.

[RFC2104] H. Krawczyk, M. Bellare and R. Canetti, IETF RFC 2104, HMAC: Keyed-Hashing for Message Authentication, February 1997.

[RFC2274] U. Blumenthal and B. Wijnen, IETF RFC 2274, User-Based Security Model (USM) for Version 3 of the Simple Network Management Protocal (SNMPv3), January 1998.

[RFC2275] B. Wijnen, R. Presuhn, and K. McCloghrie, IETF RFC 2275, View-Based Access Control Model (VACM) for the Simple Network Management Protocal (SNMPv3), January 1998.

[RFC2412] The OAKLEY Key Determination Protocol. H. Orman. November 1998. (Status: INFORMATIONAL)

[RFC2411] IP Security Document Roadmap. R. Thayer, N. Doraswamy and R. Glenn. November 1998. (Status: INFORMATIONAL)

[RFC2410] The NULL Encryption Algorithm and Its Use with IPsec. R. Glenn and S. Kent. November 1998. (Status: PROPOSED STANDARD)

[RFC2409] The Internet Key Exchange (IKE). D. Harkins and D. Carrel. November 1998. (Status: PROPOSED STANDARD)

[RFC2408] Internet Security Association and Key ManagementProtocol (ISAKMP). D. Maughan, M. Schertler, M. Schneider, and J. Turner. November 1998. (Status: PROPOSED STANDARD)

[RFC2407] The Internet IP Security Domain of Interpretation for ISAKMP. D. Piper. November 1998. (Status: PROPOSED STANDARD)

[RFC2406] IP Encapsulating Security Payload (ESP). S. Kent and R. Atkinson. November 1998. (Obsoletes RFC1827) (Status: PROPOSED STANDARD)

[RFC2405] The ESP DES-CBC Cipher Algorithm with Explicit IV. C. Madson and N. Doraswamy. November 1998. (Status: PROPOSED STANDARD)

[RFC2404] The Use of HMAC-SHA-1-96 within ESP and A.H. C. Madson and R. Glenn. November 1998. (Status: PROPOSED STANDARD)

[RFC2403] The Use of HMAC-MD5-96 within ESP and A.H. C. Madson and R. Glenn. November 1998. (Status: PROPOSED STANDARD)

[RFC2402] IP Authentication Header. S. Kent and R. Atkinson. November 1998. (Obsoletes RFC1826) (Status: PROPOSED STANDARD)

[RFC2401] Security Architecture for the Internet Protocol. S. Kent and R. Atkinson. November 1998. (Obsoletes RFC1825) (Status: PROPOSED STANDARD)

[RSA78] R. Rivest, A. Shamir, and L. M. Adelman, "A Method for Obtaining Digital Signatures and Public-Key Cryptosystems," *Communications of the ACM,* 21, no. 2 (February 1978), pp. 120–126.

[T1.210] ANSI T1.210-1993, *American National Standard for Telecommunications—Operations, Administration, Maintenance, and Provisioning (OAM&P)—Principles of Functions, Architectures, and Protocols for Telecommunications Management Network (TMN) Interfaces.*

[T1.233] ANSI T1.233-93, *American National Standard: Operations, Administration, Maintenance and Provisioning (OAM&P)—Security Framework for Telecommunications Management Network (TMN) Interfaces,* 1993.

[T1.228] ANSI T1.228-95, *American National Standard: Operations, Administration, Maintenance and Provisioning (OAM&P)—Services for Interfaces between Operations Systems across Jurisdictional Boundaries to Support Fault Management (Trouble Administration),* 1995.

[T1.243] ANSI T1.243-T1M1/95, *American National Standard: Operations, Administration, Maintenance and Provisioning (OAM&P)—Baseline Security Requirements for the Telecommunications Management Network (TMN),* 1994.

[T1.245] ANSI T1.245-1995, Directory Service for Telecommunications Management Network, 1995.

[T1.252] ANSI T1.252, *Operations, Administration, Maintenance and Provisioning (OAM&P)—* Security for the Telecommunications Management Network Directory.

[T1.254] ANSI T1.254, Application-Based Security, 1997.

[T1.259] ANSI T1.259, Security Transformations Application Service Element for Remote Operations Service Element (STASE-ROSE), 1999.

[T1.261] ANSI T1.261, *Operations, Administration, Maintenance, and Provisioning (OAM&P)—* Security for TMN Management Transactions over the TMN Q3 Interface, 1999.

[TA1080] TA-STS-001080, *Bellcore Standard Operating Environment Security Requirements,* Issue 2 (Bellcore, June 1991).

[TA1194] TA-STS-001194, *Bellcore Operations Systems Security Requirements,* Issue 1 (Bellcore, June 1991).

[TR40] T1 Technical Report No. 40 *Security Requirements for Electronic Bonding Between Two TMNs,* 1995, (also ECIC/94-001, *Security Requirements for Electronic Bonding Between Two TMNs,* 1994).

[TR815] TR-NWT-000815, *OTGR Section 2.31: Generic Requirements for Network Element and Network System Security (A Module of OTGR FR-NWT-000435),* Issue 2 (Bellcore, December 1992).

[X.200] CCITT Rec. X.200 | ISO/IEC 7498: Information Processing Systems—Open Systems Interconnection—Basic Reference Model (1984)

[X.208] CCITT Rec. X.208 (1988) | ISO/IEC 8824 : 1990, Information Technology—Open Systems Interconnection—Specification of Abstract Syntax Notation One (ASN.1).

[X.210] ITU-T Rec. X.210 | ISO/IEC 8509, Information Technology—Open Systems Interconnection—Service conventions (1987).

[X.217] CCITT Rec. X.217 | ISO/IEC 8649, Information Processing Systems—Open Systems Interconnection—Service Definition for the Association Control Service Elements, 1988.

[X.219-88] CCITT Rec. X.219—ISO/IEC 9072-1, Remote Operations: Model, Notation, and Service Definition, 1988.

[X.219-93] ITU-T Rec. X.219 | ISO/IEC 9072-1, Information Technology—Open Systems Interconnection—Remote Operations : Model, Notation and Service Definition, 1993.

[X.229] ITU-T Rec. X.229 | ISO/IEC 9072-2, Information Technology—Open Systems Interconnection—Remote Operations : Protocol Specification, 1993.

[X.25] ITU-T Rec. X.25, Interface between Data terminal Equipment (DTE) and Data Circuit-terminating Equipment (DCE) for Terminals Operating in the Packet Mode and Connected to Public Data Networks by Dedicated Circuits, 1996.

[X.273] ITU-T Rec.X.273 (1994)—ISO/IEC IS 11577: 1994(E), Information Technology—Open Systems Interconnection—Network Layer Security Protocol, November 1, 1993.

[X.500] CCITT Rec. X.500—ISO/IEC 9594-1, The Directory—Overview of the Concepts, Models and Services, 1993.

[X.501] CCITT Rec. X.501—ISO/IEC 9594-2, Information Technology—Open Systems Interconnection—The Directory—Part 2: Information Framework, 1993 (E).

[X.509] CCITT Rec. X.509—ISO/IEC 9594-8, The Directory—Authentication Framework, March 1993.

[X.680] ITU-T Rec. X.680 | ISO/IEC 8824-1, Information Technology—Abstract Syntax Notation One (ASN.1)—Specification of Basic Notation, 1994.

[X.681] ITU-T Rec. X.681 | ISO/IEC 8824-2, Information Technology—Abstract Syntax Notation One (ASN.1)—Information Object Definition, 1994.

[X.682] ITU-T Rec. X.682 | ISO/IEC 8824-3, Information Technology—Abstract Syntax Notation One (ASN.1)—Constraint Specification, 1994.

[X.683] ITU-T Rec. X.683 | ISO/IEC 8824-4, Information Technology—Abstract Syntax Notation One (ASN.1)—Parametrization of ASN.1 Specification, 1994.

[X.690] ITU-T Rec. X.690 | ISO/IEC 8825-1, Information Technology—ASN.1 Encoding Rules—Specification of Basic Encoding Rules (BER), Canonical Encoding Rules (CER) and Distinguished Encoding Rules (DER), 1994.

[X.710] ITU-T Rec. X.710 | ISO/IEC 9595, Information Technology—Open Systems Interconnection—Common Management Information Service Definition, 1991.

[X.711] ITU-T Rec. X.711 | ISO/IEC 9596, Information Technology—Open Systems Interconnection—Common Management Information Protocol Specification, 1991.

[X.720] ISO/IEC 10165-1 | ITU-T X.720, Information Technology—Open Systems Interconnection—Structure of Management Information—Part 1: Management information Model, 1992.

[X.721] ISO/IEC 10165-2 | ITU-T X.721, Information Technology—Open Systems Interconnection—Structure of Management Information—Part 2: Definition of Management Information, 1992.

[X734]ISO/IEC IS 10164-5 | ITU-T X.734, Information Technology—Open Systems Interconnection—Systems Management—Part 5: Event Report Management Function, 1993.

[X.735] ISO/IEC IS 10164-6 | ITU-T X.735, Information Technology—Open Systems Interconnection—Systems Management—Part 6: Log Control Function, 1992.

[X.736] ISO/IEC IS 10164-7 | ITU-T X.736, Information Technology—Open Systems Interconnection—Systems Management—Part 7: Security Alarm Reporting Function, August 1991.

[X.740] ISO/IEC IS 10164-8 | ITU-T X.740, Information Technology—Open Systems Interconnection—Systems Management—Part 8: Security Audit Trail Function, June 1992.

[X.741] ISO/IEC IS 10164-9 | ITU-T X.741, Information Technology—Open Systems Interconnection—Systems Management—Part 9: Objects and Attributes for Access Control, 1995.

[X.800] ITU-T Rec. X.800, Data Communication Networks: Open Systems Interconnection (OSI); Security, Structures and Applications—Security Architecture for Open System Interconnection for CCITT Applications, 1991.

[X.803] ITU-T Rec. X.803 | ISO/IEC 10745, Information Technology—Open Systems Interconnection—Upper Layers Security Model, 1993.

[X.830] ITU-T Rec. X.830 | ISO/IEC IS 11586-1, Generic Upper Layer Security—Part 1: Overview, Models and Notation, 1994.

[X.831] ITU-T Rec. X.831 | ISO/IEC IS 11586-2, Generic Upper Layer Security—Part 2: Security Exchange Service Element Service Definition, 1994.

[X.832] ITU-T Rec. X.832 | ISO/IEC IS 11586-3, Generic Upper Layer Security—Part 3: Security Exchange Service Element Protocol Specification, 1994.

[X.833] ITU-T Rec. X.833 | ISO/IEC IS 11586-4, Generic Upper Layer Security—Part 4: Protecting Transfer Syntax Specification, 1994.

[X3.106] ANSI X3.106-1983, Data Encryption Algorithm—Modes of Operation, 1983.

[X3.92] ANSI X3.92-1981, Data Encryption Algorithm, 1981.

Index

3DES (Triple DES), 64-65
key lists and, 83

A

AARE (Application association response), 108, 109, 112, 113, 114, 126, 127, 128, 129, 131
AARQ (Application association request), 108, 109, 112, 113, 114, 126, 127, 128, 129, 131, 135, 137
ABS (Application-based security), 116, 119, 124, 179
for CMIP only, 214
DES padding for, 124
main features of, 123-124
standard for, 116-117
ACC (Access control certificate), 71, 72, 83, 86, 149, 258
X.500 directory and, 83
Access control, 53, 78, 79, 112, 113, 114, 236-240, 269
authentication for, 238
and authentication scenario, 246-248
EB and, 114
initiators of, 237-238
mechanisms for, 69-73, 78
NML and, 258
management basis for, 149
MOs for, 238-240
as prevention service, 245, 252
rules of, 70-71, 237
security labels and, 73, 85-86, 237-238
security policy for, 85
SML and, 257
targets and, 236-237
Access control management, 225
Accounting management, 46
ACI (Access control information), 70, 72, 85, 236, 237, 238
initiator, 71-72, 73
request, 72-73
target, 73
ACL (Access control list), 73, 83, 85, 149, 237, 240
ACSE (Association control service element), 18-37, 76, 107, 113, 126
ACI and, 236
ASE-specific security and, 109
Q3 security and, 149
security for, 107-109
security audit trail example and, 27
STASE-ROSE and, 127, 132, 135, 136, 137, 138, 139, 140, 146, 148
TR40 for, 213
Administration functions, security, 224-226
Administrative interfaces, 88
AE (Application entity), 17, 18, 19, 126
ACSE security and, 107, 108
Agent role, 23
AH (Authentication header), 198
Alarm notification, 49
Alarms. See also Security alarms
panic mode for, 53
security, 53, 225, 233
AM (Accounting management), 8
ANSI (American National Standards Institute), 5
AP (Application process), 17, 19, 126
AEs and, 107
OSI association between, 107
title of, 108
APDU (Application protocol data unit), 140
abstract syntax definition of, 140-145

API (Application programming interface), 19, 175, 176-177
COBRA and, 29
GSS and, 89-94
local specification for, 191-193
TeNoRIOP and, 181
USM and, 204-206
Application-based security, 2670268
Application Context, 19
Application layer, 16-37
ACSE and, 18-19, 24
AE and, 17
AP and, 17
ASEs and, 16-17
DistinguishedName and, 26-27
CF and, 17
CMISE and, 20-24
FTAM and, 20
GULS and, 95
OBJECT IDENTIFIER and, 26
proper naming and, 25, 26
ROSE and, 19
security audit trail example and, 27, 28
services provided by, 16
SMASE and, 25, 26
SNMP and, 30-37
STASE-ROSE and, 19, 20
X.500 DUA and, 20
Application specific threats, 48-49
Architectural views of TMN, 7-38
communications/information architecture, 11-38
functional architecture, 7-9
physical architecture, 9-11
ASEs (Application service elements), 16-19, 107, 125
AE and, 17
AP and, 107
CMISE security and, 109
interaction of, 17
AP and, 17
security for, 107, 109
service primitives of, 17
services provided and, 132
STASE-ROSE and, 135, 136, 137, 138, 139, 147, 148
ATIS (Alliance for Telecommunications Industry Solutions), 3, 5
Attackers, potential security, 41-42
Attacks and defenses
general threats and vulnerabilities
security services
threats unique to TMN
Audit detection of security violation, containment, and recovery scenario, 262-266
Authentication and access control scenario, 246-248
Authentication protocols, 74-78
challenge-response authentication, 75-76
initiator only authentication, 87
stateful authentication, 76-78
target only authentication, 87
two-way peer entity authentication, 87

B

Basic security mechanisms, 57-80
access control mechanisms, 69-73
authentication protocols, 74-78
certificates, 68-69
Diffie-Hellman key exchange, 73-74
digital signatures, 66-68
encryption, 60-66
hashing, 57-60
mapping security services to security mechanisms, 78-80
Bell Systems in United States breakup, 3
Bellcore's
GR-1253, 224, 226
GR-2869, 8, 9, 243, 245
TR-815, 224
BER (Basic encoding rules), 16, 116, 117, 202
Biham, Eli, 115
BML (Business management layer), 8, 252
containment and recovery and, 255-256
detection and, 255
prevention and, 253, 254
security administration and, 257
Boolean, 117, 233

C

CA (Certification authority), 10, 69, 84, 85, 86, 153, 154, 225, 237, 240, 258, 268, 269
CBC (Cipher block chaining) mode, 61, 62, 63, 100, 111, 112, 113, 124, 127
access control and, 238
TLS1 and, 153
USM and, 203
CCITT (International Consultative Committee for Telephone and Telegraph), 5
CDR, 183, 184, 186, 187
Certificates, 68-69, 78, 79, 164, 238
SSL3 and, 97
STASE-ROSE and, 126, 127, 130
CF (Control function), 17
Channel management, 230-231
CipherSuite, 96
CLECs (Competitive Local Exchange Carriers), 47, 48, 49, 52, 53
access control and, 53
security alarm and, 53
Closed user groups, 88
CM (Configuration management), 8
CMIP (Common management information protocol), 22, 118, 149, 267, 268
ABS for, 214
ACI and, 236, 238
ACL and, 237
ACSE security and, 109
ASEs and, 107
-based EB applications, 115
CMISE security and, 109, 113

future for, 268
JIDM and, 267
management of security information and, 223
operations specified by, 69-70
security audit trail example and, 27
security in layers and, 89
SNMP versus, 30, 36
SSL3 and, 95
symmetric interface, 24
CMIS (Common management information service), 22, 228, 229, 230, 231, 233
CMISE (Common management information service element), 20-24, 107, 136
ACSE security and, 109
ACSE usage with, 24
addressing MOs and, 21, 22
ASEs and, 107
containment and, 20-21, 22
EB authenticator and, 110-115
EDI and, 28
filtering, 22
inheritance and, 22, 23
managers and agents and, 23-24, 25
notification service of, 23
ROSE and, 24
security, 109-124
security audit trail example and, 27
selective field protection and, 115-124
services of, 22, 23
setup for systems using, basic, 25
CO (Central office), 20, 69, 71
Column object, 32
COM (Common operations management), 9, 243, 245, 248, 252
SM and interactions between, scenario for, 246-248
Communication protocols, 75, 86, 87
anatomy of secure type of, 87-88
life cycle of secure type of, 88
out-of-stock secure type of, 89, 90
phases of, 87
Communications/information architecture, 11-38
application layer and, 16-37
data link layer and, 12, 13, 14
interoperability and, 11
lower layers of ISO model and, 11
network layer and, 14
OSI layering and, 12, 13
OSI reference model X.200 as basis for, 11
physical layer and, 13
presentation layer and, 15-16
security-related components of TMN stack and, 37-38
session layer and, 15
transport layer and, 14-15, 28
Confidentiality, 51-52
Connection access control, 50
Containment and recovery, 255-257
Context-level calls, 91, 92
Context management, 15
CORBA (Common Object Request Broker Architecture), 5, 29-30
benefit of, 29
CMIP versus, 29, 267, 268
future for, 268
GDMO versus, 29
JIDM and, 267
security in layers and, 89
SSL3 and, 95
TeNoRIOP for, 218
CORBA-based TMN security, 179-196
GIOP overview, 180
IDL syntax for NR evidence, 188-191
local API interface specification, 191-193
NR protocol machine, 195-196
TCP/IP and, 179
TeNoRIOP and, 180-188
timing of NR evidence, 193-195
COs (Central offices), 10
CUG (Closed user group), 110
connection access control and, 50
purpose of, 150
security audit trail example and, 28
X.25 and, 150
Customer private information, 49

D

Data link layer security, 13, 14
Data origin authentication, 50, 78
key lists and, 82
DCN (Data communications network), 11
DEA (Data encryption algorithm), 61
Delay/delete notifications re trouble report status/disposition, 49
Delay notification of service availability, 49
Denial of service, 44, 45, 46
DER (Distinguished encoding rules), 16, 117, 127, 129, 154, 165, 167
DES (Data encryption standard), 61, 63, 65, 74, 111, 115, 116, 127, 130, 202
access control and, 238
CBC and, 102, 112, 113
IPsec and, 197
key management and, 241
modes of, 62
padding, 114, 124, 203
SSL3 and, 98, 100
TFF and, 115
TLS1 and, 153
triple, 64-65
Detection, 254-255
Detection security services, 246
DIB (Directory information base), 84, 85
Diffie-Hellman key exchange algorithm, 73-74, 87, 97, 226, 240
Digital seals, 65, 130
Digital signatures, 66-68, 78, 79, 122
ACSE security authentication and, 108
algorithm of, 130
IA and, 177

message receipt with, 158
NR for request and, 181
protocols and, 86
for reply, 187-188
for request, 184
STASE-ROSE and, 126, 127
DistinguishedName, 26-27, 108, 150
DSA (Digital signature algorithm), 130
DSS (Digital signature standard), 69, 74

E

Eavesdropping, 42, 79
EB (Electronic bonding), 82, 149, 226, 237, 238
authenticator, 110-115
DES padding and, 114
future proofing, 115
historic background of, 110-111
key management and, 114-115
operations implications, 113-114
protocol implications, 113
replay, 112-113
vulnerabilities, 112
ECB (Electronic code book) mode, 61, 62
ECIC (Electronic Communications Implementation Committee), 5, 110
EDE (Encrypt-decrypt-encrypt) mode, 64-65, 127
EDI (Electronic Data Interchange), 5, 28-29
IA for, 217
with message integrity, 156
with NR, 156
security in layers and, 89
SSL3 and, 95
EDI-based TMN security, 153-178
COBRA versus, 179
IA client-server interaction chart and, 155
interactive agent, 154-178
message syntaxes, 158-163
TLS1 for, 153-154
EFD (Event forwarding discriminator), 233, 234
Electronic ticket, 72
EML (Element management layer), 8, 248, 249
containment and recovery and, 256-257
detection and, 255
prevention and, 253, 254
security administration and, 258-259
Encryption, 60-66, 103
ACSE security authentication and, 108
asymmetric type of, 65-66, 78, 79
digital seals and, 65
hardware for, 111
parameter components of, 134-135
protocols and, 86
public key contenders for, 65
public key distribution and, 83
STASE-ROSE and, 126, 130
symmetric key type of, 61-65, 73, 78, 79
-type values, 133
Encryption keys management, 225-226
EPROM (Electronically programmable read only memory), 84
ET (External time), 77, 78
ETSI (European Telecommunications Standards Institute), 5
External certification authorities future, 268
External entities, 47

F

F interface security future, 269
False claim of trouble report submission, 49
False inquiry, 48
False order cancellation, 48
False trouble reports, 49
Falsify testing results, 49
Fault management, 46
Filtering, 22
Flood testing capabilities, 49
FM (Fault Management), 8
FTAM (File transfer administration and maintenance), 20, 107
security in layers and, 89
Functional architecture, 7-9
LLAs and, 8
MAFs and, 7
MFAs and, 8, 9
FUs (Functional units), 25, 126, 149
ACSE security and, 108-109, 113
security audit trail example and, 27
STASE-ROSE and, 127, 135
Future enhancements to TMN security, 267-269
external certification authorities, 268
F interface security, 269
internal certification of external entities, 268
public key infrastructure, 268
related standards update, 269
secure interworking, 267
security alarm management, 268
security audit trail management, 269
X interface security, 269
FWs (fire walls), 28
connection access control and, 50
IA and, 177
IPsec and, 101
SG as, 198

G

GDMO (Guidelines for definition of managed object), 20, 29
GeneralizedTime, 112, 113, 114, 117, 126, 127, 184, 187, 199, 238
GetBulkRequest PDU, 34
GIOP (Generic Inter ORB Protocol), 29, 30
future and, 268
NR for reply and, 185
overview of, 180
request message of, 181
SECIOP and, 179
GMT (Greenwich mean time), 77, 78, 112
GSS-API (Generic security service application programming interface), 89-94, 95, 216
context level calls, 91, 92

credential management calls, 93
fatal error codes, 94
GSS administrative interfaces, 93
GSS handshaking, 90-91
GSS out-of-stack secure communication protocol, 89, 90
GSS secure transfer, 91, 92-93
informatory status codes, 94
per message calls, 93
with STASE-ROSE, 145-148
status reporting, 94
support calls, 93
Guarding, 253, 254
GUI (Graphical user interface), 11
GULS (Generic upper layer security), 94-95, 124, 125
for all OSI-compliant protocols, 210
EB security based on, 111

H

Hashing, 57-60, 121, 154
algorithms of, 58, 130
encrypted type of, 65
encryption/decryption and usage of, 203
keyed type of, 58, 59-60, 73, 78, 79, 102
MD5 and, 58, 59
requirements of, 57
SHA1 and, 184, 187
S-key and, 60
SMNPv2 and, 201
SMNPv3 and, 202
STASE-ROSE and, 126, 127
uses for, 57
HMAC, 60, 102, 127, 154, 203

I

IA (Interactive agent), 29, 154-178
ASN.1 syntax for optional receipts of, 162-163
ASN.1 syntax for status of, 162
client-server interaction chart for, 155
client specifications for, 163-165
design considerations for, 176
EDI messages and, 153
for EDI only 217
error handling/recovery and, 177-178
implementation issues and, 178
interfaces of, 175
message formatting and, 154, 155-158
message syntax definitions and, 158-163
operational concerns for, 176-177
optional message receipts and, 157-158
parsing received message by, 168-175
server specifications for, 165-175
status message detail format for, 157
IANA (Internet assigned number authority), 178
IATP (Interactive agent transfer protocol), 158, 178
IC (Interexchange Carrier), 5, 82, 110, 115
ICEC (Inter-Carrier Electronic Commerce), 47, 48
ICI (Interface control information), 12
ICV (Integrity check value), 102
IDL (Interface definition language), 29, 183, 186, 187
between client or application object and local NREvidence object, 191-193
syntax for NREvidence and, 188-191
IDUs (Interface data units), 11-12
ICI and, 12
PDU and, 12
IETF (Internet Engineering Task Force), 4, 6, 268
GSS-API by, 89
RFCs and, 268
security audit trail example and, 28
SSL3 and, 95
IIOP (Internet Inter ORB Protocol), 29, 30, 190
ILECs (Incumbent Local Exchange Carriers), 47, 48, 49, 52, 53
access control and, 53
Industry acceptance, 4
Information, management of security, 223-241
administration functions and, 224-226
information model description, 226-241
Information module, 34
InformRequest PDU, 34-35
Integrity, 51, 103
EDI with message and, 156, 158, 159-160, 167-168, 169-170
keyed hashing and, 59
Inter ORB (Object request broker) protocol (IOP), 29
Internal certification of external entities future, 268
Interoperability, 3, 4
and communications/information architecture, 11
IA and, 178
and physical architecture, 9-10
reusing same library and, 89
Invoke-ID, ROSE and, 19
IOP (Inter ORB protocol), 29
IOR (Interoperable object reference), 183, 184, 186, 191
IP (Internet protocol), 5, 11, 163, 177
IPsec for, 212
network layer and, 14, 28
PDUs and, 12
IPSec (Internet protocol security), 28, 100-102, 198
AH protocol of, 100, 101, 102
ESP (Encapsulating security payload) protocol of, 100, 101, 102
handshaking and, 100-101
ICV in, 102
for IP only, 212
message protection by, 200
protocols for, 197
SA and, 100-101, 102
secure transfer, 101-102
SNMPv1 security and, 197
ISO (International Standards Organization), 5, 15
communications/information architecture and, 11, 12-13
layering, 13-37

ITU-T (International Telecommunications Union-Telecommunications), 5, 6, 107
- ITU-T Rec. M.3400, 7-8, 243, 245
- ITU-T Rec. X.227, 136, 137
- ITU-T Rec. X.500, 20
- ITU-T Rec. X.509 and fields for a certificate, 69
- ITU-T Rec. X.741 and access control management, 149, 237, 238
- notification management and, 232, 233, 234, 235

IV (Initialization vector), 61, 111, 112, 113, 124, 131, 202
- access control and, 238
- encryption parameters and, 134
- selection, 63
- SSL3 and, 98, 100

J

JIA (Joint implementation agreement), 183, 186, 193

JIDM (Joint inter domain management), 267

K

Key distribution, 82-83
- key lists, 82-83
- public key distribution, 83

Key management, 240-241, 246

L

LAES (Lawfully authorized electronic surveillance), 52

LAN (Local area network), 89

Layers, security in, 88-89

LECs (Local exchange carriers), 3, 4, 5, 82, 110, 115

Legal review, 252, 253, 254

LLA (Logical layer architecture), 8

Local implementation freedom, 4

Login management, 224, 225, 226-232

M

MAC (Message authentication code), 27, 72, 127
- integrity and confidentiality providing and, 64
- IPsec and, 102
- MD5 and, 98
- SSL3 and, 98, 99-100

MAFs (Management Application Functions), 7, 8, 9
- NEs and, 10
- OSs and, 10

Management of security information, 223-241
- information model description, 226-241
- security administration functions, 224-226

Manager role, 23

Mapping security services to security mechanisms, 78-80

Masquerade, 43, 79

MD5, 58, 102, 127, 130, 202
- IPsec and, 197
- MAC and, 99
- operating on whole message diagram, 59
- USM and, 203

MFAs (Management Functional Areas), 8, 251
- provisioning, 45
- security risks and, 45-46
- of SM, 252-259

MIB (Management information base), 23
- access control and, 70, 236, 237
- OID and, 32
- SNMPv2 security and, 198, 199, 200
- VACM and, 207, 208

MIC (Message integrity code), 111

MIM (Management information model), 20

MIT (Management information tree), 20, 21, 22

MO (Managed object), 20, 69, 70, 118, 226
- access control and, 237, 238-240
- access control management and, 225
- addressing, 21, 22
- attributes of, 20
- channel management and, 230
- information model and, 226, 227, 228
- key management and, 240, 241
- management security of, 232
- notification management and, 232-233, 234, 235-236
- password support and, 228, 229
- security audit trail, 21
- session management and, 231

Modification of information, 43, 79

MP (Message processing), 36
- USM services to, 204

N

Negotiating of security algorithms and STASE-ROSE, 127-132
- defaults, 127, 129

NEL (Network element layer), 8, 10
- alarm detection, containment, and recovery scenario for, 248-250
- containment and recovery and, 257
- detection and, 255
- prevention and, 253, 254
- security administration and, 259

NEs (Network elements), 3, 4, 9, 10, 223, 224, 245, 246, 248
- channel management and, 230
- CMISE and, 20
- confirmation data and protected storage of, 257
- EML and, 259
- intrusion recovery of, 256
- NEL and, 259
- remote management of, 47
- revocation list administration of, 256
- ROSE and, 19
- SG and, 198

Network flooding, 43, 79

Network layer security, 14

Network management functionality, 3, 4

NMF (Network Management Forum), 5, 233

NML (Network management layer), 8, 248, 250
- containment and recovery and, 256

detection and, 255
prevention and, 254
security administration and, 258
SM and, 253
Notification management, 225, 232-236
NR (Nonrepudiation), 52-53, 78, 103, 124
ASN.1 syntax for EDI with, 160-162
COBRA and, 179-180
confidentiality for providing, 68
EDI with, 156, 168, 171-174
GIOP and, 180
IDL syntax for evidence of, 188-191
message receipt with, 158
of origin, 52, 193
as prevention service, 245, 252
protocol machine of, 195-196
public key system and, 84
of receipt, 52-53, 193-194
for reply, 185-188
for request, 181-185
SECIOP and, 179
STASE-ROSE and, 126, 147
TeNoRIOP and, 180-181
NREvidence, 181, 185, 189, 190, 191
behavior while waiting for, 194-195
timing of, 193-195

O

Object Management Group, 29
OBJECT TYPE, 30-31
access mode of, 30
parts of, 31
status of, 30
syntax of, 30
ObjectName, 31
OID (OBJECT IDENTIFIER), 26, 31, 32, 158, 159, 207, 208
MIB and, 199
PDU and, 33
OIW (Open systems environment implementers workshop) implementation agreements, 63
OMG (Object management group), 29
Open system interfaces, 47
ORB (Object request broker), 191, 192, 193
COBRA and, 179
GIOP and, 180
NREvidence and, 189, 190
TeNoRIOP and, 180, 181
Order modification, 48
Order repudiation, 49
OS (Operation system), 3, 4, 9, 10, 179, 223, 224, 245, 246
SG and, 198
TA and, 46, 110
OSF (Operation system function), 9
OSI (Open System Interconnection), 5, 15, 124, 125, 145, 199, 223, 224
ASE-to-ASE protocol and, 17
communication protocol stack of, 111
GULS and, 210
login management and, 227
lower layers of protocol stack of, 13-15
notification management and, 225
peer entity authentication in, 113
seven-layer stack of, 88, 94
stateful authentication and, 76
upper layers of protocol stack of, 15-37
OSI-based TMN protocols, security of, 107-151
ACSE security, 107-109
CMISE security, 109-124
Q3, 149
STASE-ROSE, 124-149
X.25, 150-151
X.500, 149-150
Out-of-stack secure communication protocol, 89, 90
Overview
of security, 39-103
of TMN, 1-38

P

Padding, 61, 63
DES and, 114, 124, 203
Panic mode, 53
Passwords, 130, 225
ACSE security authentication and, 108
login management and, 224
management of, 228-230
MO classes for support of, 228
S-key and, 60
PCI (Protocol control information), 12
PDU (Protocol data unit), 12, 88, 112, 113, 116, 118, 124, 126, 132, 179, 209, 268
access control authentication and, 238
ACSE security and, 108
ASE and, 17
CMIP and, 23
CMISE security and, 109
directory operations and, 150
ET and, 78
GSS-API and, 89, 148
key lists and, 82
notification management and, 233
OID and, 33
PCI and, 12
protection for, 149
ROSE type of, 125, 127, 132, 133, 140, 145
SDU and, 12
secure type of, 199-200
security audit trail example and, 27, 28
SNMP and, 32-33
SNMPv2 and, 34
SNMPv3 and, 207
STASE-ROSE and, 140
time stamp and, 77
types of, 32-33
USM and, 202-203
Peer entity authentication, 50, 78, 110
key lists and, 82
in OSI environment, 113
STASE-ROSE functionality and, 125, 126, 127, 135

strong type of, 103
Performance management, 49
Performance monitoring, 45-46
Personnel risk analysis, 253, 254
Physical access security, 253, 254
Physical architecture, 9-11
DCN and, 11
interfaces defined by, 10, 11
interoperability and, 9-10
Physical layer security, 13
PIC (Primary inter exchange carrier), 115
PM (Performance management), 8
Portraits gallery, 209-218
ABS for CMIP only, 214
GSS-API, 216
GULS for all OSI-compliant protocols, 210
IA for EDI only, 217
IPsec for IP only, 212
SNMPv2 and SNMPv3 security, 219
SSL3/TLS1 for TCP/IP only, 211
STASE-ROSE for ROSE-based protocols, 215
TeNoRIOP for COBRA only, 218
TMN security communications protocols, 209
TR40 for ACSE using protocols, 213
Potential security
attackers, 41-42
risks, 43-45
threats, 42-43
Presentation layer, 15-16, 95
Prevention, 252, 254
Prevention services, 245-246
Privileges scenario, established/change, 259-262
PROM (Programmable read only memory), 84
Protocols for security, 75, 86-89
authentication protocols, 74-78
communication protocols, 75, 86
SML and, 258
Public key encryption algorithm, 130
Public key infrastructure, 103, 268

Q

Q3
interfaces, 150
security, 149
QoP (Quality of Protection), 147, 148

R

Recovery and containment, 255-257
Remote management of NEs, 47
Replay, reroute, misroute, delete messages, 43, 79
Repudiation, 43, 79
Request/perform unnecessary intrusive testing, 49
RFCs (Request for comments), 102, 268
ROSE (Remote operations service element), 19, 135
ACSE security and, 107
PDU and, 125, 127, 132, 133, 140, 145
security audit trail example and, 27, 28
STASE-ROSE for protocols based on, 215
RRP (Registration request protocol), 150
RSA (Rivest-Shamir-Adelman) algorithm, 65, 110, 127, 154, 158
digital signatures and, 68, 184, 188
procedure example of, 66

S

SA (Security association), 100-101, 102
SAD (Security association database), 101
Salutary plug, 3-4
SC (System clock), 77, 78
ScopedPDU, 37
Scoping and filtering, 47
SDU (Service data unit), 12, 88, 89
SE (Security management). See Security management (SE)
SECIOP (Secure Inter ORB Protocol), 30
COBRA and, 179
GIOP and, 179, 268
Secure communication protocols, 87-88
Secure interworking, 267-268
Security
attackers of, potential, 41-42
in layers, 88-89
protocols for, 86-89
risks to, potential, 43-45
threats to, potential, 42-43
Security administration, 257-259
Security alarms, 53, 81-82, 103, 233-234. See also Alarms
BML and, 257
customers and, 254
detection scenario for, 248-250
EML and, 255, 259
information presented in, 81-82, 233
log for, 233, 234
management of, 268
MOs and, 233
NML and, 255, 258
notification management and, 225, 233
reporting, 246
SML and, 258
Security audit log, 81, 82
Security audit trail, 54-55, 71, 234
example of, 27, 28
management of, 269
notification management and, 225
Security engineering, 102-103
guidelines for, 103
Security functions and operations, 245-250
detection security services, 246
illustrative scenarios, 246-250
prevention services, 245-246
Security ID, 36
in layers, 88-89
Security log, 225, 246
Security management (SM), 221-241, 248
COM and interactions between, scenario for, 246-248

functions and operations for, 251-266
of information, 223-241
MAFs of, 253
scenarios for, 259-266
Security mechanisms, basic, 57-80
access control mechanisms, 69-73
authentication protocols, 74-78
certificates, 68-69
Diffie-Hellman key exchange, 73-74
digital signatures, 66-68
encryption, 60-66
hashing, 57-60
mapping security services to security mechanisms, 78-80
security services provided by, 78
Security modules and main interactions of, 38
Security MOs management, 232
Security operations, 243-269
Security overview, 39-103
Security protocols. See Protocols for security
Security risks
impacts on TMN of, 45-46
potential, 43-45
security services correlated with, 55
Security screening, 253, 254
Security services, 50-55
access control, 53
confidentiality, 51-52
connection access control, 50
data origin authentication, 51
integrity, 51
mechanisms to provide, 78
NR, 52-53
order of, partial, 80
peer entity authentication, 50
security alarm, 53
security audit trail, 54-55
security risks correlated with, 55
security threats against, 54
SecurityName, 36
Selective field confidentiality, 51-52
Selective field integrity, 51
Service ordering, 48-49
Session integrity, 51
Session layer, 15
Session management, 231-232
SessionID, 96
SG (Security gateway), 101, 197-198
SHA (Secure hashing algorithm), 98, 130
SHA1, 102, 153, 154, 184, 187, 188, 197, 202, 203
S-key, 60
SM. See Security management
SMASE (System management application service element), 25, 26
ACSE security and, 108
configuration for systems using, minimum, 25
security audit trail example and, 27
SML (Service management layer), 8, 82, 246, 247, 250
containment and recovery and, 256
detection and, 254-255
security administration and, 257-258
security screening and, 253, 254
SNMP (Simple Network Management Protocol), 4, 5, 6
application categories of, 36
application layer and, 30-37
-based TMN security, 197-208
IPsec and, 100
MIB and, 30-32, 198, 199, 200
PDU and, 32-33
relativity of time and, 199
version 1, 197-198
version 2, 33-35, 198-200, 219
version 3, 35-37, 200-208, 219
SNMP MIB, 30-31
SNMPv1 security, 197-198
SNMPv2 security, 198-200, 219
SNMPv3 (Simple Network Management Protocol Version 3), 35-37, 200-208
architecture of, 35-36
IPsec and, 200
protocol of, 37
security, 200-208, 219
security levels of, 36
user-based security model of, 201-206
VACM and, 206-208
SPD (Security policy database), 101
SPI (Security parameter index), 101, 102
SRPM (STASE-ROSE-protocol-machine), 140
SSL3 (Secure sockets layer version 3), 95-100, 154, 164, 165, 168, 175, 190
alerts for, 100
certificate and, 97
COBRA and, 30, 179
design function of, 95
EDI transactions and, 153, 170
handshake protocol of, 95-99, 100, 167, 177
message read setup and, 166
ProtocolVersion of, 95
read processing of, 167
as secure communication protocol, 87
secure transfer and, 99-100
security audit trail example and, 28
security in layers and, 88, 89
STASE-ROSE versus, 95
TeNoRIOP and, 180
TLS1 for EDI and, 153
and TLS1 for TCP/IP only, 211
ST (Security transformations), 117, 118, 119, 122, 124, 131
STASE-ROSE (Security transformations application service element for remote operations service element), 19, 20, 116, 124-149
ASE and, 125
association abort and, 138-139
association establishment and, 135-137
association release, 137-138

current status and future developments of, 148-149
data transfer and, 139-140
design features of, 148
dynamic update of security parameters and, 125, 132
encryption algorithms versus, 130
functionality of, 125
GSS API usage with, 145-148
interaction between application service elements and, 135-140
negotiation of security algorithms and, 127-132
peer entity authentication and, 125, 126, 127
-protocol-machine, 140
protocol of, 140-145
for ROSE-based protocols, 215
as secure communication protocol, 87, 88, 89
security audit trail example and, 27, 28
security transformations on ROSE PDUs and, 125-126
services of, 133-135
SSL3 versus, 95
Standard interfaces, 47
Strong data origin authentication, 59
Strong peer entity authentication, 58, 59
Support mechanisms, 81-103
directory, 83-86
GSS-API-Security in a Box, 89-94
GULS, 94-95
IPsec, 100-102
key distribution, 82-83
protocols, 87-89
security audit log, 82
security alarms, 81-82
security engineering, 102-103
SSL3, 95-100
Symmetric key encryption, 61-65
DES and, 61
error propagation and, 64
IV selection and, 63
padding and, 61, 63
triple DES and, 64-65
Syntax matching, 15

T

TA (trouble administration), 110, 111, 114, 115, 238
Tariffing/pricing, 46
TCP (Transport control protocol), 15, 30, 88, 89, 268
IIOP and, 190
SSL3 and, 95
TCP/IP (Transport control protocol/Internet protocol) , 5, 15
COBRA and, 179
connection access control and, 50
EDI and, 28-29
EDI messages and, 153, 176
port assignments and, 178
security audit trail example and, 28
SSL3 and, 95
SSL3/TLS1 for, 211
TeNoRIOP (Telecom NR Inter-ORB protocol)
API and, 181
COBRA and, 180-188
for COBRA only, 218
NR services of, 180-181
NREvidence and, 188
for reply, 185-188
for request, 181-185
Testing, 49
TFF (Technology Frontier Foundation), 115
Theft of information, 43, 45, 46
Theft of service, 44, 45, 46
Threats and vulnerabilities, 41-46
denial of service, 44
eavesdropping, 42, 78
masquerade, 43, 78
modification of information, 43, 78
network flooding, 43, 78
potential security attackers, 41-42
potential security risks, 43-45
potential security threats, 42-43
replay, reroute, misroute, delete messages, 43, 78
repudiation, 43, 78
summary of, 45
theft of information, 43
theft of service, 44
unauthorized access, 43, 79
unauthorized use of resources, 44
Threats unique to TMN, 47-49
Time stamps, 77, 108, 111, 112, 124, 149, 199
digital signature for reply and, 187, 188
digital signature for request and, 184
STASE-ROSE and, 126, 127, 130, 134, 140
UTCTime and, 150
TL1 (Transaction Language 1), 4
TLS1 (transport layer security version 1), 95, 153
for EDI, 153-154
and SSL3 for TCP/IP only, 211
TMF (Tele Management Forum), 5, 233
TMN (Telecommunications Management Network)
architectural views of,
attacks and defenses, 41-55
future enhancements to security of, 267-269
history of, brief, 3-6
management functional areas of, 8
overview of, 1-38
protocols of, security OSI-based, 107-151
securing, 105-219
security communications protocols of, 209
security functionality partitioning, 252
threats and vulnerabilities for, 41-46
threats unique to, 46-49
vulnerabilities to, generic, 47
TMN stack and security related components, 37-38
Token, 72

TP0-TP4, 15
TR40 for ACSE using protocols, 213
Traffic flow confidentiality, 52
Transfer syntax, 47
Transport layer security, 14-15, 28
Transport mode, 101
Trouble administration, 49
Tunnel mode, 101
Turing, Alan, 110

U

UDP (User datagram protocol), 100
Unauthorized access, 43, 79
Unauthorized use of resources, 44, 46
Update of related standards, 269
Usage measurement, 46
User management, 227-228
USM (user-based security model), 201-206
  APIs and, 204-206
  key items of, 201-202
  PDUs and, 202-203
  simple times of, 201

V

VACM (view-based access control model), 200, 206-208
VT (virtual time), 77, 78
Vulnerabilities
  domain-specific, 47-49
  generic, 47

W

Whole message confidentiality, 52
Whole message integrity, 51
WSs (Work Stations), 3, 10, 11, 245
  access to, 198

X

X interface security future, 269
X.25, 11, 12, 150-151
  CMIP and, 95
  CUGS and, 110, 150-151
  network layer and, 14
  security audit trail example and, 28
  security in layers and, 89
X.500 Directory, 83-86, 149-150, 226
  access control for, 85-86
  automatic registration for, 83-85
  key management and, 240
  multiple security domains and, 86
X.500 DUA (Directory user agent), 20

Z

Zero overhead padding, 63

About the Author

Moshe Rozenblit received the B.S. and M.S. degrees in physics from the University of Brussels, Belgium. He received the Ph.D. in physics from Stevens Institute of Technology, NJ, and the M.S. in computer science from Rutgers University, NJ.

Presently, Dr. Rozenblit is a senior engineer for network management and related standards at Telcordia Technologies, Inc. He is the chairman of the T1M1.5 Subworking Group on Security Management and cochairman of the ECIC Security Subcommittee. He is also a leader in information modeling, serving as the editor of ITU-T Standards on Service Level Modeling.

Much of Dr. Rozenblit's 24-year career in telecommunications has focused on network operations planning. In the early 1980s, he played a key role in the development of the Total Network Operations Plan—in particular, he developed the methodology that was used for synthesizing and evaluating alternative network operations plans. After the breakup of the Bell System, he led the effort to define the Network Services Architecture for the newly created Regional Telecoms spun off from the old Bell System. More recently, he developed network operations plans for new broadband networks for Local Exchange Carriers and private multiservice networks.

Dr. Rozenblit is the editor (or coeditor) and major contributor of several standards for Telecommunications Management Network (TMN) security, both international [ITU-T Rec. M.3016 – TMN security overview; ITU-T Rec. Q.813. Security Transformations Application Service Element for Remote Operations Service Element (STASE-ROSE)] and North American [T1.252 – Security TMN Directory; T1.254 – Application-Based Security; T1.259 – STASE-ROSE; T1.261 – Security for the Q Interface; T1 Technical Report No. 40 – Security Procedures for Electronic Bonding].

A member of the IEEE Communication Society, Dr. Rozenblit holds two U.S. patents. In 1983 while at Bell Laboratories, he was made a Distinguished Member of Technical Staff. In 1995 he received the T1M1 Leadership Award, and in 1996 he received a Certificate of Appreciation from TCIF (Telecommunications Industry Forum). Dr. Rozenblit is a frequently invited speaker on TMN at major industry forums and leading universities.